Crop Physiology

Zhenghui Liu

Crop Physiology

From Gene to Field

Springer

Zhenghui Liu
College of Agronomy
Nanjing Agricultural University
Nanjing, Jiangsu, China

ISBN 978-981-92-2959-8 ISBN 978-981-92-2960-4 (eBook)
https://doi.org/10.1007/978-981-92-2960-4

This work was supported by National Natural Science Foundation of China (31771719).

This Springer imprint is published by the registered company Springer Nature Singapore Pte Ltd.
The registered company address is: 152 Beach Road, #21-01/04 Gateway East, Singapore 189721, Singapore

Preface

Contemporary agriculture faces the challenge of producing sufficient nutritious food while caring for the environment, amid global climate change and the shifts in land use patterns resulting from urbanization and the cultivation of bioenergy crops. FAO projects that food production must increase by at least 50% compared to 2013 levels by 2050 to support global population growth. However, comprehensive analyses indicate a general slowdown in the improvement of yields for the primary field crops. Furthermore, FAO statistics reveal that 735 million people in 2023, predominantly residing in impoverished nations, suffer from daily hunger. These collectively highlight the importance of interdisciplinary efforts to synthesize critical knowledge to further enhance crop yield.

Crop physiology is such a multidisciplinary science that integrates knowledge from various disciplines, such as ecology, genomics, biochemistry, phenomics, informatics, and modelling. It investigates the physiological mechanisms underlying crop yield, quality, and resilience to abiotic stress across multiple levels of biological organization, from the cellular and tissue level to the organ, plant, population, and community levels. The aim of crop physiology is to establish a scientific basis for genetic improvement and agronomy, addressing global concerns regarding sustainable production of abundant and high-quality food, feed, fibre, and energy.

This book is structured around the author's two-decade experience in research and instruction in crop physiology, with five distinctive features as follows.

1. To be more informative to serve as a reference book, it covers the fundamental concepts and principles of crop physiology. This book focuses on the core processes that shape crop yield and quality formation, offering insights into carbon and nitrogen assimilation, source–sink interaction, seed development, plant architecture, ideal plant type, crop phenology, and canopy structure and function. These key topics are essential for comprehending modern crop science and technology and represent current hotspots and frontiers in biological research.
2. To be well structured to inspire systematic thinking, it is systematically designed, reflecting the intricate yet well-organized realm of crops by dividing the materials into eight chapters that cover the hierarchy of yield and quality formation processes in time and space. The chapters progress from the cellular metabolism of carbon and nitrogen, through organ-level embryo–endosperm interaction and plant-level source–sink relation, to canopy structure–function at the population

level. The objective is to bridge the gap between gene expression and crop performance, thereby establishing a comprehensive and integrated framework for crop scientists.

3. To be more pedagogical to fully develop cognitive ability, it is packed with features and elements designed to enhance the learning experience. The introductory section titled "Did You Know" at the beginning of each chapter lists some interesting facts, which provides insightful connections to students' daily lives or research activities. In addition, it may stimulate curiosity about the fascinating crop world around us, making crop physiology more accessible and relevant to students. "Learning Objectives" are elaborately planned based on the *Bloom's taxonomy* of cognitive domain. "Important Terms" at the end of every chapter help students assess their understanding of chapter material. "Review Questions" are set in accordance with Bloom's taxonomy to foster the integrated development of foundational cognitive abilities such as remembering, understanding, and applying, as well as higher-order skills like analysing, evaluating, and creating.
4. To be more conceptual to facilitate deep learning, it focuses on the relationship between concepts. A concept, defined as an abstract idea, serves as the fundamental building block underlying principles, thoughts, and beliefs. At the beginning of each chapter, a "Concept Map" is presented to provide mental connections between the major concepts of the material and other concepts that readers have learned, functioning as a visual display of the hierarchical organization of ideas. Some "Review Questions" are designed to develop the ability to construct a concept map using a set of related concepts, allowing instructors to assess changes and growth in students' conceptual understanding after instruction. At the end of this book, an "Idea-Based Course Design Document" is formulated to establish a conceptual foundation for the course on Crop Physiology, through which instructors and students can gain deep insights into its disciplinary characteristics. Collectively, all these elements are designed to develop logical learning and study skills by assimilating new concepts and propositions into pre-existing cognitive frameworks.
5. To serve as a linguistic foundation for crop scientists, it consistently considers the disciplinary backgrounds of scientists. For scientists from diverse disciplines to effectively collaborate in addressing the complex challenges associated with the multiple objectives of crop production, it is essential that they understand each other's terminology and methodologies. This book covers knowledge ranging from gene to field, targeting an audience of students and researchers across a range of disciplines, including crop science, molecular biology, ecology, genomics, biotechnology, phenomics, artificial intelligence, and smart agriculture. Specifically, it attempts to establish a common linguistic framework among scientists in the field of crop science, with subdisciplines such as genetics, breeding, management, and informatics. Notably, the "Abbreviations" at the beginning lists the full names of 227 common abbreviations, while the "Glossary" at the end defines 433 key terms in crop physiology, serving as a shared language among crop scientists.

The involvement in crop physiology research is an incredibly stimulating and perspective-expanding experience, often filled with excitement. In the long run, this book will serve as a starting point for young scientists who are bound to revolutionize the framework or foundations of crop physiology through their research careers. As presented in the book, crop physiology is a dynamically evolving concept and research realm. I appreciate receiving any response and feedback from readers. If you have any suggestions or comments, please do not hesitate to contact me via the email: liuzh@njau.edu.cn.

Finally, I would like to conclude with a quote from Rothamsted Research's official website, which embodies its grand vision for agriculture as the world's longest-standing agricultural experiment station. "We believe science holds the key to a world, where hunger is a distant memory, where farmers can make a fair living, where food is grown in harmony with the natural world."

Nanjing, Jiangsu, China Zhenghui Liu
May 11, 2026

Acknowledgement

Completing this book has been a long and deeply fulfilling journey. I am truly grateful to my mentors, colleagues, and students for their thoughtful insights, dedicated support, and constant encouragement. Their valuable feedback, careful suggestions, and meaningful discussions have played an essential role in shaping this work. I sincerely appreciate their generosity, collaboration, and the time they so willingly shared—each contribution has made a meaningful difference.

I would like to express my heartfelt gratitude to Professors Guoping Zhang (Zhejiang University, ZJU), Matthew Paul (Rothamsted Research, UK), Guohua Xu (Nanjing Agricultural University, NAU), Fanjiang Kong (Guangzhou University), Yanfeng Ding (NAU), Fangmin Cheng (ZJU), Gang Pan (ZJU), Si Shen (China Agricultural University), Ganghua Li (NAU), Shirong Zhou (NAU), Chengqiang Ding (NAU), Jin Xiao (NAU), Yu Jiang (NAU), and Haoyu Qian (NAU) for their thoughtful reviews and invaluable feedback. I am deeply grateful to my talented graduate students—Zhaomiao Lin, Zunxin Wang, Feng Xiao, Yang Tao, Hongfa Xu, Zongfeng Yang, Lu An, Luqi Wang, Mengmeng Sun, Hao Chen, Lei Yang, Yixian Cheng, Tang Ming, Nana Chen, Yi Li, Mengjiao Jiang, Luoluo Hu, Yanyan Ren, Xi Zhang, Junjie Lai, Ying Yang, Xueping Wang, Xuefei Yang, Tangjie Peng, Xinzhi Li and Bei Li—for their exceptional contributions in creating the illustrations. Their dedication and creativity have made this work truly special.

I would like to express my sincere gratitude to Mrs. Tabinda Tariq and Professor Haiyan Wang for their careful and thoughtful review of the manuscript. I am genuinely grateful to the committed and professional team at Springer Nature, especially Ms. Mei Hann Lee and Ms. Kavitha Jayakumar, for their steadfast support and guidance during the entire publication process. I am indebted to Ms. Lingjing Zhang from the Zhejiang University Press for her invaluable assistance during the publication process. I would also like to extend special thanks to Nanjing Agricultural University, whose funding program "Enrichment of Instructional Resources for Graduate Students" provided essential financial support, playing a meaningful role in bringing this book to completion.

Finally, I would like to express my deepest and most sincere gratitude to my beloved mentor, Professor Dajun Liu—a source of endless wisdom, constant encouragement, and lasting inspiration. His profound insights have shaped my journey, and his legacy continues to guide me with quiet strength. With great affection and respect, this book is dedicated to honouring the 100th anniversary of his birth.

Competing Interests The author has no competing interests to declare that are relevant to the content of this manuscript.

Contents

Abbreviations

2-OG	2-Oxoglutarate
2-PG	2-Phosphoglycolate
3D	Three dimensional
3-PG	3-Phosphoglycerate
AAP	Amino acid permease
ABA	Abscisic acid
AGPase	ADP-glucose pyrophosphorylase
AHA	Arabidopsis H^+-ATPase
AI	Artificial intelligence
AMP	Adenosine monophosphate
AMPK	AMP-activated protein kinase
AMT	Ammonium transporter
AS	Asparagine synthetase
Asn	Asparagine
Asp	Aspartate
AspAT	Aspartate aminotransferase
AT	Agronomic trait
ATP	Adenosine triphosphate
BC	Before Christ
BE	Starch branching enzyme
BETL	Basal endosperm transfer layer
BIT	Biotechnology–information technology
BL	Brassinolide
bp	Base pair
BRs	Brassinosteroids
bZIP	Basic leucine zipper
C	Carbon
CAM	Crassulacean acid metabolism
CAT	Cationic amino acid transporter
CBC	Calvin-Benson cycle
CC	Companion cell
CGF	Canopy gap fraction
ChlF	Chlorophyll fluorescence
CID	Carbon isotope discrimination

CKs	Cytokinins
CKX	Cytokinin oxidase/dehydrogenase
CO	CONSTANS
COV	Canopy occupation volume
CRISPR	Clustered regularly interspaced short palindromic repeat
CRY	Cryptochrome
CS	Castasterone
CT	Canopy temperature; computed tomography
CTD	Canopy temperature depression
CWIN	Cell wall invertase
CWP	Cell wall polysaccharide
DAA	Days after anthesis
DAF	Days after fertilization
DBE	Starch debranching enzyme
DH	Doubled haploid
DHAP	Dihydroxyacetone phosphate
DM	Dry matter
DP	Degree of polymerization
E4P	Erythrose 4-phosphate
ea	Actual water vapour pressure
EAA	Essential amino acid
EAS	Endosperm adjacent to scutellum
eQTL	Expression quantitative trait locus
ER	Endoplasmic reticulum
ESR	Embryo surrounding region
ET	Evapotranspiration
ETH	Ethylene
F1,6BP	Fructose 1,6-bisphosphate
F6P	Fructose 6-phosphate
FBPase	Fructose 1,6-bisphosphatase
Fd	Ferredoxin
FD	FLOWERING LOCUS D
FKF1	FLAVIN-BINDING, KELCH REPEAT, F-BOX 1
FLC	FLOWERING LOCUS C
FSPM	Functional–structural plant model
FT	FLOWERING LOCUS T
G1P	Glucose 1-phosphate
G6P	Glucose 6-phosphate
GA	Gibberellic acid
GABA	Gamma-aminobutyric acid
GAI	Green area index
GAP	Glyceraldehyde 3-phosphate
GAs	Gibberellins
GBSS	Granule-bound starch synthase
GDH	Glutamate dehydrogenase

GHG	Greenhouse gas
Gln	Glutamine
Glu	Glutamate
GN	Grain number
GOGAT	Glutamine 2-oxoglutarate aminotransferase (glutamate synthase)
GPD	Grain protein deviation
GS	Glutamine synthetase; genomic selection
Gs	Stomatal conductance (*G* for conductance, *s* for stomata)
Gt	Gigaton
GW	Grain weight
GWAS	Genome-wide association study
GWD	Glucan water dikinase
HI	Harvest index
HK	Hexokinase
HMW	High-molecular-weight
HTP	High-throughput phenotyping
IAA	Indole-3-acetic acid
InDel	Insertion-deletion
INV	Invertase
iP	Isopentenyl adenosine
IPA1	Ideal Plant Architecture 1
IPT	Isopentenyltransferase
JA	Jasmonic acid
JAs	Jasmonates
K	Potassium
kDa	Kilodalton
Km	Substrate concentration at half Vmax (substrate-saturated enzyme activity)
LAI	Leaf area index
LAT	Laser ablation tomography
LC-MS/MS	Liquid chromatography–tandem mass spectrometry
LDP	Long day plant
LiDAR	Light detection and ranging, also Lidar, LIDAR
LM	Lateral meristem
LPR	Leaf to panicle ratio
LR	Long-wave radiation
M	Molarity (mol/L); million
MAGIC	Multiparent advanced generation inter-cross
MAS	Marker-assisted selection
Mb	Megabase
MRI	Magnetic resonance imaging
mRNA	Messenger ribonucleic acid
MST	Monosaccharide transporter
MW	Molecular weight
N	Nitrogen

NADH	Nicotinamide adenine dinucleotide
NADPH	Nicotinamide adenine dinucleotide phosphate
NDVI	Normalized difference vegetation index
NIL	Near-isogenic line
NIR	Near infrared
NiR	Nitrite reductase
NMR	Nuclear magnetic resonance
NPF	Nitrate transporter 1/peptide transporter family
NPT	New plant type
NR	Nitrate reductase
NRT1	Nitrate transporter 1
NSC	Non-structural carbohydrate
NUE	Nitrogen use efficiency
P	Phosphorus
PAR	Photosynthetically active radiation; PHYTOCHROME RAPIDLY REGULATED
PAT	Polar auxin transport
PBF1	Prolamin-box binding factor 1
PCD	Programmed cell death
PEPC	Phosphoenolpyruvate carboxylase
PFK	Phosphofructokinase
Pg	Gross photosynthesis
PGR	Plant growth regulator
PHY	Phytochrome
PIN	PIN-FORMED
PM	Physiological maturity; plasma membrane; plasmalemma
Pn	Net photosynthesis
PPFD	Photosynthetic photon flux density
ppm	Parts per million
PPP	Pentose phosphate pathway
PSV	Protein storage vacuole
PT	Physiological trait
PUE	Phosphorus use efficiency
PWD	Phosphoglucan water dikinase
QTL	Quantitative trait locus
R	Respiration
RAM	Root apical meristem
RBCS	Ribulose 1,5-bisphosphate carboxylase/oxygenase small subunit
RCA	Ribulose 1,5-bisphosphate carboxylase/oxygenase activase
RFOs	Raffinose family oligosaccharides
Rg	Growth respiration
RGB	Red-green-blue
RIL	Recombinant inbred line
Rm	Maintenance respiration
RNAP	RNA polymerase

ROS	Reactive oxygen species
RS	Resistant starch
RTM	Radiative transfer model
Rubisco	Ribulose bisphosphate carboxylase/oxygenase
RuBP	Ribulose 1,5-bisphosphate
RUE	Radiation use efficiency
S	Sulphur
SA	Salicylic acid
SAG	Senescence-associated gene
SAM	Shoot apical meristem
SAS	Shade-avoidance syndrome
SBPase	Sedoheptulose 1,7-bisphosphate phosphatase
SCN	Stem cell niche
scRNA-seq	Single-cell RNA sequencing
SDP	Short day plant
SEM	Scanning electron microscope
SIF	Sun-induced chlorophyll fluorescence
SLA	Specific leaf area
SLs	Strigolactones
SM	Spikelet meristem
SNP	Single nucleotide polymorphism
SNR	Signal-to-noise ratio
SnRK1	SUCROSE NON-FERMENTING 1 (SNF1)-RELATED KINASE 1
SOC1	SUPPRESSOR OF OVEREXPRESSION OF CONSTANS 1
SPL	SQUAMOSA PROMOTER-BINDING PROTEIN-LIKE
SPP	Sucrose-phosphate phosphatase
SPS	Sucrose-phosphate synthase
SR	Short-wave radiation
SRR	Shoot to root ratio
SS	Starch synthase
SSP	Seed storage protein
SSS	Soluble starch synthase
STE	Sieve tube element
Stereo-seq	Spatially enhanced resolution omics sequencing
STP	Sugar transport protein
SUC	Sucrose carrier
SUS	Sucrose synthase
SUT	Sucrose transporter
SWEET	Sugars will eventually be exported transporter
T6P	Trehalose 6-phosphate
TAG	Triacylglycerol
TB1	Teosinte Branched 1
TC	Transfer cell
TCA	Tricarboxylic acid
TF	Transcription factor

TMD	Transmembrane domain
TOC	Top of canopy
TOF	Time to flowering; time-of-flight
TOR	Target of rapamycin kinase
TST	Tonoplast sugar transporter
UAV	Unmanned aerial vehicle
UDP	Uridine diphosphate
UGPase	UDP-glucose pyrophosphorylase
UmamiT	Usually multiple acids move in and out transporter
UPS	Ureide permease
UV	Ultraviolet
VPD	Vapor pressure deficit
W	Watt
WT	Wild type
WUE	Water use efficiency

Introduction to Crop Physiology

1

Bridging the Knowledge Gap Between Genotype and Phenotype

Contents

Z. Liu, *Crop Physiology*, https://doi.org/10.1007/978-981-92-2960-4_1

Abstract

Crop physiology conducts a comprehensive and in-depth investigation into the functions of crop plants, specifically the mechanisms underlying yield formation, quality development, and resistance to abiotic stress across multiple levels of biological organization. It serves as a crucial link between the genome and phenome, playing a fundamental role in crop science by facilitating advancements in both breeding and management. This chapter commences by providing an overview of crop domestication, emphasizing the importance of crop production and the fundamental principles of crop science. Next, the concepts of crop physiology are interpreted in detail, highlighting its distinctions from plant physiology. Subsequently, the challenges and opportunities facing crop physiology in the era of Omics are outlined. This chapter concludes with an emphasis on the significance of integrating a physiological perspective into crop improvement strategies.

Keywords

Physiology · Crop physiology · Crop production system · Biological organization · Omics

Did You Know

- The domestication of wild plants into crops, known as the Neolithic Revolution, occurred approximately 10,000–12,000 years ago. It was a crucial turning point in human history that paved the way for urbanization and civilization.
- Structures of interest in plant biology span 11 orders of magnitude in size, ranging from the world's tallest tree (116.07 m, a specimen of *Sequoia sempervirens* in Redwood National Park in California, USA) to the thickness of a biological membrane (5–7 nm).
- The Nobel Prize in Physiology or Medicine honours groundbreaking discoveries that significantly benefit humanity. From 1901 to 2025, it has been awarded 116 times to 232 laureates.
- Tu Youyou was awarded the 2015 Nobel Prize in Physiology or Medicine for her discovery of artemisinin, a novel therapy against malaria. Her recognition underscored the successful integration of traditional herbal medicine with modern pharmacological research.

Learning Objectives

After reading this chapter, students will be able to:

- Appreciate the significance of crop production and perceive the challenges it encounters.
- Understand the key principles of crop science.
- Comprehend the characteristics and significance of crop physiology.
- Summarize the history of crop physiology and identify the lessons learned from it.
- Justify the genetic strategy of enhancing crop performance through the modification of a single gene or cellular process.
- Evaluate the challenges and opportunities facing crop physiology.

Introductory Remarks

Crops are domesticated wild plants that serve as essential sources of staple foods for humans, feed for livestock, and raw materials for various industries. To meet the demands of a burgeoning population and sustainable development, the objectives of the crop industry have been reoriented towards efficiently producing abundant, nutritious, and healthy food while minimizing ecological footprints. To achieve this, it is necessary to implement an interdisciplinary approach that integrates all branches of agricultural science.

The current research focus in crop science is mainly on genomics, which provides deep insights into the molecular and genetic aspects of crop yield, quality, and resilience. In contrast, traditional cognate disciplines like crop physiology and agronomy have been marginalized, as evidenced by the decreasing allocation of human resources and funding. It is imperative to reintegrate crop science to establish a cohesive framework for crop breeding and management.

Crop physiology examines the functions and their underlying mechanisms within crop plants across multiple hierarchical levels, from cell and tissue, through organ and plant, to community and ecosystem. It employs the principles of systems thinking and emphasizes the interactions occurring at each level of biological organization. The emergence of cutting-edge technologies, particularly artificial intelligence (AI), is revolutionizing the identification and characterization of physiological traits (phenotypes), thereby revitalizing traditional whole-plant-level physiological studies in crops. By establishing a robust connection between genotype (gene) and phenotype (trait), crop physiology is playing an indispensable role in facilitating the integration of various subdisciplines within crop science.

This chapter focuses on the core principles of crop physiology, emphasizing its unique features in comparison to plant physiology and its essential contribution to crop science. It argues that, in the era of genomics, crop physiology will become increasingly vital for elucidating the complex regulatory mechanisms that govern crop yield and quality, as well as for understanding how crops respond to dynamic environmental conditions and advancing agronomic technologies.

Concept Map

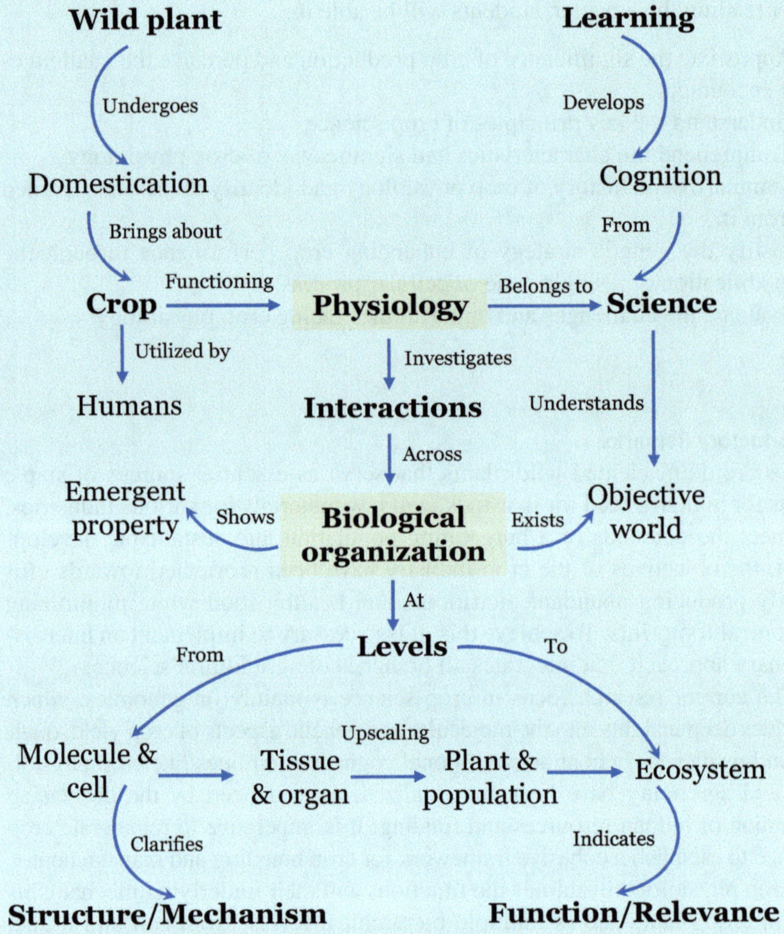

Annotation: Science is the knowledge system that covers general truths or the operations of fundamental laws governing the physical world and its phenomena. Cognition refers to the mental states and processes involved in acquiring knowledge through perception, recognition, conception, and reasoning. The primary objective of college education is to cultivate cognitive abilities through the study of science. Crop physiology, as a branch of agricultural science, focuses on the functional processes of crop plants that have been domesticated from wild species for agricultural use. It primarily investigates interactions across multiple levels of biological organization, ranging from molecules to organs and ultimately to entire ecosystems. At any given level, reduction to lower levels of organization helps clarify structural and mechanistic details, whereas upgrading to higher levels enhances the understanding of function and contextual relevance. The emergent properties of biological systems add complexity to the study of crop physiology, making it particularly challenging.

1 Introduction to Crop Production

Crop production is a fundamental component of agriculture, forming the basis for economic stability, social well-being, and environmental sustainability. Economically, it provides employment opportunities and generates income for farmers and related industries. Furthermore, it plays a vital role in international trade, as agricultural commodities represent a significant share of global exports. Socially, crop production ensures food security and sustenance, particularly in regions facing food shortages, and contributes to poverty alleviation and the enhancement of rural livelihoods. Environmentally, sustainable crop management practices contribute to carbon sequestration, enhance soil health, and support the conservation of biodiversity.

1.1 Crop Domestication

The Neolithic Revolution, also referred to as the Agricultural Revolution, represents a pivotal transformation in human history. This significant transition took place roughly 10,000 to 12,000 years ago across multiple regions, including the Fertile Crescent, China, and the Americas. It was characterized by a fundamental shift from nomadic hunting and gathering practices to settled agricultural communities. As a pivotal moment in history, the advent of agriculture and the establishment of permanent settlements facilitated population growth and the development of intricate social hierarchies, thereby laying the groundwork for human civilization.

The key aspect of the Neolithic Revolution involved adopting a selection of plant species and wild animals as sources of proteins, carbohydrates, oils, and fibre, which later became the cultivated crops and livestock through human *domestication*. Domestication is a complex and lengthy process that occurs along a continuum, involving humans, plants, and animals, and is jointly driven by ecological, biological, and cultural factors (Larson et al. 2014). It emerged independently in multiple regions where local ecosystems harboured the wild ancestors of important domesticated crops and animals, such as wheat, barley, rice, maize, potatoes, sheep, goats, and pigs. Today, ten typical centres of plant domestication are widely acknowledged, among which the Near East, Mesoamerica, South America, China, and India stand out as the earliest and most significant (Table 1.1).

Over the past approximately 12,000 years of agricultural development, humans have domesticated wild plant species, transforming them into cultivated crops that possess more valuable traits than their wild counterparts. There is a well-documented suite of traits, known as the "*domestication syndrome*", that distinguishes most cultivated crops from their wild progenitors (Doebley et al. 2006). Domestication syndrome refers to a consistent set of phenotypic traits that emerge repeatedly as wild plant species are transformed into cultivated crops through human selection, either intentional or unconscious. Core traits include reduced seed shattering, increased size of harvestable organ, compact plant architecture, determinate growth habit, improved nutritional or sensory quality, and enhanced ease of harvesting. The

Table 1.1 Ten major centres of crop domestication

Major centres	Domesticated crops
Central Asia	Apple; Apricot; Carrot; Garlic; Onion; Peach; Pear; Pistachio; Various melon species
China	Adzuki bean; Broomcorn millet; Buckwheat; Foxnut; Foxtail millet; Ginger; Litchi; Rice; Soybean; Taro; Tea; Yam (*D. polystachya*)
Ethiopia	Coffee; Finger millet; Sorghum; Teff
India	Cotton (*G. arboreum*); Cucumber; Cucurbit; Eggplant; Mango; Mung bean; Pigeon pea; Radish; Rice; Sesame; Tree cotton; Urd bean
Meso-America	Avocado; Cacao; Chayote; Chili pepper; Common bean; Cotton (*G. hirsutum*); Maize; Tobacco; Tomato; Vanilla
Northeast America	Squash (*C. pepo* ssp. *ovifera var. ovifera*); Sunflower
Papua New Guinea	Banana; Sugarcane; Taro; Yam (*D. alata*)
South America	Cacao; Cassava; Chilis; Common bean; Cotton (*G. barbadense*); Lima bean; Peanut; Potato; Quinoa; Rubber plant; Squash (*C. maxima*); Squash (*C. moschata*); Strawberries; Sweet potato; Yam (*D. trifida*)
Sub-Saharan Africa	African rice; Coffee; Cotton (*G. herbaceum*); Cowpea; Oil palm; Okra; Pearl millet; Sorghum; Yam (*D. dumetorum*)
The Near East	Barley; Chickpea; Common wheat (*T. aestivum*); Date-palm; Einkorn wheat; Emmer wheat; Field pea (*P. sativum*); Fig; Flax; Grape; Lentil; Lettuce; Olive; Watermelon

Note: The domesticated crops from these centres are listed based on Abbo et al. (2017), with some modifications. The origins of cotton are listed with reference to Snider et al. (2021). Sugarcane is classified as originating from Papua New Guinea (Singels et al. 2021). The origin of the watermelon is widely believed to be in the region of the Near East, particularly associated with ancient Egypt. In addition, Europe is considered to be the origin centre of canola (*Brassica napus*) and sugar beet (*Beta vulgaris*) (Sadras and Calderini 2021); however, it is not included among the ten major origin centres

partitioning of assimilates is perhaps the main difference between crop species and their wild relatives. For instance, cultivated potato plants allocate approximately 80% of the dry matter to tubers, which is ten times more than that in their wild relatives (Halford 2010).

Many of the domesticated traits are governed by single genes or a limited number of genetic loci. Notable examples include *TB1* (*TEOSINTE BRANCHED 1*), which inhibits lateral shoot development in maize; *sh4* (*shattering 4*), which reduces seed shattering in rice; and *fw2.2* (*fruit weight 2.2*), which governs fruit size and weight in tomato. These single genes often exert major phenotypic effects. For instance, *tb1*, *sh4*, and *fw2.2* account for more than 20% to 30% of the phenotypic variation associated with key morphological changes in their respective crops. These differences are substantial enough to enable direct human selection during the early stages of domestication. In addition, homologous genes are often independently targeted during the domestication of distantly related crops, resulting in convergent evolution of key traits, such as analogous shattering-resistant or branching genes across cereal crops. The identification of such critical domestication genes not only deepens our understanding of crop evolution but also enables strategies such as de novo domestication of wild species or the precise introgression of favourable alleles into existing crops, thus speeding up the development of contemporary breeding initiatives.

1.2 Classification of Crops

Crops are the domesticated wild plants that can be extensively cultivated and harvested for profit or subsistence. Globally, there are approximately 382,000 species of vascular plants, with slightly over 6000 being domesticated as crops for human consumption (FAO 2019). *Field crops* are grown extensively on a large scale and directly or indirectly provide a human benefit, such as food, feed, fibre, oil, biofuels, or other materials. As of 2014, fewer than 200 crop species had significant global production levels, with only nine crops (wheat, rice, maize, barley, sorghum, potatoes, sweet potatoes, sugarcane, and soybeans) covering more than 75% of human consumption (FAO 2019).

Field crops can be classified into five typical categories based on their botanical and agronomic attributes. *Cereals* are primarily annual grasses cultivated predominantly for their grain, with wheat, rice, and maize collectively accounting for over 60% of human dietary intake. *Legumes*, belonging to the family Fabaceae, are characterized by their pod-shaped fruits, and include examples such as peanuts, soybeans, field beans, field pea, and cowpea. They can convert atmospheric nitrogen into organic amino acids, thereby enhancing soil fertility. *Root and tuber crops*, such as sweet potato, cassava, and potato, exhibit a high level of diversity in terms of their phylogeny. They are primarily characterized by the fact that the part harvested is typically located underground, although it may not always be the actual root. *Industrial crops*, such as sugarcane, sugar beet, and cotton, are cultivated to produce oil, fibre, sugar, rubber, medicinal compounds, and other raw materials used across various industries. *Forage crops*, including alfalfa, turnip, and milk vetch, are specifically grown to serve as primary sources of animal feed. Among the five types of crops, cereals and legumes are widely cultivated worldwide, serving as vital sources of carbohydrates (cereals) and proteins (legumes) for both feed and food purposes. Moreover, cereals are also a significant source of proteins in many countries, and legumes have been acknowledged as an important source of micronutrients.

2 Fundamentals of Crop Science

Crop science is an interdisciplinary field dedicated to understanding and improving cultivated plants throughout their entire life cycle. It integrates core knowledge from plant biology, genetics, physiology, ecology, and informatics to enhance crop productivity. When conducting research in crop science, it is essential to consider the following key shared principles and concepts.

2.1 The Essence of Crop Production Is the Chemical Transformation of Solar Energy

Crop production can be regarded as a complex system consisting of natural and economic reproduction components (Fig. 1.1). The inputs into the system include natural resources such as solar energy, temperature, precipitation, and soil, as well as economic resources like seeds, *agrichemicals* (e.g. fertilizers, herbicides, plant growth regulators), irrigation systems, and machinery. Its outputs include a variety of products that serve as food, feed, and industrial materials. At the heart of this system lies its core element: the crop population or stand.

This artificial system is designed to capture solar energy through the crop canopy, which subsequently converts it via photosynthesis into chemical energy stored as starch, sucrose, proteins, fibre, and lipids. Among all chemical inputs, nitrogen stands out as the most critical fertilizer because it is an indispensable component of proteins that carry out essential functions in plant physiology. Specifically, the leaf enzyme Rubisco plays a pivotal role in assimilating atmospheric CO_2. From a physicochemical perspective, crop production can be defined as the process in which nitrogen fertilizer is applied to the soil to promote CO_2 assimilation via light-driven

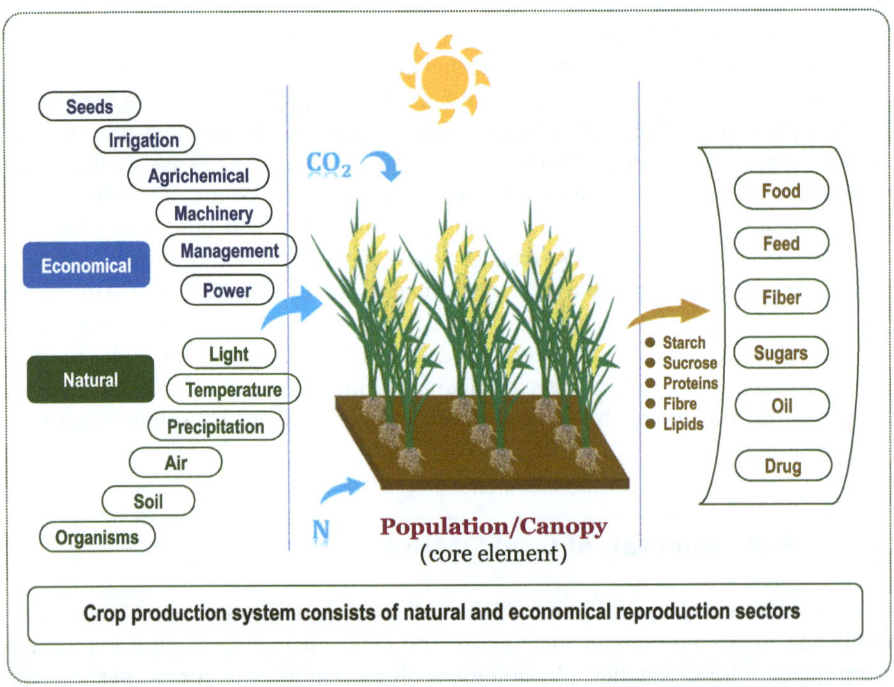

Fig. 1.1 The input, output, and core elements of the crop production system. Crop cultivation is a complex biological process through which solar energy is converted into chemical energy stored in the harvested plant organs in the form of starch, sucrose, proteins, fibre, and lipids

photosynthesis in leaves. Subsequently, the resulting photosynthates (sugars) are converted into diverse crop products for human consumption.

Crop yield formation depends on three key processes: (1) the capacity of the canopy to capture CO_2 and solar energy, and the ability of roots to acquire water, nitrogen, and other mineral nutrients from the soil; (2) the efficiency of converting the captured resources into biomass; and (3) the allocation of biomass to harvestable organs, such as seeds, grains, tubers, and roots. All these processes are critically dependent on the interception of solar energy within the specific wavelength range of 400–700 nm, commonly known as *photosynthetically active radiation* (*PAR*). In the crop stand, the transformation of solar radiation into chemical energy in the form of carbohydrates occurs through three sequential processes (Hay and Porter 2006): (1) interception of incident solar radiation by the canopy, (2) conversion of the harvested radiation to chemical energy by photosynthesis, and (3) allocation of the photosynthates from source to sink organs.

To simplify, the yield potential of a given genotype is determined by the product of the solar radiation received over the growing season per unit area of land (Q) and the efficiencies with which the crop intercepts that radiation (I), converts it into biomass energy (ε), and then allocates the biomass to the harvestable organs (H). The yield (Y) of a crop, per unit area, over a given period can, therefore, be expressed by the equation:

$$\mathrm{Y} = Q \times I \times \varepsilon \times H$$

where Q is the total quantity of incident solar radiation received over the period; I is the fraction of Q intercepted by the canopy; ε represents the conversion efficiency of radiative energy into chemical energy, which is commonly defined as the quantity of dry matter produced per unit of intercepted radiation; and H is the harvest index. Ultimately, crop yield can be influenced by multiple factors during this energy-transformation process, such as photosynthetic pathways (C3 or C4), source–sink relationships, and canopy structure. In addition, environmental conditions (e.g. light intensity, temperature, and rainfall) and agronomic practices (e.g. the amount of irrigation and fertilization) strongly influence crop yield by modifying this process.

This equation can also be expressed in an alternative form as follows (Araus et al. 2021):

$$\mathrm{Y} = RU \times \mathrm{RUE} \times HI$$

where RU (radiation uptake) represents the cumulative radiation intercepted by the canopy over the entire crop life cycle; RUE (radiation use efficiency) denotes the efficiency with which the intercepted radiation is converted into biomass through photosynthesis; and HI (harvest index) indicates the proportion of total biomass allocated to the harvestable portion. Compared with the preceding equation, RU is the product of Q and I; RUE is equivalent to ε, and HI is identical to H. For more information about RUE, please refer to Sect. 3.1.3 in Chap. 6.

2.2 The Crop Production System Possesses Emergent Property

According to *systems theory*, a *system* is a construct composed of elements that interact with one another (Box 1.1). *Interactions* among system components give rise to *organized complexity*, a property absent in individual elements when considered in isolation. This characteristic is also referred to as the *emergent property*, indicating a phenomenon that arises from the interactions and collective behaviour of individual components within a complex system. Emergent properties are characteristics that do not manifest in the individual components of a system but become apparent only when these components interact as a whole. Therefore, they can also be considered unexpected phenomena arising from such interactions. For instance, within the field of neuroscience, the concept of emergent properties can be observed in operation when a collective system, such as the brain composed of numerous neurons, exhibits the capacity to think. This capacity for thought isn't found in any single neuron, but rather emerges from the complex interactions within the *neural network*. In short, emergent properties highlight the complexity of systems and the emergent behaviours that arise from interactions, emphasizing the importance of holistic and interdisciplinary approaches to understanding the world. As such, systems thinkers generally hold to the adage that *the whole is greater than the sum of its parts*.

Conventionally, the theory of *reductionism* believes that a complicated system can be explained by considering it as a combination of simple parts. By contrast, the *holism* view of systems theory shifts from isolated parts to the whole organization of parts. It is obvious that no matter how exact the study of single elements may be *in vitro*, it does not permit an understanding of the system characteristics. To achieve a comprehensive understanding of the yield formation of the cropping system, it is imperative to move beyond mere reductionism at the molecular level. Only through complementary research on the characteristics of the cropping system *in vivo*, where organized complexity is preserved, can we ensure that reductions to the molecular level remain physiologically relevant.

Emergence, which refers to the phenomenon in which complex systems display properties that cannot be predicted from their individual components, emphasizes the indispensable value of research at every biological scale. Although molecular studies lie at the heart of modern plant biology, they are unable to fully explain the emergent behaviours observed at higher levels, such as crop performance in field populations. Therefore, physiological- and systems-level research is not only complementary but also essential for translating mechanistic insights at the molecular level into tangible crop performance under field conditions.

Box 1.1: The Systems Theory

Systems theory states that the universe consists of open, closed, and isolated systems at various scales. Central to its interdisciplinary study is the principle that systems exhibit certain structural and behavioural similarities and commonalities. Systems theory integrates principles derived from the defining characteristics of systems irrespective of their field of origin (e.g., neurology or ecology) or scale of analysis (e.g., microscopic or macroscopic). Consequently, it enables insights gained from one system to be applied to the understanding, management, or construction of another. For example, a more comprehensive understanding of neural network mechanisms in the brain has provided valuable insights into the design of machines that mimic human cognitive functions, such as learning and problem-solving, which is referred to as artificial intelligence (AI).

Systems theory has its historical roots in the pursuit of a general framework applicable to all systems. This intellectual movement originated in the 1950s, coinciding with the development of other related theories such as information theory, game theory, cybernetics, chaos theory, autopoiesis theory, complexity theory, and dynamic systems theory. These theories collectively explore the nature and characteristics of systems across various domains. Systems theory has exerted a profound influence on a wide array of disciplines, including engineering, biology, ecology, geography, sociology, psychiatry, and neurology. This influence has reshaped the scholarly understanding of how human societies operate, how living organisms function, and how the natural world evolves and interacts.

2.3 Understanding the Structure–Function Relationship Is Fundamental to Deciphering the Mechanisms Underlying Crop Productivity

Biological organization, also called the *hierarchy* of life or levels of organization, is the hierarchy of complex biological structures and systems that define life using a reductionistic approach. The traditional hierarchy spans from atoms to biospheres. In crop science, relevant *levels* of organization include molecule, organelle, cell, tissue, organ, plant, population (crop), community, and ecosystem. The hierarchy is structured in a way that each level represents a progressive elevation in organizational complexity. The fundamental principle underlying the organization is the concept of emergence, which refers to the phenomenon where the properties and functions observed at a hierarchical level are absent and irrelevant at the lower levels.

The levels consist of families of *entities*, typically of comparable size and dynamic properties, which interact primarily with each other in a characteristic manner. For example, at the molecular level, DNA molecules primarily interact with protein molecules such as *transcription factors* and RNA polymerase II (RNAPII), thereby establishing the fundamental machinery for transcription initiation. At the plant level, source organs such as leaves, bracts, and glumes interact with the sink organ, the grain, to coordinate the allocation of photosynthates. In addition, the rates of key processes decrease as they progress from lower to higher levels, partly because it takes more time for causal effects to spread over longer distances (Wimsatt 1994).

When investigating the *mechanism* of a biological system, insights are often revealed at levels beneath the observable occurrence. By dissecting the system into its constituent parts, we uncover its *structure* and examine the chemical and physical *processes* driven by the interactions between these parts, thereby gaining deeper insights into how the system operates. On the other hand, if we want to interpret the *function* or significance of that system, it is usually clarified at higher levels by viewing the system as a whole and examining its interactions with the environment in terms of exchanges of *matter*, *energy*, and *information* (or *signals*). For instance, we pose inquiries at the canopy or population level: how does the canopy assimilate CO_2? Therefore, we should identify the underlying mechanisms at the lower level of the leaf by investigating its anatomy and photosynthetic processes. In addition, we can seek the significance of canopy photosynthesis at higher levels to elucidate its effect on the carbon footprint in the ecosystem. In this regard, the crop level (population) acts as a reference for inferring significance at lower organizational levels (ranging from plant to molecule) and serves as a basis for clarifying mechanisms at higher organizational levels (such as the community and beyond).

2.4 Crop Science Investigates Various Processes Across Levels of Biological Organization

As a comprehensive and applied discipline, crop science integrates a diverse array of knowledge spanning from molecular biology to ecosystem biology, which is crucial for dissecting the intricate processes underlying crop yield formation, quality development, and resilience.

2.4.1 At the Molecular Level

Living organisms are composed of four primary compounds, namely carbohydrates, lipids, proteins, and nucleic acids. *Carbohydrates* function as energy molecules and structural components. *Lipids* serve as vital energy reserves in plant cells and play a crucial role in forming the waxy, protective cuticles that coat the surfaces of leaves, stems, and roots. The primary role of *proteins* is to store materials (e.g. nitrogen), build structures, perform movements, and transport substances. However, their most critical function is to serve as catalytic *enzymes* for essential biochemical reactions.

The development of plants and their ability to respond to changes in the environment depend largely on the readout of the genetic information. The instructions for all the structures and functions of a plant are encoded in and translated by nucleic acids. There are two types of nucleic acids: deoxyribonucleic acid (*DNA*) and ribonucleic acid (*RNA*). DNA is the largest molecule in the cell and contains genetic information organized into units called *genes*. RNA decodes the information stored in DNA and translates it into proteins. A gene is a basic unit of heredity and a sequence of nucleotides in DNA or RNA that encodes the synthesis of a gene product, either RNA or protein. In most cases, genes consist of DNA, which serves as the primary genetic material due to its ability to store genetic information stably and long term. However, in RNA viruses, genes are RNA segments that either directly encode viral proteins or perform regulatory functions.

Genes are made up of promoter regions and alternating regions of introns (non-coding sequences) and exons (coding sequences; Box 1.2). *Gene expression* is the process by which information from a gene is transformed into a functional gene product, such as proteins, transfer RNA (tRNA), or small nuclear RNA (snRNA). The production of a functional protein encompasses the *transcription* of the gene from DNA into RNA, the excision of introns and ligation of exons, the *translation* of the spliced RNA into an amino acid chain, and the post-translational modification of the protein molecule (Fig. 1.2). This unidirectional flow of genetic information, proceeding from DNA to RNA and then to protein, is known as the Central Dogma, which serves as the cornerstone of modern biology.

Box 1.2: Structure and Organization of a Eukaryotic Gene

The majority of genes, up to 90% in any given cell, remain inactive until the cellular environment requires the synthesis of specific enzymes encoded by those genes. Each gene has a structural organization that enables recognition by regulatory molecules, facilitates binding interactions, and ensures precise activation at the appropriate time. The structure of a eukaryotic gene comprises several essential components.

1. Exons: These are the sequences that encode proteins or functional RNA molecules. Following splicing, exons are joined together to form a continuous coding sequence in mature mRNA. In many cases, exonic regions account for less than half of a gene's overall structural composition.
2. Introns: These regions are interspersed between exons and do not code for proteins due to being excised by splicing during post-transcriptional processing. They are often involved in regulatory functions, such as enhancing gene expression or serving as sites for non-coding RNA production.
3. Regulatory elements: These include promoters, which control transcription initiation; enhancers, which enhance transcription; and other regulatory sequences. The first approximately 1000 bp in the 5′ flanking region is called the gene promoter, as it contains essential sequence motifs for initiating transcription. For example, the TATA box is a short sequence of about six to eight base pairs rich in A and T. It helps recruit RNA polymerase II (RNAPII) to start transcription. The CAAT box is a conserved promoter element in eukaryotic genomes, typically characterized by the consensus sequence CCAAT and located approximately 70–80 base pairs upstream of the transcription start site. This cis-regulatory element functions by binding transcription factors, facilitating the recruitment of RNAPII, and enhancing transcriptional efficiency. Once RNAPII binds to the promoter, it moves downstream toward the 3′ end of the DNA strand and into the transcribed region.
4. Untranslated regions (UTRs): The 5′ and 3′ UTRs of mRNA play essential roles in post-transcriptional gene regulation, influencing processes such as mRNA stability, translation efficiency, and subcellular localization. In the 3′ UTR, there exists the poly(A) signal, such as the conserved hexamer AAUAAA. This signal directs the cleavage of the pre-mRNA, followed by the addition of a poly(A) tail to the 3′ end of the mRNA to further stabilize the transcript and assist in its nuclear export to the cytoplasm (Fig. 1.3).

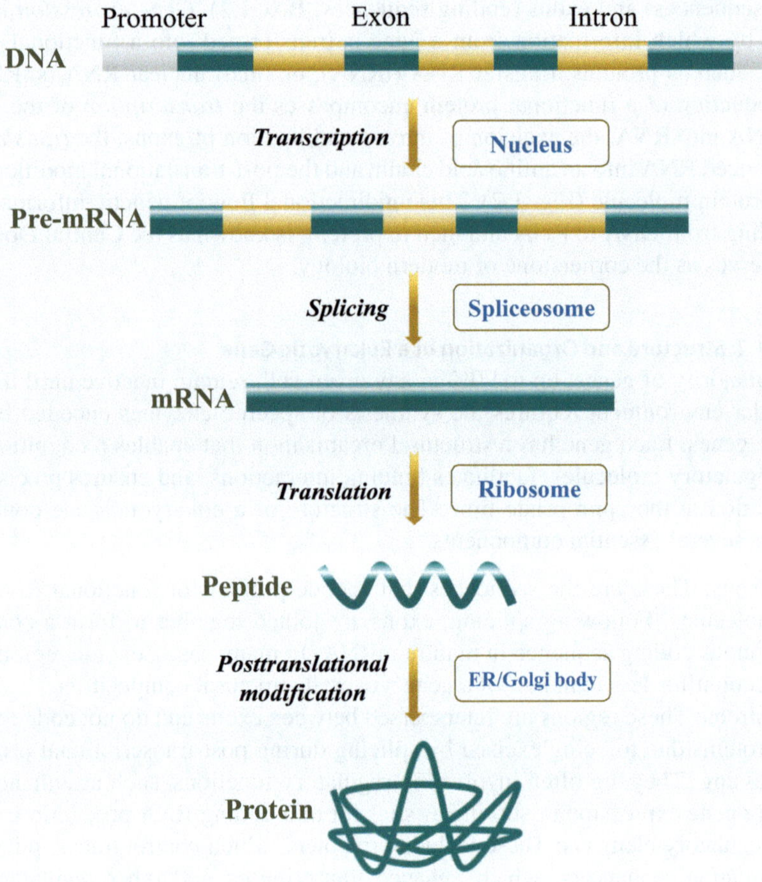

Fig. 1.2 The central dogma of molecular biology. Transcription refers to the synthesis of a complementary RNA molecule (mRNA) using a DNA template, a reaction catalysed by the enzyme RNA polymerase. Exons are regions of DNA that are transcribed into RNA and retained in the mature mRNA after splicing, encoding functional proteins or regulatory elements. Introns are non-coding sequences within genes that are transcribed into RNA but not translated into protein. These sequences are removed during the processing of precursor mRNA (pre-mRNA) through a process known as splicing, performed by the spliceosome in the nucleus. Translation is the process by which the nucleotide sequence of mRNA is decoded to synthesize a specific polypeptide chain. It involves key cellular components such as ribosomes, transfer RNA (tRNA), and ribosomal RNA (rRNA), which collectively interpret mRNA codons and link amino acids in the correct sequence. Together, transcription and translation constitute the core mechanisms of the Central Dogma of molecular biology, which describes the unidirectional flow of genetic information from DNA to RNA to protein. First articulated by Francis Crick in 1958, the Central Dogma asserts that once genetic information has been converted into protein, it cannot be reverse-translated or reverse-transcribed back into nucleic acid sequences under typical cellular conditions. While exceptions do exist, such as reverse transcription mediated by retroviruses and RNA-dependent RNA polymerase activity, the core framework of DNA → RNA → protein remains a foundational concept in molecular biology. ER, endoplasmic reticulum. (Redrawn from Encyclopaedia Britannica. https://www.britannica.com)

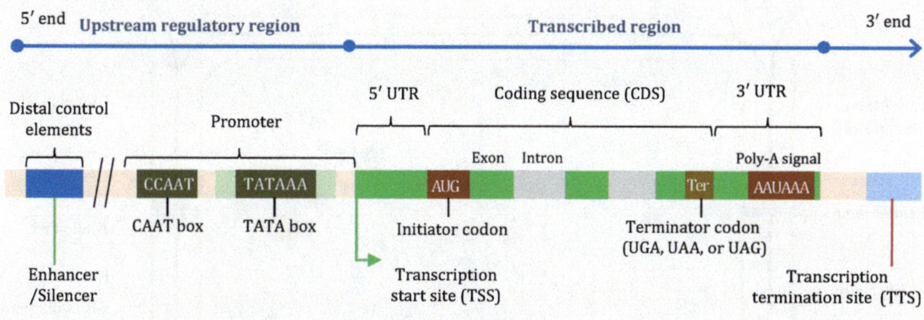

Fig. 1.3 The structural organization of a eukaryotic gene. (The diagram is not drawn to scale)

2.4.2 At the Cellular Level

Every living organism is composed of microscopic units referred to as cells. A plant cell consists of a box-like *cell wall* that surrounds a mass of *protoplasm* (the protoplasm of a single cell is called its *protoplast*). The protoplasm is composed of the *plasmalemma*, the *nucleus*, and the *cytoplasm*. Most of the volume of cytoplasm is a clear substance called *cytosol*. It primarily consists of water, enzymes, and numerous metabolites and products of enzymatic reactions. Inside the cytoplasm exist a wide variety of *organelles*, such as the chloroplasts, ribosomes, and mitochondria (Fig. 1.4).

1. *The Cell Wall*

Cell walls are primarily composed of polysaccharides, proteins, and aromatic compounds. The cell wall contains signalling molecules that facilitate both intercellular communication and wall-nucleus interactions. When pathogens launch an attack, the polysaccharides of the cell wall are degraded into fragments that serve as alarm signals, which in turn prompt the plant to secrete defence molecules. In response, the cell wall is fortified through the deposition of defence-related proteins and lignin, thus forming a more robust barrier against subsequent fungal and bacterial invasions. Additionally, cell wall surface molecules enable plant cells to recognize compatible partners during pollen–style interactions. Approximately 10% of the plant genome is dedicated to cell wall biogenesis, which involves substrate synthesis, polymer assembly, and reorganization during growth. In agriculture, cell walls contribute significantly to the texture of fresh fruits and vegetables and serve as essential dietary fibres in human nutrition. They also provide a major source of calories for ruminant animals. Moreover, cell walls represent a key source of lignocellulosic biomass, which is expected to become an important substrate for biofuel production.

2. *The Plasmalemma*

The cytoplasm and its organelles are enclosed by membranes that consist of a phospholipid bilayer and associated proteins, typically in a ratio of 40% lipid and 60% protein. Most proteins possess extensive hydrophilic domains, enabling their

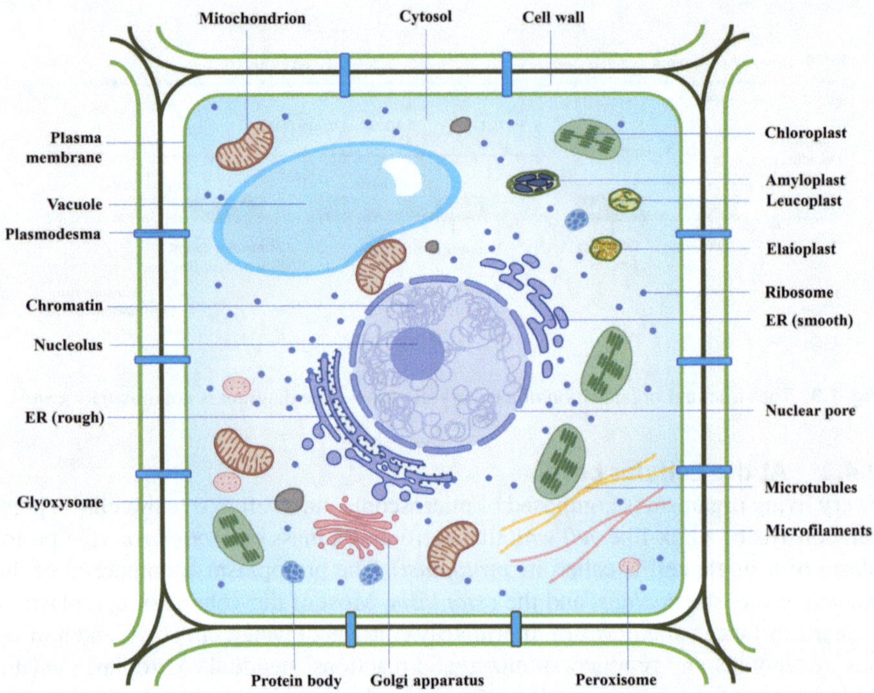

Fig. 1.4 A generalized plant cell. Plant cells demonstrate considerable diversity in the quantity and types of organelles they possess. This diagram provides an idealized representation of a typical plant parenchyma cell, illustrating all organelles in an organized manner for clarity. The three essential components are: the cell wall, the plasma membrane, and the cytoplasm, accommodating a wide variety of organelles. ER, endoplasmic reticulum

association primarily with phospholipid head groups and water. Additionally, many proteins contain substantial hydrophobic regions that facilitate their integration into the lipid bilayer and interaction with fatty acid chains. Membranes fulfil several critical functions in cellular metabolism: regulating the transport of molecules into and out of cells and organelles; compartmentalizing the cell into distinct regions, each with specialized metabolic activities; and serving as scaffolding platforms for enzyme attachment.

The plasma membrane is a selectively permeable barrier that completely envelops the cytoplasm. As the outermost layer of the cell, it must prevent the entry of harmful substances while allowing the passage of beneficial ones. Molecular pumps within the plasma membrane actively transport essential molecules into the cell and expel others for secretion. Given that one side of the plasma membrane interfaces with the external environment and the other side contacts the cytoplasm, the two surfaces exhibit significant differences, particularly in their protein components.

3. *The Nucleus*

The *nucleus* is a membrane-bound organelle that functions as the central regulatory hub. It is enclosed by a double-layered nuclear envelope, consisting of inner and outer membranes that are continuous with the endoplasmic reticulum. Embedded within this envelope are nuclear pore complexes, which facilitate the selective transport of macromolecules between the nucleus and the cytoplasm. Within the nucleus, the nucleoplasm houses chromatin fibres that consist of DNA wrapped around histone proteins. Chromatin fibres can be differentiated into *euchromatin* (transcriptionally active regions) and *heterochromatin* (highly condensed and largely inactive regions). The nucleolus, a prominent subnuclear structure, forms around nucleolar organizer regions and serves as the primary site for ribosomal RNA (rRNA) transcription and ribosomal subunit assembly. Additional structural components include the nuclear matrix, which provides a structural scaffold for chromatin organization, and various nuclear bodies involved in RNA processing and regulation.

The nucleus contains the genetic material in the form of DNA organized into chromosomes and governs gene expression, thereby coordinating cellular growth, development, and responses to environmental stimuli. The nucleus performs several critical functions in plant cells. First, it serves as the site of genetic information storage: nuclear DNA contains all the hereditary instructions required for the plant's development and life cycle. Second, it regulates transcription by recruiting transcription factors, RNA polymerases, and co-activators to gene promoters, thereby precisely controlling when and which proteins are expressed. Third, the nucleus is responsible for RNA processing, during which pre-mRNA transcripts undergo 5′ capping, splicing, and 3′ polyadenylation before being exported to the cytoplasm. The spliceosome, which is the core molecular machine responsible for splicing out introns from pre-mRNA and ligating exons, is mainly located in specific subnuclear compartments within the nucleus. Fourth, during interphase, the nucleus coordinates DNA replication, ensuring accurate duplication of the genome so that each daughter cell receives a complete and faithful copy of the chromosomes. Fifth, it governs cell cycle progression through nuclear checkpoints that monitor DNA integrity and prevent the propagation of mutations. Finally, the nucleus integrates internal and external signals via nuclear receptors and transcription factors, converting these cues into appropriate gene expression responses.

4. *The Organelles*

Plant cells possess a diverse array of membrane-bound organelles with specific functions. *Vacuoles* play several metabolic roles in addition to promoting cell expansion, such as storage, digestion, pH and ionic homeostasis, defence against microbial pathogens and herbivores, sequestration of toxic compounds, and pigmentation. *Mitochondria* are organelles that house enzymes and electron transfer chain components that produce ATP by ways of the *Krebs cycle*. *Plastids*, including *chloroplasts*, *leucoplasts*, *elaioplasts*, and *amyloplasts*, are major organelles found only in plant cells, carry out photosynthesis, synthesize and store starch and lipids, and provide substrates to other biosynthetic pathways (refer to Sect. 1.2 in Chap. 2).

The *endoplasmic reticulum* (ER) is responsible for synthesizing, processing, and sorting proteins targeted to membranes, vacuoles, or the secretory pathway, as well as synthesizing a diverse array of lipids. It is broadly classified into two distinct types based on the presence or absence of ribosomes on its membrane surface: the rough endoplasmic reticulum (RER), which features ribosomes attached to its membrane, giving it a granular appearance and primarily functioning in protein synthesis and post-translational modification; and the smooth endoplasmic reticulum (SER), which lacks membrane-bound ribosomes and exhibits a smooth morphology, mainly participating in lipid biosynthesis, carbohydrate metabolism, and intracellular calcium ion storage.

The *Golgi apparatus* is responsible for transporting, modifying, and packaging proteins and lipids into vesicles for delivery to targeted destinations. *Microbodies* have two classes: *peroxisomes* and *glyoxysomes*. Peroxisomes are involved in detoxifying certain by-products of photorespiration and are found closely associated with chloroplasts. Glyoxysomes are involved in converting stored fats into sugars. They are important during the germination of fat-rich, oily seeds such as peanuts, sunflowers, and coconuts.

Plant cells also possess non-membranous organelles. *Ribosomes* are responsible for protein synthesis. They are complex aggregates composed of three molecules of RNA (ribosomal RNA) and approximately 50 types of protein, which associate to form two subunits. *Microtubules* and *microfilaments* are constructed through the assembly of globular proteins, providing a skeletal structure and facilitating organelle movement. Microtubules are primarily composed of heterodimeric tubulin subunits, while microfilaments are mainly constituted of actin subunits. The extensive network of microtubules and microfilaments within the cytosol is referred to as the *cytoskeleton*. Its primary functions include maintaining cell shape, organizing the cytoplasmic architecture, facilitating the intracellular transport of structures such as vesicles and chromosomes, and enabling cell motility.

2.4.3 At the Tissue and Organ Level

Plant *organs* are composed of *tissues* in which cells are interlinked by narrow cytoplasmic channels known as plasmodesmata. These microscopic structures traverse the cell walls of adjacent cells and function as critical routes for the exchange of metabolites, proteins, and signalling agents. By establishing a continuous cytoplasmic network, plasmodesmata create *symplastic continuity* to facilitate intercellular communication.

There are over a dozen types of vegetative plant tissues alone, along with tissues specialized for reproduction. Tissues can be organized into tissue systems, which are composed of multiple cell types that work synergistically to perform specific functions. The plant is composed of three fundamental tissue systems: the dermal tissue, the ground tissue, and the vascular tissue (Fig. 1.5).

The *dermal tissues* cover the surfaces of plants, comprising the epidermis and the periderm. The *epidermis*, consisting of multiple cell types such as pavement cells, guard cells, and trichomes, is the primary outer protective covering of the plant

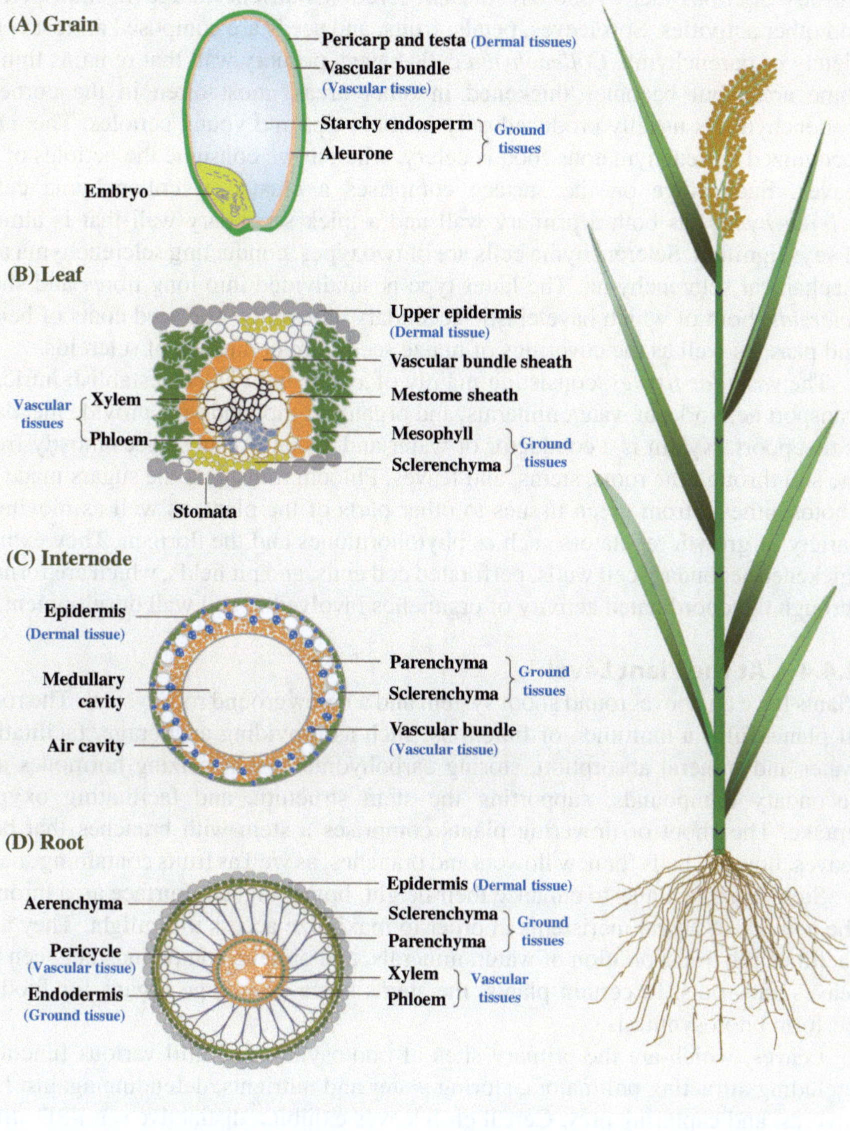

Fig. 1.5 Diagram representation of tissues of the crop plant (rice as an example). This diagram illustrates the three fundamental tissue systems, namely dermal, ground, and vascular, and their diverse structural forms across seeds, leaves, internodes, and roots

body. The *periderm*, which is the protective tissue replacing the epidermis, is primarily found in plants that experience a secondary increase in thickness.

The *ground tissues* form the bodies of plants, consisting of three simple tissues: parenchyma, collenchyma, and sclerenchyma. *Parenchyma* is typically composed of living cells with thin walls and a simple structure, enabling them to adapt to various functions such as photosynthesis, secretion, nutrient storage (in endosperm), and other activities. Soft leaves, petals, fruits, and seeds are composed almost completely of parenchyma. *Collenchyma* cells have a primary wall that remains thin in some areas but becomes thickened in other areas, most often in the corners. Collenchyma is usually produced only in shoot tips and young petioles. The most recognized collenchymatous food is celery, wherein we consume the petioles of its leaves. Each ridge on the surface comprises a cluster of collenchyma cells. *Sclerenchyma* has both a primary wall and a thick secondary wall that is almost always lignified. Sclerenchyma cells are of two types: conducting sclerenchyma and mechanical sclerenchyma. The latter type is subdivided into long fibres and short *sclereids*, both of which have elastic secondary cell walls. The seed coats of beans and peas, as well as the coverings of maize seeds, are composed of sclereids.

The *vascular tissues*, consisting mainly of *xylem* and *phloem*, establish intricate transport networks of water, minerals, and organic compounds and provide mechanical support. Xylem is a conductor of water and dissolved substances mostly from the soil through the roots, stems, and leaves. Phloem transports the sugars made by photosynthesis from green tissues to other parts of the plant, as well as moving a variety of growth regulators such as phytohormones and the florigen. They exhibit thickened secondary cell walls, perforated cell ends, and pit fields, which are formed through the coordinated activity of organelles involved in cell wall development.

2.4.4 At the Plant Level

Plants have an aboveground shoot system and a belowground root system. The roots of plants fulfil a multitude of functions, such as providing anchorage, facilitating water and mineral absorption, storing carbohydrates, synthesizing hormones and secondary compounds, supporting the stem structure, and facilitating oxygen uptake. The shoot of flowering plants comprises a stem with branches that bear leaves, flowers, buds for new flowers and branches, as well as fruits containing seeds.

Stems enable plants to enhance their height, branching, and surface area through the activity of apical meristems in order to maximize access to sunlight. They also facilitate the transportation of water, minerals, and organic compounds between the leaves and roots. In certain plants, the stems serve as storage organs for food or perform photosynthesis.

Leaves, which are the primary sites of photosynthesis, fulfil various functions including attracting pollinators, storing water and nutrients, defending against herbivores, and capturing prey. Cereal crop leaves exhibit a distinctive semi-3D structure (McAllister et al. 2025): the basal sheath envelops younger leaves and the stem, providing mechanical support and accounting for the majority of shoot elongation during vegetative growth; the leaf blade extends laterally to maximize light harvest; the ligule, located at the junction between the sheath and blade, functions as a protective barrier against water ingress and pathogen invasion; and a pair of auricles modulates blade orientation to optimize light capture.

The structures of flowers, buds, fruits, and seeds are more intricate as they are composed of multiple basic organs. Flowers are also sophisticated shoots that bear leaf-derived organs known as sepals, petals, stamens, and carpels. Seeds are complex structures because they contain dormant embryos that possess primordia of shoots and roots, enclosed within a protective seed coat derived from the maternal plant. It should be mentioned that several foliage leaves are present in mature embryos of cereals, including rice, maize, and wheat. This fact contrasts with the situation in other monocot and eudicot embryos, where foliage leaves are not formed before dormancy and germination. This phenotype indicates that in grasses, early vegetative stages are incorporated into embryos before dormancy occurs.

2.4.5 At the Population/Crop Level

In crop production, a *population* is comprised of individual plants belonging to the same genotype or cultivar. While interactions among individuals within plant populations are often weak, the structure and function of the population transcend the mere sum of its constituents. In species characterized by dioecy or self-incompatibility, sexual reproduction is successful at the population level, yet fundamentally impossible for a single individual. When pollination relies on animal vectors, an isolated plant typically lacks the ability to produce sufficient nectar to sustain even a single pollinator over an extended period. In contrast, a plant population can collectively support a pollinator community. Nevertheless, dense populations increase the risk of disease transmission: spores from pathogenic fungi have a high probability of encountering and infecting at least one susceptible host, thereby promoting rapid spread among neighbouring individuals.

Plant stands in contemporary crop production differ from those in natural ecosystems. In dense, mixed vegetation typical of natural habitats, plants must compete for essential resources such as light, water, and mineral nutrients to enhance their survival and reproductive success. Traits such as a loose structure and greater height may provide competitive advantages over adjacent plants, thus enhancing individual fitness. In contrast, modern agricultural systems aim to maximize productivity at the population level, favouring cooperative interactions that enhance overall yield. Throughout the history of domestication, humans may have unintentionally favoured plants that were stronger competitors rather than more effective cooperators by selecting those with greater potential for propagation and growth. This preference for competitive traits has resulted in an overemphasis on characteristics, such as plant height and branching, that enhance individual light interception but ultimately reduce RUE at the population level (Slattery and Ort 2021). The Green Revolution of the 1960s addressed this issue by promoting the cultivation of more cooperative cultivars through the development of semidwarf plant types.

2.4.6 At the Community and Ecosystem Level

A population does not exist in isolation; rather, it coexists with diverse populations of other plant species, as well as animals, fungi, bacteria, and viruses. From an ecological perspective, the plants inhabiting the fields form a *community* of interdependent organisms. The genetic content of a community is determined by the species and cultivar, while its physical structure is defined by multiple factors such as planting density and spacing pattern. Crop communities can be described in simple terms. Monoculture refers to the exclusive cultivation of a single crop species in a field at any given time, whereas polyculture or *intercropping* involves the coexistence of two or more crop species. Crops selected for intercropping typically exhibit complementary resource use abilities, which can result in yield advantages and greater system stability compared to monocultures in low-input agricultural systems. The adoption of high-yielding cultivars, increased mechanization, and intensified use of fertilizers and pesticides have largely displaced intercropping from most modern farming systems. However, maize and cowpea intercropping remains common in certain regions of Africa and Asia (Asseng et al. 2014).

The community, together with its chemical and physical environments, forms a fundamental ecological unit known as an *ecosystem*. Crop production systems depend on *ecosystem services* provided by natural ecosystems, such as pollination, biological pest control, maintenance of soil structure and fertility, nutrient cycling, and hydrological regulation. Preliminary studies suggest that the value of these services to crop productivity is substantial and frequently underestimated. In turn, sustainable management practices can generate a variety of ecosystem services to the natural world, such as improvements in soil and water quality, carbon sequestration, and the preservation of biodiversity. By contrast, a crude management pattern may lead to ecosystem disservices, including the loss of wildlife habitats, nutrient runoff, sedimentation of waterways, greenhouse gas emissions, and pesticide-related risks to human health and non-target species. Therefore, it is essential to design cropping systems within the framework of ecosystem ecology to maximize the provision of ecosystem services while minimizing negative environmental and social impacts (Power 2010).

3 Concepts of Crop Physiology

3.1 The Meaning of Physiology in Biology

According to the Encyclopaedia Britannica, *physiology* is a scientific discipline that investigates the functioning of living organisms, whether animal, plant, or microbe, and of their constituent tissues or cells (https://www.britannica.com). As such, human physiology is defined as the integrated study of the normal functioning of the human body (Feher 2017), whereas microbial physiology focuses on the structure–function relationships in microorganisms, specifically their responses to environmental cues (Cole 2018).

As a specialized branch of biology, physiology delves into the structure–function relationships by dissecting living systems into their constituent parts and investigating the interactions among these components that drive chemical and physical processes. Through this approach, physiologists can elucidate the mechanisms by which organisms, organ systems, organs, cells, and biomolecules perform their essential functions within a living system.

The fundamental principles of human physiology, such as homeostasis, structure–function relationships, and hierarchical organization, provide a foundational framework for understanding the nature and essence of physiology. As delineated by Feher (2017), these principles are as follows: (1) cells serve as the fundamental functional units of human life; (2) physiological systems endeavour to maintain *homeostasis*; (3) all causal mechanisms within the body conform to physical and chemical laws; (4) our bodies are the product of evolutionary processes from ancestral forms; (5) our bodies transform matter and energy into bodily materials and carry out physiological work; (6) the function of body parts is determined by their structure; (7) integrated control of body systems requires signalling across all levels of organization; (8) control systems employ negative feedback, anticipatory (feed-forward) control, or developmental (threshold) control mechanisms to regulate bodily response and behaviour; and (9) physiology is intrinsically quantitative.

3.2 The Difference Between Crop and Plant Physiology

As a branch of physiology, *plant physiology* is a purely basic science that focuses on the functioning of the plant organism and its constituent organs, tissues, and cells. More precisely, it is the scientific investigation of the complex biochemical and biophysical processes that govern plant growth, development, and responses to abiotic and biotic factors. Plant physiology conducts investigations across various scales of size and time, spanning from the molecular interactions implicated in gene expression to the intricate processes of ontogenesis. It has several key research fields, including phytochemistry, plant morphology and anatomy, molecular and cell biology, functional genomics, and environmental physiology.

Crop physiology, derived from plant physiology, is an applied basic science that examines the structure–function relationships underlying the biochemical and biophysical processes involved in yield and quality formation within crop stands, as well as their responses to environmental signals and management practices. Crop stands, whether monocultures or mixed cropping systems, serve as the core focus of crop physiology. They should be managed to minimize *competition* among individual plants or plant species and to promote *cooperation* in resource sharing and utilization, which significantly differs from natural plant populations (Struik 2023).

Crop physiology is situated at the intersection of genomics and phenomics (Fig. 1.6). It employs two complementary analytical approaches (Struik 2023): (1) Scaling down—agronomically significant problems observed at the crop level are translated into targeted research questions at lower levels of biological organization, such as the cellular or tissue levels, enabling in-depth investigation and elucidation of the regulatory mechanisms. (2) Scaling up—advances in mechanistic understanding at molecular or cellular levels are integrated into broader conceptual frameworks or crop models at higher levels, thereby demonstrating their relevance and significance for realistic crop production. The ultimate goal of crop physiology is to provide a scientific foundation for genetic improvement and agronomy, addressing global challenges related to the sustainable production of abundant, high-quality food, feed, fibre, and energy. As such, it holds considerable appeal for scientists across diverse agricultural disciplines, from molecular sciences to natural resources management.

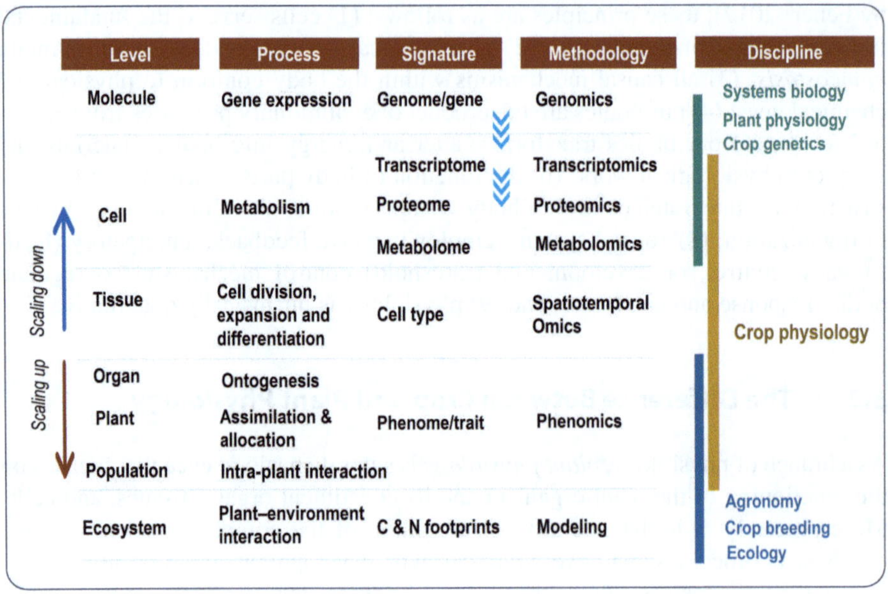

Fig. 1.6 Crop physiology: the key interplay between genomics and phenomics. It functions as a bridge that connects fundamental sciences, such as systems biology, plant physiology, and genetics, at the cellular level with applied disciplines, including agronomy and crop breeding, at the population level and beyond. Dark green arrows indicate the unidirectional flow of genetic information, known as the Central Dogma. C, carbon; N, nitrogen

Just as the concept of *plant* differs from that of *crop*, crop physiology has developed into a distinct discipline, separate from plant physiology. The focus of crop physiology is on the whole plant and plant community rather than that of contemporary plant physiology and genomics, which emphasizes cells and tissues, since most yield-controlling processes occur at the crop level in close interaction with the environment. In addition, crop physiology focuses on the examination of the complex interaction networks that underlie the fundamental life processes of crop plants at multiple levels of biological organization (Struik 2023). Therefore, crop physiology covers both fundamental research, driven by scientific curiosity (similar to plant physiology), and applied research, aimed at addressing practical challenges, to deepen our understanding of crop functioning.

In summary, plant physiology offers a fundamental and mechanistic understanding of plant functions, whereas crop physiology applies this knowledge to meet the specific requirements of agricultural systems. It emphasizes the operation of physiological processes at both the individual plant and canopy levels, with the overarching objective of enhancing yield, quality, resource use efficiency, and stress resilience. The two disciplines are closely interconnected: crop physiology draws upon foundational mechanisms elucidated by plant physiology, such as photosynthesis and hormonal regulation, to interpret crop performance. Conversely, insights derived from crop physiology, particularly the structure–function relationships at the organ, plant, and population levels, can serve as a foundation for novel research directions in plant physiology, thereby promoting deeper investigation into the underlying biological mechanisms.

3.3 Interdisciplinary Approaches Focusing on the Key Interactions

Crop physiology is an interdisciplinary science that integrates knowledge from various disciplines, such as soil science, ecology, biochemistry, biophysics, genomics, phenomics, informatics, and modelling. It has broad research areas, spanning from the molecular mechanisms of plant development to the ecological processes of greenhouse gas (GHG) emissions from the crop stand. Therefore, crop physiology adopts multiple approaches to fully uncover the functioning of crop plants. These include (1) systems biology methodologies such as functional genomics, transcriptomics, proteomics, and metabolomics; (2) biochemical analyses of carbon and nitrogen metabolism; (3) histochemical identification and characterization of metabolites in vegetative and reproductive tissues; (4) anatomical and morphological studies of organogenesis (the formation of tissue or organs); (5) biophysical analysis of plant structure and its response to environmental stimuli; (6) ecophysiological methods for quantifying gas fluxes (CO_2, O_2, H_2O, and GHGs) between crops and the environment; (7) phenotyping of physiological traits across various levels, especially under field conditions; and (8) modelling approaches that link plant or canopy structure with corresponding functional attributes.

At the heart of each level of biological organization are the *interactions* or relationships between constituent parts that have comparable size or dynamics. Interactions at the specific level give rise to its organized complexity or emergent property. The intricate interactions between carbon and nitrogen metabolisms, source and sink organs, embryo and endosperm, and canopy structure and function play a dominant role in physiological processes at their respective levels, thereby being crucial for the formation of crop yield, quality, and resilience (Fig. 1.7). The subsequent chapters will offer a thorough analysis and clarification of the mechanism and relevance of these crucial interactions (Table 1.2).

4 History of Crop Physiology

Interest in the physiology of crop yield emerged during the mechanization phase of the agricultural intensification revolution, beginning in the early twentieth century. It subsequently evolved through periods of increased consumption of chemicals and energy in developed countries (1940s and 1950s), significantly influencing the conceptual framework and interpretation of the Green Revolution (1960s and 1970s). From the mid-1960s to the 1990s, crop physiologists made significant contributions to understanding yield physiology at the whole-plant level. Significant advancements in crop physiology over a span of 50 years (1955–2005) were

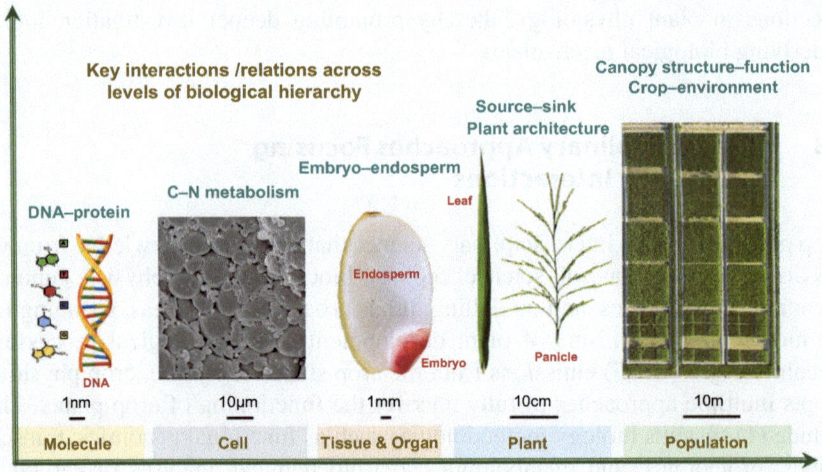

Fig. 1.7 A visual representation illustrating key interactions across organizational levels. At the cellular level, the SEM image of rice endosperm reveals a loose arrangement of large starch granules surrounded by small protein bodies, which are the primary end products of carbon (C) and nitrogen (N) metabolism, respectively. Scale bars indicate the relative magnitude at different levels, which may not precisely represent the actual dimensions

Table 1.2 Key interactions/relationships for crop performance across organizational levels

Levels	Interactions/relations	Relevance	Chapter
Molecule	The macromolecules of proteins and nucleic acids	Transcription of the genetic blueprint, marking the initiation of the crop life cycle; sharing a linguistic foundation with molecular biologists	1
Cell	Carbon and nitrogen metabolism	Establishing the biochemical foundation for crop productivity and quality	2
Tissue and organ	Embryo and endosperm	The most prominent relationship in the seed, affecting the process of grain-filling (sink activity) and seed variability	3
Plant	Spatial arrangement of vegetative and reproductive organs (plant architecture)	Optimizing light interception and utilization to achieve yield potential at favourable conditions; the foundation of the canopy structure	4
	Source and sink organs	The primary factor influencing the allocation of photosynthates to harvested organs; the foundation of canopy function	5
Population and community	Canopy structure and function	Underpinning the productivity and environmental impact of the crop production system	6
Ecosystem	Crop and the environments (crop phenology)	The fitness of crop development for local conditions, being crucial for a successful harvest; sharing a linguistic foundation with crop ecologists	7
All levels	All the abovementioned (physiological traits)	Identifying novel physiological traits suggestive of crop performance for breeding and management; linking the fundamental genomic research with applied crop science	8

comprehensively summarized by Boote and Sinclair (2006). These include C4 photosynthesis and photorespiration, upscaling of leaf-level carbon assimilation to the canopy level, phloem loading and unloading of sucrose, seed development and sink strength, water relations, nitrogen metabolism and symbiotic N_2 fixation, and the impacts of global climate change and air pollutants, as well as environmental influences on crop yield.

The period from 1960 to 1990 marked the peak of crop physiology, but since then, it has played a less prominent role in plant and crop science, except for its application in crop models. Currently, the focus of plant physiology research is to identify and understand gene regulation in plant development and growth. Unfortunately, this genomic perspective has resulted in significant gaps in the traditional understanding of plant physiology as a partnership between fundamental and applied research aimed at improving crop productivity (Boote and Sinclair 2006). One of the gaps lies in the lack of consideration for practical applications when organizing and conducting molecular or genomic research aimed at improving crop performance under real-world conditions. Many genetic analyses rely on phenotypic data obtained from very limited or controlled environments, which may not effectively identify genes that can enhance breeding populations.

In addition, the regulation of plant systems has proved to be complex and redundant. A single physiological process, when altered, is typically compensated for or dampened by the plant, resulting in minimal impact on overall crop performance and yield. Likewise, altering one cellular process could negatively impact another. For example, in contrast to single-gene traits such as herbicide tolerance or Bt toxin resistance, abiotic stress tolerance is a complex, polygenic trait regulated by a network of genes involved in metabolism, development, and resilience under diverse environmental conditions. To tackle this challenge, scientists have focused on modifying entire stress-resilience gene sets and multi-functional pleiotropic genes that help plants withstand environmental stresses such as drought, salinity, and extreme temperatures. Advances in synthetic biology and genome editing now make such coordinated genetic strategies feasible, offering more durable and practical solutions beyond single-gene approaches. However, genes conferring resistance to one stress frequently result in a reduction in performance or yield under other stresses or normal conditions. For example, water-saving traits like stomatal closure can improve drought tolerance but hinder photosynthesis and biomass growth (Lawson and Leakey 2024). Such physiological trade-offs complicate breeding efforts in the real world, thereby impeding the development of high-yielding, stress-resistant crops and decelerating their commercial adoption (Ganie and Azevedo 2025).

Boote and Sinclair (2006) summarized that crop physiology research has come full circle, shifting from a focus on specific processes such as nitrate reductase in the 1960s to whole-plant physiology during the 1970s–1990s, and later returning to specific processes such as genes and enzymes. They therefore formulated a strong hypothesis that this cyclical pattern will inevitably recur in the future and that the molecular-level perspective will ultimately meet the challenge of integrating knowledge at the crop level.

5 Challenges and Opportunities

5.1 In the Era of Cellular Omics

The omics disciplines, such as *genomics*, *transcriptomics*, *proteomics*, *metabolomics*, and *phenomics*, encompass various branches of biology that focus on characterizing and quantifying biological molecules or components to gain insights into the structure, function, and dynamics of organisms. At the heart of omics sciences lies *genomics*, which is the study of an organism's complete genetic makeup. The 2000 publication of the *Arabidopsis thaliana* genome marked a significant milestone in plant genomics. From 2000 to 2020, the genomes of 782 plant species were sequenced, totalling 1144 assemblies. Between 2021 and 2023, an additional 2373 genomes from 1031 plant species, including 793 newly sequenced species, were assembled, marking a significant leap forward in plant genomics (Song et al. 2023). These advances have revolutionized plant biology and enabled groundbreaking discoveries in the mechanisms controlling growth and development.

Since the completion of the rice genome sequencing project in 2005, crop science has entered an era of genomics, consisting of three distinct subdisciplines: structural genomics, functional genomics, and quantitative genomics (Wang and Han 2022). *Structural genomics* refers primarily to genome sequencing for the construction of a complete map of the plant genome sequence. This is an essential undertaking for conducting research in crop genetics and molecular biology. *Functional genomics* aims to unravel the functions of genes in crop plants. *Quantitative genomics* is large-scale sequence- and statistics-based research to characterize the genetic features and quantitative traits of crop populations.

Tremendous progress has been achieved in crop genomics research. The most significant outcomes include the cloning and functional characterization of a large number of genes, many of which are directly associated with important agronomic traits such as disease resistance, response to environmental stress, crop yield, grain quality, and nitrogen use efficiency (Li et al. 2018). The significant advancements have greatly enhanced our understanding of the genetic and molecular mechanisms, as well as the related biological processes that regulate crop performance, thereby reshaping the field of plant biology and crop breeding.

Crop breeding has traditionally relied on conventional cross-breeding methods, where genotype selection is based solely on pedigree and phenotype (Alemu et al. 2024). With the rapid advancement of genome sequencing technologies and the development of high-density genome-wide marker systems, *genomic selection* (GS) has emerged as a powerful tool in plant breeding, particularly following the widespread availability of genome-wide single-nucleotide polymorphisms (SNPs). GS is considered to have a greater advantage over pedigree breeding and marker-assisted selection (MAS) as it can improve complex traits such as yield in a shorter period, by reducing the number of breeding cycles (Krishnappa et al. 2021). The ongoing rapid advancement of next-generation sequencing technology, enabling the generation of dense genome-wide SNP markers, combined with significant cost reductions for *genotyping* in various crops, renders GS an indispensable method in breeding programs (Alemu et al. 2024).

5.2 Beyond the Cellular Level

The advancements in molecular biology during the twentieth century, particularly the establishment of the Central Dogma, elucidated how genetic information is transcribed into messenger RNA (mRNA) and subsequently translated into proteins (Fig. 1.2). This discovery uncovered the fundamental processes driving development and growth. Building on this foundation, the Human Genome Project (HGP) was launched in 1990 to identify and map the three billion DNA base pairs and about 25,000 genes in the human genome, marking the beginning of a new era in understanding how genetic information controls biological functions. However, one of the principal challenges confronting the HGP is elucidating how genomic sequences are translated into complex biological systems. This challenge is exemplified by the process through which a fertilized egg, containing a complete genome, develops into a fully formed organism with diverse cell types organized into functionally distinct tissues via gene expression regulation (Liu et al. 2024).

The advances in sequencing technologies, particularly the massively parallel sequencing, have enabled large-scale, genome-wide multi-omics analyses of tissues. However, bulk tissue-based multi-omics analyses primarily capture average-level signals, which limits their ability to fully characterize the heterogeneity of different cell types within tissues (Liu et al. 2024). Recent advancements in spatial omics technologies, encompassing both imaging- and sequencing-based approaches, have provided novel insights into the regulatory mechanisms of genomic elements, as well as the interplay among gene expression, cellular functions, and cell fate determination.

Traditional imaging techniques, including X-rays, computed tomography (CT) scans, and magnetic resonance imaging (MRI) scans, have enabled the visualization of the 3D structure of tissues and organs; however, they are deficient in molecular and cellular resolution. Immunohistochemistry (IHC) and in situ hybridization (ISH) facilitate the spatial localization of specific genes or proteins but can only detect a limited number of targets. Recent advancements in spatial omics technologies have overcome these limitations by acquiring high-content omics data from tissues while retaining spatial localization information. Notably, techniques such as single-cell RNA sequencing (scRNA-seq) and spatially enhanced resolution omics sequencing (Stereo-seq) enable the detection of spatial and temporal variations in gene expression, as well as cell–cell interactions, at subcellular resolution (~500 nm) across large tissue areas (up to centimetre-scale dimensions).

Yao et al. (2024) integrated Stereo-seq with scRNA-seq to map the spatially resolved transcriptomes of individual cells in germinating rice embryos. To meet the demands of high-precision cellular annotation, they developed a deep learning framework for automated cell segmentation. Transcriptomic data acquired at four critical time points, 6, 24, 36, and 48 h after imbibition, enabled the identification of two distinct scutellum subtypes. These findings were rigorously validated using spatial transcript detection via in situ hybridization and functional annotation of cell-type-specific marker genes. Time-resolved expression analysis revealed stage- and cell-type-dependent transcriptional dynamics, identifying key regulators involved in nutrient mobilization, metabolic pathway activation, and phytohormone responses, each exhibiting cell-context-specific reprogramming during germination. Collectively, this study establishes the first high-resolution spatiotemporal atlas of the rice embryonic transcriptome and introduces an innovative analytical pipeline for dissecting the functional heterogeneity of embryonic cell lineages during seed germination.

5.3 The Knowledge Gap Between Genomics and Breeding

With the significant advancements in plant genomics, a wealth of knowledge has been acquired regarding the genetics of crop function, including both the heritability of genetic traits and the genetic regulation of physiological and biochemical processes. Numerous studies published in prestigious journals such as NATURE, CELL, and SCIENCE have demonstrated that modifying one or a few genes can lead to substantial increases in crop yields, ranging from 10% to 68%, across crops like rice, maize, tobacco, and soybean. These studies have provided significant insights into molecular biology and gene discovery. However, the findings have rarely translated into yield improvements in actual farming environments, as most yield tests, using outdated and non-elite plant lines, are conducted under controlled conditions in greenhouses or small-scale field trials that differ significantly from those in commercial crop production systems (Khaipho-Burch et al. 2023). Although there are certain exceptions, especially in traits related to disease resistance, this discrepancy raises a crucial concern about the applicability of findings from basic plant science research to real-world crop improvement. As a result, it is essential to not only enhance our understanding of individual genes, pathways, and regulatory elements at the cellular level from a reductionist perspective but also to explore how these components interact collectively to perform specific functions at the crop level from a holistic viewpoint.

Over the past century, breeders have selected crop variants that tolerate high planting densities. However, in many small-scale trials, plants are often spaced at low densities that are not commercially relevant. In some cases, yield is measured from individual plants rather than from entire plots. Owing to the emergent property, plant-level performance cannot be extrapolated to crop-level performance. For example, low-density planting may result in larger spikes and more tillers in individual wheat plants, but it could reduce the overall grain yield at the population level. The emergent nature of yield from physiological processes demands that all components contributing to yield be considered. As a result, selecting for individual plant yields is generally ineffective in breeding at the population level. The consensus among breeders supports the idea that there is often little to no noticeable correlation between crop yield and the yield components of individual plants. Accordingly, breeders emphasize the synergistic effect of individual traits at the population level. In contrast, the mainstream of functional genomics and molecular biology focuses on examining plant-level phenotypes, which may not necessarily be extrapolated to the population level (Fig. 1.8; Liu 2024).

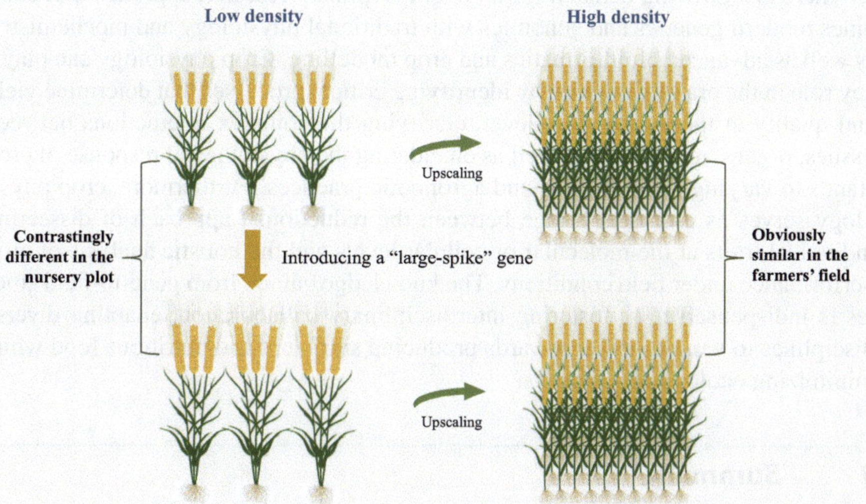

Fig. 1.8 Illustration of the density effect on wheat spike size. A low planting density usually promotes the development of larger spikes. The introduction of a "large-spike" gene into a recipient parent can lead to larger spikes under low planting density. However, it may have minimal effect under high planting density due to intense intraplant competition within the dense population. (Redrawn from Liu (2024))

6 Concluding Remarks and Perspectives

According to the report by FAO et al. (2024), it is estimated that between 713 and 757 million people may have experienced hunger in 2023, equating to one out of every 11 individuals worldwide, and one out of five in Africa. The world's population is expected to grow to nearly 10 billion by 2050, increasing food demand by approximately 50% compared to 2013 (FAO 2017). Achieving this requires an annual crop yield growth rate of 2.4%. However, current rates for maize, rice, wheat, and soybean fall significantly short of this target. In fact, they are increasing at less than half the required rate, reaching a plateau in many major producing regions. It is even more difficult to achieve this rate of genetic progress, given the challenges posed by soil degradation, dwindling water reserves, soaring energy prices, and climate change.

Most yield and agronomic traits are governed by multiple genes and strongly influenced by environmental factors, rendering them difficult to enhance using conventional breeding methods. While genetic modification or genome editing of single traits can enhance plant-level biomass, these improvements often do not translate into consistent increases in field-level grain yield across diverse agroclimatic environments. This is largely because yield formation depends on complex, interdependent physiological processes (Burgess et al. 2022). The high expectations from genomics and molecular approaches are now shifting towards recognizing the importance of physiology for realistic improvements in crop productivity.

There is a growing demand for an interdisciplinary research approach that combines modern genetics and genomics with traditional physiology and biochemistry, as well as advanced bioinformatics and crop modelling. Crop physiology can play a key role in the era of genomics by identifying critical processes that determine yield and quality at the whole plant level, clarifying the complex interactions between tissues, organs, and plants, as well as elucidating the physiological response of crop stands to varying environments and agronomic practices. Furthermore, crop physiology serves as a critical bridge between the reductionist approach of dissecting individual traits at the molecular or cellular levels and the holistic analysis of crop performance under field conditions. The knowledge gained from gene-to-field studies is indispensable for fostering interdisciplinary collaboration, enabling diverse disciplines to work together towards producing sufficient and nutritious food while minimizing ecological footprints.

7 Summary

1. Domestication syndrome refers to a consistent set of phenotypic traits that repeatedly emerge during the transformation of wild plant species into cultivated crops, primarily including reduced seed shattering, increased size of harvestable organs, compact plant architecture, determinate growth habit, improved nutritional or sensory quality, and enhanced harvest convenience.
2. When studying crops, it is essential to consider several key principles: crop production fundamentally relies on the biochemical conversion of solar energy into chemical forms that are stored in the harvested organs; crop production systems display emergent properties; structure, function, and interaction constitute the fundamental conceptual framework of crop science; and crop science examines a range of processes across multiple levels of biological organization.
3. Crop physiology is the scientific study of how crop plants function, focusing on the structure–function relationship to elucidate the mechanisms underlying yield formation, quality development, and tolerance to abiotic stresses across multiple biological levels.
4. Crop physiology has evolved from plant physiology; however, it shows a significant difference, concentrating on the interactions at different organizational levels.
5. The period between 1960 and 1990 witnessed the peak of crop physiology, but since then, it has played a less prominent role in plant and crop science. The prevalence of a genomic perspective on plant behaviour has resulted in significant gaps between fundamental and applied research aimed at improving crop production.
6. Crop physiology should occupy a middle ground between the reductionism of genomics and the holistic approach of agronomy, aiming to establish a crucial link between genotype and phenotype.

Important Terms

Biological organization; Crop level; Crop physiology; Domestication; Emergent property; Function; Genomics; Hierarchy; Holism; Interaction; Level; Mechanism; Physiology; Plant physiology; Reductionism; Structure; System

Review Questions

Multiple Choice

1. The essential of crop production is the chemical transformation of solar energy. In the crop stand, this transformation involves three processes in sequence, EXCEPT for the (   ):

   (a) Interception of incident solar radiation by the canopy
   (b) Conversion of the intercepted radiant energy to chemical potential energy
   (c) Allocation of dry matter production between the harvestable organs and non-harvestable organs
   (d) Production of a functional protein by gene expression

2. According to the report by FAO et al. (2024), it is estimated that (   ) people may have experienced hunger in 2023.

   (a) Between 713 and 757 million
   (b) Between 71.3 and 75.7 million
   (c) Between 7.13 and 7.57 million
   (d) Between 7.13 and 7.57 billion

3. Globally, there are approximately 382,000 species of vascular plants, with about (   ) crop species having significant global production levels.

   (a) 6000
   (b) 200
   (c) 9
   (d) 5

4. Regarding the statements on the hierarchy of life or levels of biological organization, (   ) is NOT correct.

   (a) Each level in the hierarchy represents an increase in organizational complexity.
   (b) The concept of emergence refers to the properties and functions found at a hierarchical level that are present and relevant at the lower levels.
   (c) The levels consist of families of entities, typically of comparable size and dynamic properties, which interact primarily with each other in a characteristic manner.
   (d) Rates of key processes also decrease as they progress from lower to higher levels, partly because it takes more time for causal effects to spread over longer distances.

5. The endosperm tissue of cereal grains belongs to (   ).
   (a) Epidermis tissue
   (b) Parenchyma
   (c) Sclerenchyma
   (d) Collenchyma
6. The process by which information in a DNA molecule is converted to information in an RNA molecule is called (   ).
   (a) Translation
   (b) Transcription
   (c) Gene expression
   (d) Genotyping
7. The set of all traits expressed by a cell, tissue, organ, organism, or species is collectively called (   ).
   (a) Genome
   (b) Proteome
   (c) Transcriptome
   (d) Phenome

True or False

1. Gene is a basic unit of heredity and a sequence of nucleotides in DNA or RNA that encodes the synthesis of a gene product, either RNA or protein (   ).
2. Although originating from plant physiology, crop physiology has emerged as a distinct discipline (   ).
3. Crop physiologists use both fundamental and applied research to better understand how crop plants function (   ).
4. Systems comprise interacting components that collectively exhibit emergent behaviours not seen in isolated parts (   ).

Blank Filling

1. Emergent property refers to a phenomenon that arises from the____________ and collective behaviour of individual components within a complex system.
2. The genomic perspective of the current study has resulted in significant gaps in our traditional understanding of plant physiology as a _____________ between fundamental and applied research aimed at improving crop production.

Short Answer

1. The relevance of crop physiology to crop breeding and management.
2. Challenges and opportunities for crop physiology in the era of genomics.
3. The relations between the disciplines of crop and plant physiology.
4. Key principles and concepts when dealing with the study of crop sciences.
5. Major centres of crop domestication, with at least two representative crops listed for each centre.

Concept Map

1. Develop a concept map to elucidate the meaning of SYSTEM, utilizing the following terms: *System*; *Parts*; *Structure*; *Function*; *Environment*; *Mechanism*; *Level*; *Relevance*; *Interaction*; *Emergent Property.*
2. Please complete the concept map provided below.

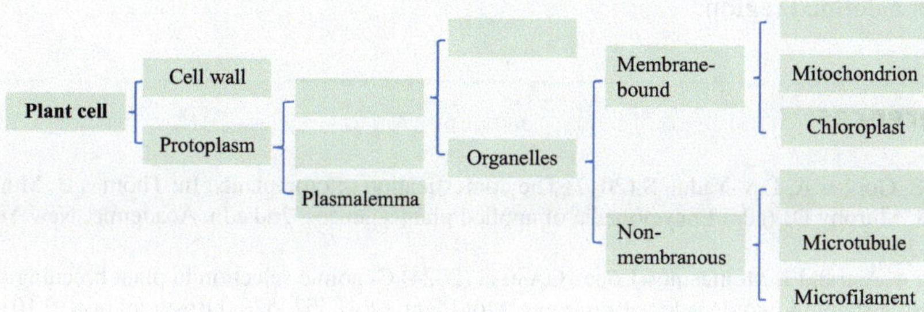

Essay Question

1. After a thorough review of the critical historical events and significant discoveries in crop physiology research over the past 50 years since the establishment of the Crop Science Society of America, Boote and Sinclair concluded that "crop physiology research, … … has made a full circle, from emphasis on specific processes such as nitrate reductase, to whole-plant physiology, back to specific processes (genes to enzymes). We believe the present phase of the circle is in danger of losing focus on the whole-plant response. We also strongly suspect that the cycle will repeat itself, and that there will again be a recognition of the need to put together understanding of individual processes into a whole-plant perspective." (Crop Sci, 2006, 46: 2270–2277) *Do you agree with them that crop physiology will soon focus on studying at the whole-plant level? What does this proposition imply for your research? So, where do you see crop physiology headed in the future?*

2. Biological organization is the hierarchy of complex biological structures and systems that define life using a reductionistic approach. The traditional hierarchy extends from molecules to cells, tissues, plants, populations, communities, and even biospheres. The basic principle behind this organization is the concept of emergence—the properties and functions found at a higher hierarchical level are not present or relevant at lower levels. According to the theory of biological organization, when we look for the mechanism of a biological phenomenon, it is usually found at levels below where it occurs. On the other hand, if we want to understand the significance of that phenomenon, it is typically explained at higher levels. *Please specify the level of organization in your research work and explain how to clarify its mechanism and significance.*

Authentic Performance Task

Assemble a multidisciplinary team composed of experts in crop breeding, genetics, physiology, informatics, and agronomy to develop an integrated, site-specific strategy designed to enhance high-yielding crop production for a specific crop species within a defined region.

References

Abbo S, Gopher A, Lev-Yadun S (2017) The domestication of crop plants. In: Thomas B, Murray BG, Murphy DJ (eds) Encyclopedia of applied plant sciences, 2nd edn. Academic, New York, pp 50–54

Alemu A, Åstrand J, Montesinos-López OA et al (2024) Genomic selection in plant breeding: key factors shaping two decades of progress. Mol Plant 17(4):552–578. https://doi.org/10.1016/j.molp.2024.03.007

Araus JL, Sanchez-Bragado R, Vicente R (2021) Improving crop yield and resilience through optimization of photosynthesis: panacea or pipe dream? J Exp Bot 72:3936–3955. https://doi.org/10.1093/jxb/erab097

Asseng S, Zhu Y, Basso B et al (2014) Simulation modeling: applications in cropping systems. In: Van Alfen NK (ed) Encyclopedia of agriculture and food systems. Academic, New York, pp 102–112

Boote KJ, Sinclair TR (2006) Crop physiology: significant discoveries and our changing perspective on research. Crop Sci 46(5):2270–2277. https://acsess.onlinelibrary.wiley.com/doi/abs/10.2135/cropsci2006.01.0039gas

Burgess AJ, Masclaux-Daubresse C, al Strittmatter G e (2022) Improving crop yield potential: underlying biological processes and future prospects. Food Energy Secur 12(1):e435. https://doi.org/10.1002/fes3.435

Cole JA (2018) Anaerobic bacterial response to nitrosative stress. In: Poole RK (ed) Advances in microbial physiology. Academic, New York, pp 193–237

Doebley JF, Gaut BS, Smith BD (2006) The molecular genetics of crop domestication. Cell 127(7):1309–1321. https://doi.org/10.1016/j.cell.2006.12.006

FAO (2017) The future of food and agriculture–trends and challenges, Rome

FAO (2019) In: Bélanger J, Pilling D (eds) The state of the world's biodiversity for food and agriculture, FAO commission on genetic resources for food and agriculture assessments, Rome

FAO, IFAD, UNICEF, WFP, WHO (2024) The state of food security and nutrition in the world–financing to end hunger, food insecurity, and malnutrition in all its forms, Rome
Feher J (2017) The core principles of physiology. In: Feher J (ed) Quantitative human physiology, 2nd edn. Academic, New York, pp 3–14
Ganie SA, Azevedo RA (2025) Why stress–resistant crops remain a scientific promise rather than a farming reality? Bridging the gap between genetic discovery and agricultural impact. Ann Appl Biol:1–5. https://doi.org/10.1111/aab.70050
Halford NG (2010) Photosynthate partitioning. In: Pua EC, Davey MR (eds) Plant developmental biology-biotechnological perspectives. Springer
Hay RKM, Porter JR (2006) The physiology of crop yield, 2nd edn. Wiley-Blackwell, Hoboken
Khaipho-Burch M, Cooper M, Crossa J et al (2023) Genetic modification can improve crop yields–but stop overselling it. Nature 621(7979):470–473. https://doi.org/10.1038/d41586-023-02895-w
Krishnappa G, Savadi S, Tyagi BS et al (2021) Integrated genomic selection for rapid improvement of crops. Genomics 113(3):1070–1086. https://doi.org/10.1016/j.ygeno.2021.02.007
Larson G, Piperno DR, Allaby RG et al (2014) Current perspectives and the future of domestication studies. Proc Natl Acad Sci USA 111(17):6139–6146. https://doi.org/10.1073/pnas.1323964111
Lawson T, Leakey AD (2024) Stomata: custodians of leaf gaseous exchange. J Exp Bot 75(21):6677–6682. https://doi.org/10.1093/jxb/erae425
Li Y, Xiao JH, Chen LL et al (2018) Rice functional genomics research: past decade and future. Mol Plant 11:359–380. https://doi.org/10.1016/j.molp.2018.01.007
Liu DC (2024) Towards cultivar-oriented gene discovery for better crops. Crop J 12(3):670–675. https://doi.org/10.1016/j.cj.2024.01.008
Liu L, Chen A, Li Y et al (2024) Spatiotemporal omics for biology and medicine. Cell 187(17):4488–4519. https://doi.org/10.1016/j.cell.2024.07.040
McAllister T, Nelissen H, Strable J et al (2025) Unlocking grass leaf development: foundations for tunable cereal design. New Phytol 248(4):1655–1676. https://doi.org/10.1111/nph.70477
Power AG (2010) Ecosystem services and agriculture: Tradeoffs and synergies. Philos Trans R Soc Lond Ser B Biol Sci 365(1554):2959–2971. https://doi.org/10.1098/rstb.2010.0143
Sadras VO, Calderini DF (eds) (2021) Crop physiology case histories for major crops. Academic, New York
Singels A, Jackson P, Inman-Bamber G (2021) Sugarcane. In: Sadras VO, Calderini DF (eds) Crop physiology case histories for major crops. Academic, New York, pp 674–713
Slattery RA, Ort DR (2021) Perspectives on improving light distribution and light use efficiency in crop canopies. Plant Physiol 185(1):34–48. https://doi.org/10.1093/plphys/kiaa006
Snider J, Bange M, Heitholt J (2021) Cotton. In: Sadras VO, Calderini DF (eds) Crop physiology case histories for major crops. Academic, New York, pp 714–746
Struik PC (2023) From photons to farmers: consilience in crop physiology. Wageningen University & Research. https://doi.org/10.18174/640011
Wang CS, Han B (2022) Twenty years of rice genomics research: from sequencing and functional genomics to quantitative genomics. Mol Plant 15(4):593–619. https://doi.org/10.1016/j.molp.2022.03.009
Wimsatt WC (1994) The ontology of complex systems: levels of organization, perspectives and causal thickets. Can J Philos Supp 20:207–274. https://doi.org/10.1080/00455091.1994.10717400
Yao J, Chu Q, al Guo X e (2024) Spatiotemporal transcriptomic landscape of rice embryonic cells during seed germination. Dev Cell 59(17):2320–2332. https://doi.org/10.1016/j.devcel.2024.05.016

Carbon and Nitrogen Metabolism

2

Primary Cellular Metabolic Processes Underlying Crop Yield and Quality

Contents

Z. Liu, *Crop Physiology*, https://doi.org/10.1007/978-981-92-2960-4_2

Abstract

Carbon (C) and nitrogen (N) metabolism are the primary cellular processes governing plant growth and development. Their products, including starch, lipids, and proteins, dominate seed chemical composition and provide the fundamental basis for crop yield and quality. Interactions between C and N metabolism occur at multiple points within plant metabolic pathways, posing challenges when attempting to discern their individual contributions to crop productivity. This chapter begins by exploring the fundamental principles of plant metabolism, followed by an analysis of the key processes involved in C and N assimilation and metabolism. The focus is on elucidating the major biochemical components that have agronomic importance, namely sucrose, starch, and storage proteins. Subsequently, three typical scenarios are presented to interpret the interplay between C and N metabolism, with two case studies that exemplify its significance for crop production. This chapter concludes by providing a perspective on future studies regarding the C–N interaction.

Keywords

Assimilation · Metabolism · Carbon · Nitrogen · Interaction · Crop quality

Did You Know

- Photosynthesis is responsible for producing virtually all the O_2 in the atmosphere and for fixing about 10^{11} tonnes of carbon from CO_2 into organic compounds annually.
- Rubisco, the principal enzyme responsible for C fixation of CO_2, is the most abundant enzyme on earth. A rudimentary calculation estimates that there is about 0.8–1.1 Gt (10^9 t) of Rubisco within the 610 Gt of C in the terrestrial biosphere. In addition, it typically constitutes 10%–30% of the total nitrogen content in leaves, while this proportion can reach as high as 50% under certain conditions.
- Carnivorous plants typically grow in nitrogen-poor soils and have evolved specialized leaf structures to capture and digest insects, thereby supplementing their nutrient intake. For instance, the Venus flytrap (*Dionaea muscipula*) possesses hinged leaf lobes that rapidly close when trigger hairs on the inner surface are stimulated, effectively trapping prey. In contrast, pitcher plants (genus *Nepenthes*) develop deep, tubular leaves filled with digestive fluids that serve as pitfall traps. Once insects fall into these chambers and drown, enzymes break down their bodies, releasing nitrogen and other essential nutrients.
- Natural processes fix around 190 M tonnes of nitrogen each year, with lightning accounting for ~8%, photochemical reactions between NO and O_3 contributing ~2%, and biological nitrogen fixation responsible for the remaining 90%.
- Melvin Calvin, along with his collaborators Andrew Benson and James Bassham, elucidated the biochemical pathway by which plants fix CO_2 during photosynthesis, the Calvin-Benson cycle, and was awarded the Nobel Prize in Chemistry in 1961 for this groundbreaking work.

Learning Objectives

After reading this chapter, students will be able to:

- Understand basic concepts of C and N assimilation.
- Describe the main storage carbohydrates in plants, how they are synthesized, and how they are degraded to release energy and carbon.
- Explain the process of assimilating inorganic N into amino acids and describe how plants store N.
- Summarize how C and N interact at the cellular level.
- Interpret the difficulty in simultaneously improving grain yield and quality from the perspective of C–N interaction.

Introductory Remarks

The majority of the organic matter on Earth is produced by plants through the assimilation of inorganic carbon (C) and nitrogen (N) from the environment into organic molecules, processes powered by solar energy. These processes, namely photosynthesis and N assimilation, are crucial not only for plants themselves but also for nearly all other forms of life, including animals, fungi, and most bacteria, which obtain their C and N nutrition exclusively from plant-derived compounds. In this regard, the essence of crop production lies in harvesting the C assimilates (e.g. starch, lipids, and fibre)[1] and N assimilates (e.g. proteins) produced by crops to supply humans with adequate nutrients, energy, and industrial materials.

Plants utilize light energy and transform it into chemical energy via photosynthesis, synthesizing simple carbohydrates from water and CO_2. They possess an outstanding ability to concentrate atmospheric CO_2 (at a concentration of approximately 430.5 ppm in 2025) into carbohydrates (about 40% by dry mass), representing an enrichment of roughly 800- to 1000-fold. These carbohydrates are subsequently converted into more chemically reduced compounds, primarily proteins and lipids, which together form the foundation of food chains. The biochemical and physiological processes governing C uptake and release are critical determinants of crop yields. Moreover, plants play a crucial role in regulating atmospheric composition and Earth's energy balance. Each year, they assimilate approximately 100 billion tonnes of C, roughly 15% of the total C contained in atmospheric CO_2. Meanwhile, respiration by plants and heterotrophic organisms returns nearly the same amount of C from organic compounds back to the atmosphere as CO_2.

(continued)

[1] Cotton fibres are composed of approximately 95% cellulose, which is derived from the extremely thick cell walls of epidermal hair cells, known as lint, that surround cotton seeds.

Following C, N is the element that plants require in the largest quantities. Plants store N as a short-term buffer to cope with temporary imbalances between N demand and supply, and also as a long-term reserve in storage organs such as seeds, roots, and tubers. Approximately 1%–5% of the total dry matter of plants is composed of N, an essential component of proteins, nucleic acids, chlorophyll, coenzymes, phytohormones, and secondary metabolites. Proteins are large biomolecules composed of one or more long chains of amino acid residues. They perform a wide range of crucial functions within organisms, including catalysing metabolic reactions, facilitating DNA replication, and mediating the transmembrane flux of biological molecules. Moreover, proteins serve as reserves for N and energy in the form of storage proteins.

The availability of N to plant roots is a crucial determinant of plant growth and crop productivity. To meet crop requirements and boost agricultural productivity, over 110 M tonnes of N fertilizers are applied globally on an annual basis. Although these fertilizers notably enhance yields, they also incur significant environmental and economic costs (Liu et al. 2025). Approximately 60% of total N fertilizer is dedicated to cereal crops. Typically, only 30%–50% of the applied N is effectively taken up by crops, while the remainder is either emitted into the atmosphere as nitrous oxide, a potent greenhouse gas that contributes to climate change, or leached into water bodies, leading to eutrophication and the formation of hypoxic dead zones in aquatic ecosystems.

Comprehending the intricate mechanisms governing C and N metabolism is pivotal for enhancing crop yield and quality in an environmentally sustainable manner. However, these two fundamental metabolic processes are closely intertwined at the cellular, tissue, organ, and plant levels. For instance, at the cellular level, N metabolism, such as amino acid biosynthesis, predominantly relies on C skeletons derived from glycolysis, the *Calvin-Benson cycle* (CBC), the *Krebs cycle* (or the tricarboxylic acid cycle, TCA cycle), and the *pentose phosphate pathway* (PPP). At the plant level, nitrates absorbed by roots can be locally reduced to ammonium, which is either utilized in root N metabolism or transported to the leaves for synthesizing photosynthetic enzymes, like Rubisco. Above-ground, C is assimilated through photosynthesis and transformed into sugars. These sugars either facilitate N assimilation and amino acid biosynthesis in the leaves or are transported as sucrose to the roots, where they provide energy and C skeletons for N uptake and assimilation. The interdependence of C and N metabolism across various organizational levels poses challenges when attempting to delineate an independent contribution of either C or N to physiological processes.

From a metabolic perspective, the life of crop plants is fundamentally shaped by continuous exchanges and interactions between C and N metabolism. This dynamic relationship operates across multiple spatial scales, from biochemical pathways within cells to whole-plant physiological processes.

(continued)

Temporally, it spans from rapid enzymatic reactions occurring within seconds to circadian rhythms, developmental transitions, and the entire crop lifecycle. Understanding these dynamic spatiotemporal interactions is essential for optimizing crop productivity, resource use efficiency, and sustainable agricultural practices. This chapter focuses on the interaction between C and N metabolism and its significance for crop production, aiming to establish a common biochemical basis for comprehending the formation of crop productivity and resilience.

Concept Map

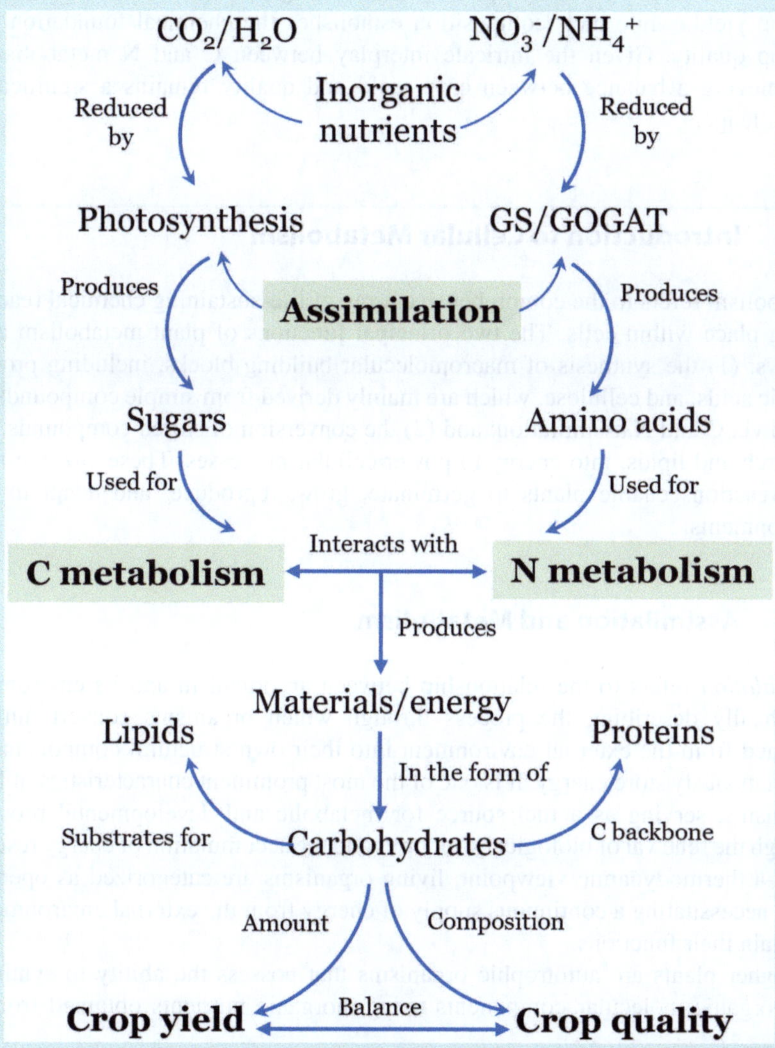

(continued)

Annotation: This map illustrates the key physiological processes at the cellular level and highlights the interaction between the two primary metabolic pathways: C and N metabolism. Assimilation refers to the incorporation of inorganic compounds from the environment into plant cells, whereas metabolism is defined as the set of biochemical reactions occurring within plant cells. Sugars, which are products of C assimilation via photosynthesis, and amino acids, which result from N assimilation through the GS/GOGAT pathway, serve as the primary sources of materials and energy for plant metabolism. C and N metabolisms interact at multiple levels, ranging from cellular processes to tissue and whole-plant levels. The resulting carbohydrates, such as starch and sucrose, storage proteins, and lipids, form the physicochemical basis of harvested crop organs. The quantity of these organic compounds determines crop yield, while their composition establishes the chemical foundation of crop quality. Given the intricate interplay between C and N metabolism, achieving a balance between crop yield and quality remains a significant challenge.

1 Introduction to Cellular Metabolism

Metabolism refers to the comprehensive array of life-sustaining chemical reactions taking place within cells. The two principal functions of plant metabolism are as follows: (1) the synthesis of macromolecular building blocks, including proteins, nucleic acids, and cellulose, which are mainly derived from simple compounds produced via C and N assimilation; and (2) the conversion of stored compounds, such as starch and lipids, into energy to power cellular processes. These enzyme-mediated reactions enable plants to germinate, grow, reproduce, and adapt to local environments.

1.1 Assimilation and Metabolism

Assimilation refers to the relationship between an organism and its environment, specifically describing the process through which organisms convert nutrients obtained from the external environment into their own structural components and simultaneously store energy. It is one of the most prominent characteristics of living organisms, serving as a fuel source for metabolic and developmental processes through the renewal of biological structures and the accumulation of energy reserves. From a thermodynamic viewpoint, living organisms are categorized as open systems, necessitating a continuous supply of energy from the external environment to maintain their functions.

Higher plants are autotrophic organisms that possess the ability to synthesize their organic molecular components using inorganic nutrients obtained from the environment. The assimilation of inorganic nutrients involves their absorption by roots from the soil (e.g. N, P, and K), by leaves from the atmosphere (carbon dioxide), and subsequent incorporation into vital organic compounds (e.g. sucrose and amino acids) to support growth and development. In crop plants, photosynthetic C assimilation in leaves and N assimilation in roots and leaves are crucial determinants of grain yield and quality.

Metabolism is defined as the biochemical processes occurring within living organisms. It encompasses the aggregate of intracellular chemical reactions, which provide energy for vital physiological functions and supply substrates for the synthesis of novel organic compounds. Metabolic processes entail the transformation of nutrients into energy and building materials for growth via anabolic and catabolic pathways. *Anabolism* encompasses all reactions that convert simpler molecules into more complex ones, and thus is characterized as divergent. It requires an input of energy and raw materials for the synthesis of macromolecules, including proteins, carbohydrates, lipids, and nucleic acids. *Catabolism* involves the transformation of complex substances into simpler molecules. For instance, during glycolysis, glucose is broken down into pyruvate, a process associated with the release of energy. Catabolic pathways are convergent because numerous metabolic reactions ultimately merge into common pathways to produce simpler molecules. Energy transfer within these pathways mainly occurs via two crucial high-energy molecules: NADH, the reduced form of nicotinamide adenine dinucleotide, and ATP, a high-energy phosphate compound.

1.2 The Plastids and Metabolic Compartmentalization

Plastids are believed to have originated from free-living photosynthetic cyanobacteria via endosymbiotic associations with non-photosynthetic eukaryotic hosts. This endosymbiotic event resulted in the stable incorporation of photosynthetic prokaryotes into host cells, ultimately leading to the development of plastids. Throughout evolution, while the majority of plastid genes have been transferred to the host nucleus, a number of essential functions still remain within the plastids. The preservation of these functions highlights the significance of intracellular compartmentalization in plant metabolic processes.

Plastids are responsible for performing photosynthesis, synthesizing and storing starch and oils, and contributing to other biosynthetic pathways. C metabolism primarily occurs within various types of plastids, such as *chloroplasts*, *leucoplasts*, *elaioplasts*, and *amyloplasts*. Chloroplasts, the photosynthetic plastids responsible for energy capture in vascular plants, typically exhibit a hemispherical or lens-shaped morphology. Leucoplasts are colourless plastids involved in the synthesis of monoterpenes and other volatile compounds found in essential oils. Elaioplasts are small, round plastids that serve as sites for oil storage in the form of membrane-associated droplets. Amyloplasts are unpigmented plastids that serve as the primary site for the synthesis and storage of starch.

Metabolic pathways, including starch synthesis, are found in both the cytosol and plastids, with specific enzyme isoforms in each compartment (see Sect. 2.4.2). This functional redundancy enables parallel metabolic fluxes and enhances metabolic flexibility. The plastid envelope membrane, functioning as a selective barrier between the plastid and the cytosol, serves as a vital interface for metabolic integration by dynamically coordinating plastid metabolism with the broader cellular metabolic network. Transporters embedded in the plastid membranes facilitate metabolite exchange between plastids and the cytosol, enabling selective transport of specific metabolites, ions, and essential energy carriers such as ATP and reducing equivalents (e.g. NAD(P)H). In addition to their transport functions, these proteins serve as signals that integrate metabolic information across cellular compartments, thereby playing a crucial role in coordinating various metabolic pathways.

Consequently, the *compartmentalization* of metabolic pathways between the cytosol and plastids, along with the activity of plastid envelope transporters, facilitates efficient spatial division and coordinated regulation of crucial processes, including photosynthesis, photorespiration, starch and lipid biosynthesis, and secondary metabolism. Moreover, this compartmentalization not only enhances the resilience to varying intracellular environments and maintains metabolic homeostasis but also offers a unique "cell factory" platform for metabolic engineering.

1.3 Modulation of Enzyme Activities

The chemical reactions of metabolism are systematically organized into *metabolic pathways*, in which one chemical is gradually transformed via a series of enzymatic steps into another chemical. *Enzymes* are biological catalysts synthesized by living cells, mainly composed of proteins that accelerate biochemical reactions without undergoing any permanent chemical alteration themselves. These complex proteins promote highly specific reactions by interacting with substrates at their active sites, thus lowering the activation energy required for the reactions to take place.

Enzyme activity is regulated by two distinct mechanisms, specifically coarse control and fine control (Smith et al. 2009). Generally, coarse control regulates the quantity of enzymes by modulating gene expression and operates on a relatively slower time scale. In contrast, fine control modifies the activity of existing enzyme molecules through mechanisms such as reversible binding, covalent modification, proteolytic cleavage, and protein–protein interactions, enabling cells to respond promptly to metabolic signals. Alterations in enzyme levels resulting from coarse control typically occur over a period of hours or days, whereas fine control acts more rapidly to adjust enzyme activities within the limits set by coarse control.

Coarse control is achieved through the modulation of the rates of enzyme protein synthesis and turnover. It can be regarded as establishing the cellular capacity for catalysing specific reactions or metabolic pathways, thereby governing the cellular abundance of enzyme proteins. Developmentally programmed metabolic changes and responses to environmental shifts frequently involve coarse control mechanisms. These mechanisms can regulate entire pathways, individual enzymes, or groups of enzymes within a pathway. For example, the levels of all enzymes involved in the CBC increase as a young leaf develops and acquires photosynthetic capacity, whereas these enzyme levels decline during leaf senescence.

Fine control is accomplished either via covalent modification of specific amino acid residues within the protein or through non-covalent interactions between the enzyme protein and specific metabolites. The two principal forms of covalent modification for regulating enzyme activity are *phosphorylation*/dephosphorylation of specific amino acid residues, frequently serine residues, and the oxidation/reduction of the sulfhydryl (–SH) groups of cysteine residues. Non-covalent interaction, referred to as allosteric regulation, is mediated by an *allosteric effector*, a small molecule that binds to regulatory sites distinct from the active site. This binding can induce conformational changes that substantially modulate enzyme activity. A classic illustration is the interaction between ATP and the enzyme phosphofructokinase (PFK) within the negative feedback loop that governs glycolysis. During this process, PFK catalyses the phosphorylation of F6P by ATP to F1,6BP (Smith et al. 2009). When ATP levels are elevated, ATP binds to an allosteric site on PFK, inducing a conformational change in the enzyme. This structural modification decreases the enzyme's affinity for its substrates, F6P and ATP, at the active site, thereby inhibiting its activity. In this way, ATP acts as a negative allosteric effector of PFK, even though it also serves as one of its substrates.

2 Carbon Assimilation and Metabolism

C assimilation, a plant process that transforms atmospheric CO_2 into organic compounds by harnessing solar energy, is a fundamental component of terrestrial ecosystems and the Earth's climate system. As primary producers, plants perform photosynthesis to synthesize sucrose, a precursor for essential biomolecules including starch, cellulose, proteins, and lipids. More than 90% of plant dry mass originates from photosynthetic assimilation, underscoring its critical role in determining crop productivity. Terrestrial vegetation sequesters approximately 123 billion metric tons of C annually, thereby offsetting roughly 30% of anthropogenic CO_2 emissions and making a significant contribution to mitigating climate change. Consequently, plant C assimilation and associated metabolic processes hold profound global significance. Although certain aspects of C metabolism are conserved across organisms, plants have developed unique biochemical pathways that differ from those of other life forms.

2.1 Photosynthetic Carbon Assimilation

Life on Earth is ultimately dependent on solar energy. Photosynthesis is the sole biological process capable of harvesting this energy. This process comprises two primary stages: the light-dependent reactions and the light-independent reactions, also referred to as the Calvin-Benson cycle (CBC) or the dark reactions. During the light-dependent reactions, chlorophyll and other carotenoid pigments located in the thylakoid membranes capture light energy, which initiates the photolysis of water and the transport of electrons, thereby generating a proton gradient used for ATP synthesis. Nicotinamide adenine dinucleotide phosphate ($NADP^+$) is subsequently reduced to NADPH, a key *reducing agent* in the CBC. Additionally, NADPH and ATP contribute to N and S assimilation, as well as other biosynthetic pathways.

The CBC is composed of three phases: carboxylation, reduction, and regeneration. Overall, producing one molecule of triose phosphate from three molecules of CO_2 consumes six molecules of NADPH and nine of ATP. The net glyceraldehyde 3-phosphate (GAP) is either metabolized immediately or transformed into storage carbohydrates, such as starch in the chloroplast or sucrose in the cytosol. Synchronization of CBC activity with the supply of ATP and NADPH is crucial for plants to efficiently assimilate C. Fluctuations in energy supply caused by changing light conditions necessitate precise regulation of specific CBC enzymes. The primary mechanism regulating CBC enzymes involves covalent modification mediated by ferredoxin and thioredoxin. Through this fine control mechanism, the activities of multiple enzymes are tightly linked to the electron flux through the photosystems, thereby ensuring a balanced supply and demand of NADPH and ATP for C assimilation.

2.2 The Hexose Phosphate Pool: A Major Intersection

The hexose phosphate pool is composed of three interconvertible metabolites, glucose 6-phosphate (G6P), glucose 1-phosphate (G1P), and fructose 6-phosphate (F6P), as illustrated in Fig. 2.1. Its composition is sustained in dynamic equilibrium by two key enzymes: *phosphoglucomutase* and *glucose 6-phosphate isomerase*. Both enzymes regulate readily reversible reactions that remain near equilibrium in vivo. Metabolites may enter this pool via the transformation of triose phosphates into hexose phosphates during photosynthesis. Conversely, they may exit the pool via synthesis of starch, sucrose, trehalose, and structural components of cell walls (Fig. 2.1). *Gluconeogenesis* represents an alternative pathway in which stored lipids are converted into glucose in germinating oilseeds. During the night or within heterotrophic tissues, C re-enters the pool via phosphorylation of free hexoses derived from the catabolism of sucrose or transient starch. In the absence of photosynthesis, glycolysis and the pentose phosphate pathway serve as the major metabolic drains, depleting the hexose phosphate pool to generate cellular energy and reducing equivalents.

2.3 Sucrose Metabolism

Sucrose, the principal product of photosynthesis, functions as the primary compound for long-distance transport of photosynthates in most plants and serves as a storage compound in certain plants, such as sugar beet, sugar cane, and carrot.

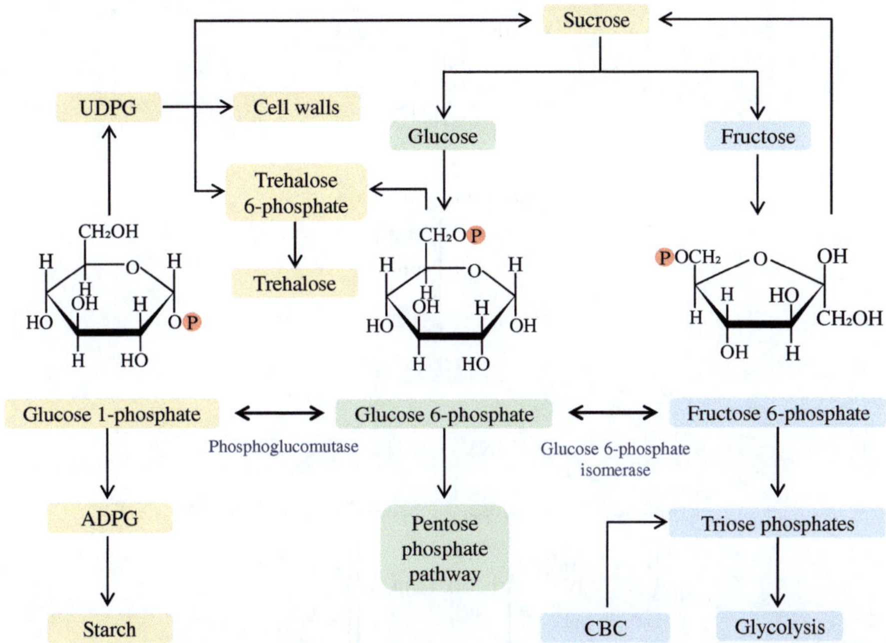

Fig. 2.1 The hexose phosphate pool and its role in cellular metabolism. The hexose phosphates serve as a crucial source of intermediates for glycolysis, the oxidative pentose phosphate pathway, and diverse biosynthetic processes. The intermediates of the hexose phosphate pool are interconverted by two enzymes: phosphoglucomutase and glucose 6-phosphate isomerase. The pathway of trehalose synthesis resembles that of sucrose synthesis, as both occur in the cytosol and utilize UDP-glucose and hexose phosphate as substrates. However, trehalose synthesis specifically uses glucose 6-phosphate, whereas sucrose biosynthesis employs fructose 6-phosphate. In the majority of plant tissues examined, trehalose 6-phosphate levels are typically very low but generally correlate with sucrose levels, potentially serving as a proxy for sucrose. It is now evident that most, if not all, plants possess the capacity to synthesize and degrade trehalose, with trehalose 6-phosphate, the biosynthetic intermediate, acting as the biologically active signalling molecule rather than trehalose itself. ADPG, ADP-glucose; CBC, Calvin-Benson cycle; UDPG, UDP-glucose. (Redrawn from Buchanan et al. (2015))

Sucrose forms a stable glycosidic bond between the anomeric carbons of glucose and fructose, effectively protecting these reactive sites from non-enzymatic oxidation by cellular components. This structural feature underlies its classification as a non-reducing sugar, a property that makes it highly suitable for transport.

2.3.1 Sucrose Synthesis

Sucrose synthesis commences with the formation of UDP-glucose from G1P and UTP, a reaction catalysed by UDP-glucose pyrophosphorylase (UGPase). Subsequently, UDP-glucose is condensed with F6P through sequential reactions catalysed by two key enzymes: sucrose-phosphate synthase (SPS) and sucrose-phosphate phosphatase (SPP; Fig. 2.2). This process is confined to the cytosol and

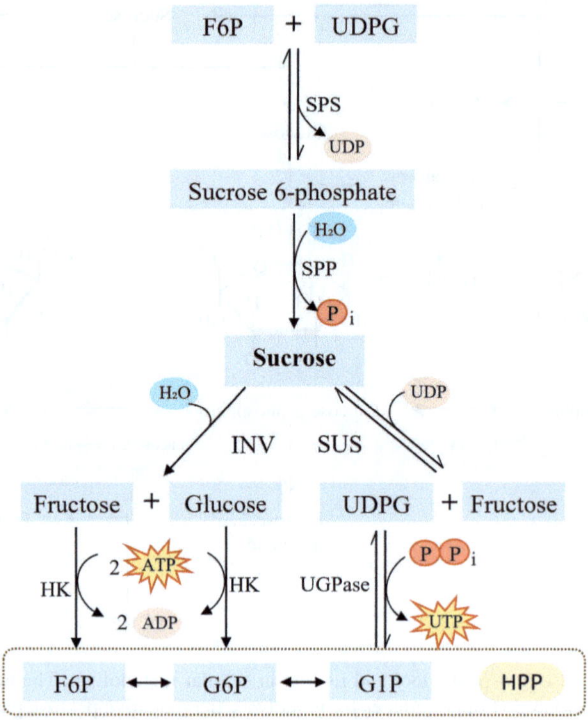

Fig. 2.2 An overview of sucrose metabolism and its relationship to the hexose phosphate pool. Sucrose synthesis is catalysed by sucrose-phosphate synthase (SPS) and sucrose-phosphate phosphatase (SPP), whereas its degradation is carried out by sucrose synthase (SUS) and invertase (INV). A key distinction between these two sucrose-degrading enzymes lies in the energy status of their products relative to the cytosolic hexose phosphate pool (HPP). Invertase catalyses the hydrolysis of sucrose into free hexoses, glucose and fructose, which subsequently are phosphorylated by hexokinase (HK) at the expense of ATP, producing G6P and F6P. In contrast, sucrose synthase catalyses the formation of UDP-glucose (UDPG), which can then undergo a reaction with pyrophosphate mediated by UDP-glucose pyrophosphorylase (UGPase) to yield G1P and UTP via an ATP-independent metabolic pathway. (Redrawn from Buchanan et al. (2015))

utilizes C from the hexose phosphate pool, whose composition varies between day and night.

During the day, CO_2 fixed by Rubisco enters the CBC and is converted into triose phosphates, specifically glyceraldehyde 3-phosphate (GAP) and dihydroxyacetone phosphate (DHAP). These triose phosphates are transported from the chloroplast to the cytosol, serving as precursors for the synthesis of fructose 1,6-bisphosphate (F1,6BP). In the cytosol, fructose 1,6-bisphosphatase (FBPase) catalyses the degradation of F1,6BP to form F6P, which is subsequently isomerized to G6P. The resultant F6P and G6P enter the hexose phosphate pool, providing essential substrates for sucrose biosynthesis.

During the night, photosynthetic C fixation ceases, resulting in the termination of triose phosphate production via the CBC. To maintain sucrose synthesis, transient starch that accumulates during daylight hours is hydrolysed by starch-phosphorylating enzymes, such as β-amylase and starch phosphorylase, in the chloroplasts. The resultant breakdown products are released directly into the chloroplast stroma as G6P, thereby bypassing the CBC. Subsequently, G6P is transported to the cytosol, replenishing the hexose phosphate pool for sucrose biosynthesis throughout the night. Consequently, nocturnal sucrose production primarily relies on the catabolism of transient starch stored in the chloroplasts during daytime. This dual mechanism ensures continuous sucrose synthesis and subsequent supply to the developing sinks, despite the diurnal shift between light-dependent CO_2 fixation and the utilization of stored C reserves.

2.3.2 Sucrose Degradation

Sucrose can be degraded by either *sucrose synthase* (SUS) or *invertase* (INV) (Fig. 2.2). The reaction catalysed by invertase is irreversible and results exclusively in sucrose degradation. The term "sucrose synthase" may be misleading: although the enzyme catalyses both the synthesis and cleavage of sucrose, its predominant physiological role is sucrose degradation. SPS, rather than SUS, serves as the primary enzyme accountable for sucrose biosynthesis in source tissues, where it demonstrates high activity. In contrast, SUS activity is significantly higher in sink tissues, which import and metabolize sucrose, than in source tissues. This tissue-specific activity pattern strongly supports the notion that SUS primarily functions in sucrose cleavage to supply hexose substrates for energy production and biosynthetic processes.

The precise roles of invertase and SUS in sucrose breakdown remain unclear. Both enzymes commonly coexist in the same tissues. A key distinction between them lies in how their reaction products interface with the cytosolic hexose phosphate pool: invertase yields free hexoses, which require ATP-dependent phosphorylation; in contrast, SUS produces UDP-glucose, which can react with pyrophosphate to yield G1P and UTP. Thus, in concert with UGPase, SUS enables an ATP-independent pathway for hexose phosphorylation (Fig. 2.2).

Both SUS and invertase are present as multiple isoenzymes, each with a specific function in particular tissues or cellular compartments. In potato tubers, the activity of invertase is inadequate to decompose sucrose at the requisite rate, suggesting a primary role of SUS in this process. Suppression of the major SUS genes expressed in tubers leads to a reduction in starch content and an inhibition of tuber growth. Similar findings in pea and maize support the notion that SUS activity is generally associated with the utilization of sucrose for the biosynthesis of *non-structural carbohydrates* (NSCs) such as starch, as well as structural carbohydrate polymers like cellulose and callose.

Invertases can be broadly classified into two major classes: acid invertases, which include cell-wall invertases (CWINs) and vacuolar invertases (VINs), and neutral/alkaline cytosolic invertases (CINs) that are situated in the cytoplasm. These different forms function in distinct cellular compartments. The cytosolic CINs play a crucial role in plant growth, particularly in root development, as evidenced by studies on *Arabidopsis* and rice mutants. VINs presumably catalyse the hydrolysis of stored sucrose and are usually found in tissues that accumulate hexoses, such as developing fruits. The apoplastic invertase isoform, CWINs, is firmly bound to the cell wall matrix and facilitates the hydrolysis of sucrose transported to sink tissues through the apoplast. This function is of particular significance in enhancing sink activities for cereal grain filling. *GRAIN INCOMPLETE FILLING 1* (*GIF1*), which encodes the rice CWIN, regulates the allocation of photosynthates during the early stage of grain filling and affects grain yield (Wang et al. 2008). In maize, loss of function of *MINIATURE 1* (*Mn1*), the *ortholog* of *GIF1*, leads to a typical miniature-seed phenotype and approximately a 70% reduction in grain yield (Cheng et al. 1996).

2.4 Starch Biosynthesis

Starch plays a crucial role in the carbon economy of most plants, with its functions varying across developmental stages and in response to environmental cues. Globally, it is estimated that cereal crops alone produce over 1 Gt of starch annually, which significantly contributes to the world's food security.

2.4.1 The Chemical Properties of Starch

Starch is predominantly composed of two glucose polymers, *amylose* and *amylopectin*, that form large, insoluble granules with a semi-crystalline structure. Starch granules vary in size, ranging from less than 1 μm to more than 100 μm in diameter. Transient starch granules in leaves are generally smaller than those in storage organs such as tubers and seeds, reflecting their synthesis over a shorter period during the day and subsequent degradation at night. By linking thousands of glucose monomers into a relatively small number of partially crystalline molecules, starch functions as a compact and stable form of carbohydrate storage. If the plastid were to accumulate an equivalent amount of sucrose, it would cause water to flow into the plastid from the cytosol via osmosis, leading to swelling and eventual rupture.

Both amylose and amylopectin are homopolymers composed of α-1,4-linked glucose units. Amylopectin typically consists of more than 100,000 glucose residues and constitutes 70% or more of the total starch content. It is a branched molecule, with approximately 4%–5% of its glycosidic linkages being α-1,6 bonds, which serve as branch points. In contrast, amylose generally contains around 1000 glucose residues and accounts for 30% or less of the starch, with significantly fewer branch points. The physicochemical properties of extracted starch are determined by the ratio of amylose to amylopectin, as well as the morphology of starch granules. These characteristics vary depending on the plant species and the storage organ. Archaeologists can identify the types of crops used by ancient civilizations by analysing residual starch grains adhering to their cooking utensils. In addition, starch properties can be modified through biotechnology by engineering plants to express altered levels of starch biosynthetic enzymes.

Amylopectin, the primary polymer responsible for the granular structure, exhibits a non-random branching pattern that results in the formation of clusters of unbranched chain segments. These linear regions are connected by left-handed double helices, each consisting of six glucose residues per turn. In contrast, amylose is predominantly linear or only slightly branched and tends to adopt single-helical conformations. The double helices of amylopectin further assemble into ordered crystalline lamellae, alternating with amorphous regions at intervals of approximately 9 nm. This hierarchical organization manifests on a larger scale as a periodic structure with a repeat distance of hundreds of nanometres, similar to the growth rings observed in trees. Nevertheless, the ring-like features within starch granules merely reflect the intrinsic structural organization, rather than developmental dynamics.

2.4.2 Starch Synthesis in the Chloroplast and Amyloplast

Starch synthesis in plants commences with ADP-glucose. In leaf chloroplasts, triose phosphates from the CBC enter the stromal hexose phosphate pool and are subsequently channelled into starch biosynthesis. In contrast, starch synthesis in amyloplasts depends entirely on sucrose imported from photosynthetic leaves (Fig. 2.3). In the amyloplasts of cereal endosperm, the G6P used as the substrate for ADP-glucose synthesis is typically transported from the cytosol. However, a second isoform of *ADP-glucose pyrophosphorylase* (AGPase) exists in the cytosol, in addition to the one located in the amyloplast. A specific transporter, namely, the Brittle 1 (BT1) protein situated at the plastid envelope, facilitates the direct influx of ADP-glucose into amyloplasts via an exchange mechanism with ADP. Empirical evidence indicates that this cytosolic pathway, rather than the plastidic pathway, serves as the predominant source of ADP-glucose in cereal endosperm. Mutations affecting either the cytosolic isoform of AGPase or the ADP-glucose transporter are associated with a 75% reduction in starch content.

2.4.3 Key Enzymes for Starch Synthesis

Starch biosynthesis proceeds via ADP-glucose as a substrate and requires the coordinated action of three distinct enzymes (Fig. 2.3): *starch synthases* (SSs), *starch branching enzymes* (BEs), and *starch debranching enzymes* (DBEs).

Five distinct isoforms of starch synthase (SS) are present in plants, each possessing a unique biochemical function. Granule-bound starch synthase (GBSS) differs from the other isoforms in that it exclusively binds to starch and becomes incorporated into the growing granule. GBSS is responsible for amylose synthesis; mutants lacking this enzyme produce waxy starch devoid of amylose. GBSS synthesizes long glucan chains through repeated action on the same substrate, ADP-glucose. It diffuses into the matrix of the starch granule, a phenomenon that may be crucial for amylose production. This is because nascent amylose chains can thus be shielded from the branching enzymes present in the plastid stroma. The remaining four SS isoforms are soluble within the plastid stroma and are jointly referred to as soluble starch synthases (SSS). They are partially associated with the surface of the starch granule and contribute to the biosynthesis of amylopectin by preferentially

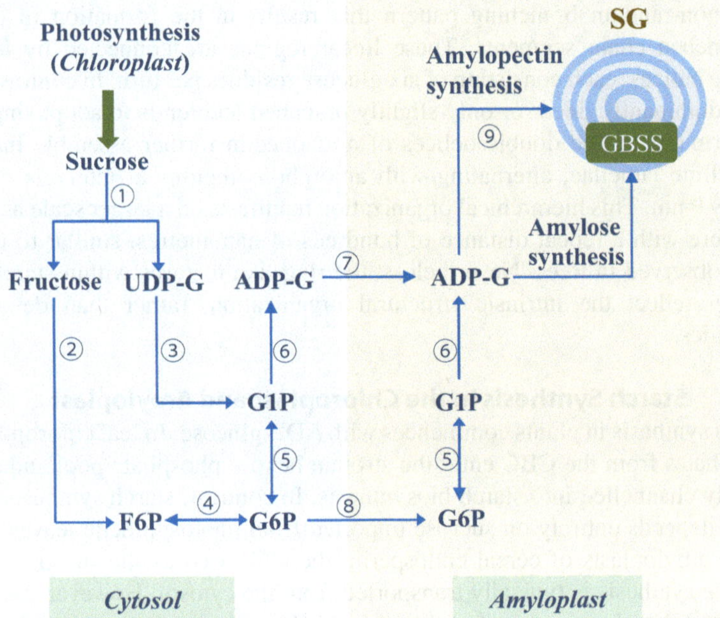

Fig. 2.3 The compartmentalization of starch biosynthesis in cereal endosperm. The energy and substrates required for starch synthesis in leaves are derived from photosynthesis, whereas in heterotrophic endosperm tissues, they are obtained through the metabolism of imported sucrose from the leaves. Cereal crops possess the unique capacity to synthesize ADP-glucose in the cytosol and subsequently transport it into the amyloplast, in addition to utilizing the plastid-localized biosynthetic pathway. In the amyloplast, starch synthesis involves four isoenzymes of soluble starch synthases, two branching enzymes, and a specialized debranching enzyme working together to produce branched amylopectin on the granule surface. The resulting amylopectin forms a semicrystalline matrix, trapping granule-bound starch synthase (GBSS) that synthesizes amylose using ADP-glucose diffusing into the amylopectin matrix. Enzymes or transporters: ① sucrose synthase, ② hexokinase, ③ UDP-glucose pyrophosphorylase, ④ glucose 6-phosphate isomerase, ⑤ phosphoglucomutase, ⑥ ADP-glucose pyrophosphorylase, ⑦ ADP-glucose transporter (restricted to the Poaceae family), ⑧ hexose transporter, and ⑨ soluble starch synthases 1–4, branching enzymes 1–2, and debranching enzyme. ADP-G, ADP-glucose; GBSS, granule-bound starch synthase; SG, starch granule. (Redrawn from Buchanan et al. (2015))

elongating glucan chains of diverse lengths. Single-isoform mutants are still capable of synthesizing starch, yet they typically display an altered amylopectin structure.

The α-1,6 branches in starch are synthesized by BEs. The majority of plants possess two isoforms of this enzyme, class I and class II, which differ in substrate specificity. These two isoforms exhibit distinct substrate preferences: class I enzymes transfer longer glucan chains than class II enzymes. Amylopectin biosynthesis generally involves partial debranching. Mutants lacking *isoamylase 1*, encoding a branching enzyme (DBE), accumulate a highly branched, water-soluble polymer that resembles *glycogen* more closely than amylopectin. Debranching promotes crystallization by removing misplaced branches that impede the formation of higher-order lamellar structures within amylopectin. This represents an intriguing evolutionary innovation: an enzyme ancestrally involved in starch degradation has been co-opted for a biosynthetic role through gene duplication and functional specialization. Other plant debranching enzymes catalyse branch-point hydrolysis during starch degradation.

2.5 Starch Degradation

2.5.1 In the Germinating Seeds

The understanding of starch degradation primarily derives from biochemical analyses conducted on germinating cereal grains, such as barley and wheat. This is driven by the importance of starch breakdown in industrial malt and flour production. The remobilization of starch from the endosperm of cereal grains represents an atypical instance of starch breakdown in plants. During this process, gibberellins (GAs) released by the embryo stimulate the *aleurone* layer, a living tissue surrounding the *starchy endosperm* and *scutellum*, to secrete hydrolases. Three types of starch hydrolases are secreted: *α-amylase*, *limit dextrinase* (a debranching enzyme), and *α-glucosidase* (*maltase*). In addition, secreted proteases activate a stored *β-amylase* in the endosperm via limited proteolysis at its carboxyl terminus. Subsequently, the embryo absorbs the released glucose to support seedling establishment and growth.

It is likely that GAs interact with a specific outward-facing receptor on the plasma membrane of aleurone cells. The signalling cascade initiated upon GA perception leads to transcriptional regulation of genes encoding α-amylase. The promoters of these α-amylase genes contain GA-responsive elements (GA-boxes), which are responsible for conferring GA-induced transcriptional activation. In barley, GAMYB has been identified as a key transcription factor that binds to the GA-box and regulates the expression of α-amylase genes. The semi-crystalline lamellar structure of the starch granules, which is a consequence of the packing of double helices formed by adjacent amylopectin chains, poses a challenge for starch degradation enzymes to access and hydrolyse. In the endosperm, α-amylase is believed to initiate granule degradation by targeting specific surface sites, namely, pores on the granule surface, resulting in the characteristic pitting morphology of the granules.

2.5.2 In the Chloroplast of Leaves

The transient starch stored within leaf chloroplasts serves as the primary energy and C source for plant metabolism and growth during the nocturnal period. In chloroplasts, glucan phosphorylation, catalysed by glucan water dikinase (GWD) and phosphoglucan water dikinase (PWD), modifies the surface properties of starch granules. The initial step of starch degradation in the chloroplast is the phosphorylation of a small fraction of glucosyl residues in amylopectin by GWD and PWD. GWD binds to the surface and introduces phosphate groups at the C-6 position of glucosyl residues, whereas PWD acts on pre-phosphorylated starch and adds phosphate groups at the C-3 position of adjacent glucosyl residues. Subsequent starch degradation involves several enzymes, such as α-amylase, β-amylase, debranching enzymes, *disproportionating enzyme*, and *starch phosphorylase*, yielding a complex reaction network in chloroplasts that ultimately generates maltose, glucose, and G1P.

2.5.3 In the Developing Seeds

In the developing *Arabidopsis* seed, the endosperm undergoes programmed cell death (PCD) and is degraded after fulfilling its nourishing function. The GA-induced expression of *AMY1*, encoding α-amylase, is closely associated with nutrient remobilization and PCD initiation at an early stage (Sreenivasulu et al. 2010). In barley with a persistent endosperm, genes encoding hydrolases are expressed in the maternal nucellus 2–3 days after fertilization (DAF), in the pericarp 8–12 DAF, and in the nucellar projection at 8 DAF. Nevertheless, they are not expressed during endosperm PCD in the maturing seed because of the expression of α-amylase inhibitors (Sreenivasulu et al. 2010). Sreenivasulu and Wobus (2013) proposed that the GA–α--amylase–PCD network may be activated in a highly time-specific manner during seed development across various species. In *Arabidopsis*, where the endosperm has a short lifespan, the GA–AMY1 pathway directly triggers PCD. In barley, however, the persistent endosperm requires additional regulatory mechanisms: hydrolase activity is spatially compartmentalized and suppressed to avoid premature degradation, thereby ensuring that PCD is initiated only when the seed is fully developed.

Starch granules within the developing rice endosperm are susceptible to partial degradation under conditions of pre-harvest sprouting or heat stress. In the *PHS8/ISA1* mutant exhibiting pre-harvest sprouting, elevated isoamylase activity has been attributed to the down-regulation of *OsABI3* and *OsABI5*, resulting in diminished ABA sensitivity (Du et al. 2018). Significantly, no evidence was found for GA involvement in starch hydrolysis under these conditions. Likewise, Hakata et al. (2012) proposed that α-amylase activation during heat stress may result from decreased ABA levels rather than GA induction. Collectively, these findings suggest that distinct hormonal regulatory mechanisms govern starch degradation in cereal endosperm: ABA predominantly suppresses degradation during grain development, whereas GA promotes it during germination.

3 Nitrogen Assimilation and Metabolism

N is one of the most essential mineral nutrients for plant growth and development, as it plays a key role in the synthesis of essential macromolecules, such as amino acids, proteins, nucleic acids, chlorophyll, and flavonoids. Given its critical importance to agricultural productivity, extensive research has been conducted on the biochemical and physiological mechanisms underlying N assimilation and metabolism (Krapp 2015).

3.1 Nitrogen Assimilation

3.1.1 Nitrate and Ammonium

Nitrate (NO_3^-) and ammonium (NH_4^+) are the primary forms of inorganic N absorbed by crop plants. NH_4^+ exhibits lower mobility in soil and persists in the vicinity of the fertilized root zone, thereby reducing leaching losses. However, it can induce toxicity at high concentrations, particularly in alkaline soils. In contrast, NO_3^- is highly soluble, rendering it readily available to plants but also susceptible to leaching, especially in sandy soils or during intense rain events. Fine-tuning the NH_4^+: NO_3^- ratio is a practical, evidence-based strategy to enhance NUE while reducing nitrate leaching and greenhouse gas emissions. For crops such as wheat and maize that thrive in well-aerated soils, a higher proportion of NO_3^- is beneficial. For rice paddies or greenhouse crops, a higher ratio of NH_4^+ can be advantageous.

Urea, the most extensively utilized N fertilizer, undergoes a two-step transformation in soil prior to being absorbed by crops (Fig. 2.4). Right after application, soil urease enzymes catalyse the breakdown of urea, resulting in the formation of ammonium carbonate, which rapidly dissociates into NH_4^+ and carbon dioxide. Subsequently, soil nitrifying bacteria convert NH_4^+ into NO_3^- through two distinct stages. Firstly, ammonia-oxidizing bacteria (such as Nitrosomonas) oxidize NH_4^+ to nitrite (NO_2^-). Secondly, nitrite-oxidizing bacteria (such as Nitrobacter) convert NO_2^- to NO_3^-. Although NH_4^+ is absorbed rapidly but has lower mobility, NO_3^- is highly mobile in soil water and can be efficiently taken up during the later growth stages.

Soil NO_3^- is taken up into the root cell via specific transporters and subsequently reduced to NO_2^- by nitrate reductase (NR) in the cytoplasm. NO_2^- is then transported into the plastids, where it is further reduced to NH_4^+ by nitrite reductase (NiR) (Fig. 2.4). The plastids responsible for nitrite reduction and its subsequent assimilation into amino acids are proplastids and amyloplasts in roots and chloroplasts in leaves. In many plants, when the roots receive small amounts of nitrate, nitrate is primarily reduced in the roots. As the absorption increases, a greater proportion of the nitrate is translocated to the shoot and assimilated therein. Even under similar conditions of nitrate supply, the balance between root and shoot nitrate metabolism varies from species to species, as evidenced by NR activity in both organs or by the NH_4^+: NO_3^- ratio in the xylem sap. In plants like cocklebur (*Xanthium strumarium*), NO_3^- metabolism is limited to the shoot, whereas in other plants like white lupine (*Lupinus albus*), most NO_3^- is metabolized in the roots.

3.1.2 The GS/GOGAT Cycle

Ammonium serves as a central intermediate in plant N assimilation, which comprises two interconnected processes: *primary* and *secondary assimilation*. Primary assimilation refers to the uptake of inorganic N from external sources, such as NH_4^+ and NO_3^-, and its direct conversion into organic compounds within the plant. Secondary assimilation, in contrast, involves the recapture of NH_4^+ generated internally through metabolic processes like photorespiration and protein degradation, thereby recycling N within plant tissues.

During primary N assimilation, NH_4^+ is acquired by plant roots either from the external soil environment, through the reduction of absorbed nitrate, or via symbiotic N fixation in legumes and other N-fixing plants. This "primary" NH_4^+ is rapidly incorporated into organic molecules through the *glutamine synthetase/glutamine 2-oxoglutarate aminotransferase* (GS/GOGAT) pathway, predominantly occurring in the root cytosol or leaf chloroplasts. The GS/GOGAT cycle constitutes the principal route for NH_4^+ assimilation (Fig. 2.4). In this metabolic pathway, GS catalyses

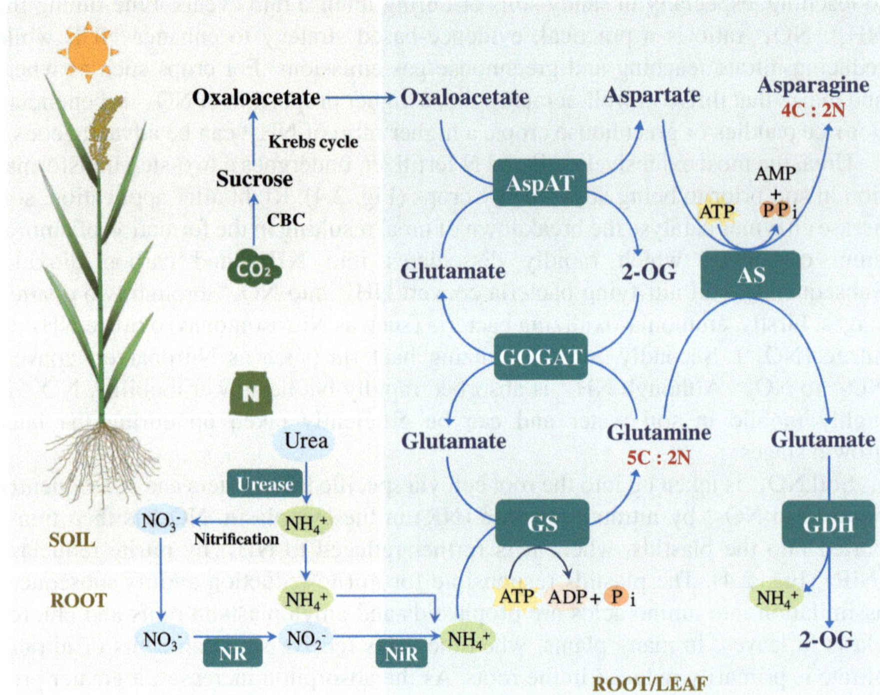

Fig. 2.4 Overview of nitrogen assimilation. Urea added to soil is rapidly hydrolysed by urease, producing ammonia and carbon dioxide. The ammonia quickly protonates, forming ammonium (NH_4^+), which plants can absorb directly; alternatively, soil microbes nitrify ammonium into nitrate (NO_3^-), another readily taken-up N form. Nitrate and ammonium are the main forms of inorganic N for plant nutrition. Nitrate is taken up by the cell and subsequently reduced to nitrite by nitrate reductase (NR) in the cytoplasm. Nitrite then enters plastids, such as proplastids and amyloplasts in roots, or chloroplasts in leaves, where nitrite reductase (NiR) converts it to ammonium. Ammonium is assimilated into glutamine and glutamate via the GS/GOGAT cycle, using 2-oxoglutarate (2-OG) from the Krebs cycle as a C source. Subsequently, aminotransferase reactions mediate the transfer of N-containing groups from glutamate and glutamine to asparagine and aspartate, which function as key N transport molecules in plants. Note that asparagine represents a more efficient use of reduced C than glutamine does. Both amino acids bind two N atoms, but glutamine contains five C atoms, while asparagine contains four. AS, asparagine synthetase; AspAT, aspartate aminotransferase; CBC, Calvin-Benson cycle; GDH, glutamate dehydrogenase; GOGAT, glutamine 2-oxoglutarate aminotransferase; GS, glutamine synthetase; NiR, nitrite reductase; NR, nitrate reductase; 2-OG, 2-oxoglutarate. (Redrawn from Smith et al. (2009) and Buchanan et al. (2015))

the ATP-dependent amination of glutamate (Glu) with NH_4^+ to form glutamine (Gln). Subsequently, GOGAT, also known as *glutamate synthase*, mediates the transfer of the amide group from Gln to 2-oxoglutarate (2-OG), yielding two molecules of Glu.

During secondary N assimilation, NH_4^+ is released intracellularly as a result of various metabolic processes within the plant. Internal sources of NH_4^+ include deamination of amino acids during protein turnover during leaf senescence, photorespiration-derived NH_4^+ production, and N remobilization during seed germination or under stress conditions. The released NH_4^+ is subsequently reassimilated via the GS/GOGAT pathway. This process is termed "secondary" because the N originates from endogenous organic compounds rather than from external environmental uptake. In contrast to animals, plants do not possess excretory mechanisms for nitrogenous waste; therefore, liberated NH_4^+ must be efficiently recaptured and reassimilated to sustain growth and development. Notably, NH_4^+ release through photorespiration can exceed primary N assimilation by up to tenfold, posing a significant risk of NH_4^+ toxicity in plants with impaired recycling capacity. To prevent such toxicity, plant cells rapidly convert the generated NH_4^+ into amino acids through the GS/GOGAT cycle.

Distinct isoenzymes of GS and GOGAT have been identified in all plant species examined (Liu et al. 2022). GS1 and GS2, the two isoforms of GS, are localized to the cytosol and chloroplasts, respectively. Cytosolic GS1 enzymes play a role in primary N assimilation in roots, particularly under high-nitrate conditions. In mature leaves, GS1 isoforms are mainly found in companion cells to promote phloem loading for N translocation to sinks. During leaf senescence, GS1 activity is also induced in mesophyll cells to re-assimilate NH_4^+ derived from amino acid catabolism. The chloroplastic GS2 primarily contributes to primary N assimilation and secondary assimilation of photorespiratory NH_4^+ in leaf mesophyll cells.

The two major classes of plant GOGAT enzymes are ferredoxin-dependent GOGAT (Fd-GOGAT) and NADH-dependent GOGAT (NADH-GOGAT), both of which are localized in plastids. Fd-GOGAT, mainly located in the chloroplasts, is the primary isoform involved in the GS/GOGAT cycle, which mediates secondary NH_4^+ assimilation associated with photorespiration. In contrast, NADH-GOGAT primarily functions in non-photosynthetic plastids. In leaves, it is localized in the plastids of companion cells within vascular bundles; in roots, it is found in analogous non-green plastids, thus supporting N assimilation in tissues devoid of photosynthetic activity. Consequently, the two GOGAT isoforms exhibit complementary roles: Fd-GOGAT mediates N metabolism in photosynthetically active mesophyll cells, whereas NADH-GOGAT fulfils this function in transport-associated companion cells of leaves and in root tissues (Tegeder and Masclaux-Daubresse 2018).

3.1.3 The AS/AspAT Cycle

The initial products of NH_4^+ assimilation, primarily glutamate and glutamine, are rapidly utilized in protein biosynthesis. These compounds also participate in N transport and serve as key precursors for the synthesis of amino acids, nucleobases, and other N-containing metabolites. Especially, assimilated N can be channelled into the synthesis of aspartate and asparagine. *Asparagine synthetase* (AS) catalyses the transfer of the amide group from Gln to aspartate (Asp), yielding asparagine (Asn) and Glu. *Aspartate aminotransferase* (AspAT) transfers the amide group from Glu to oxaloacetate, producing Asp and 2-OG. Aspartate is metabolically versatile and acts as an important N donor in aminotransferase reactions, as well as a precursor for a wide range of amino acids. In contrast, asparagine mainly functions in the transport and storage of N within the cell.

Additionally, *glutamate dehydrogenase* (GDH) plays a dual role by catalysing both the amination of 2-OG with NH_4^+ to generate Glu and the deamination of Glu to produce NH_4^+ and 2-OG. GDH has a high *Km* for NH_4^+ (10–80 mM), while tissue NH_4^+ concentrations typically range from 0.2 to 1.0 mM. It is widely accepted that in vivo, GDH primarily functions in glutamate catabolism, such as providing C skeletons for the Krebs cycle during darkness or supporting amino acid catabolism in germinating seeds and senescing leaves. Collectively, the enzymes GS, GOGAT, AS, AspAT, and GDH are central to the primary assimilation of inorganic N from soil and secondary assimilation of free NH_4^+ within plants (Fig. 2.4).

3.2 Amino Acid Biosynthesis

Amino acids are organic compounds that contain both an amino group and a carboxyl group. There are twenty proteinogenic amino acids that are incorporated into proteins during translation. The protein quality of crop products depends on *essential amino acid* (*EAA*) content. Animals cannot synthesize EAAs easily, so they must be supplied through diet. If one EAA is insufficient, others are broken down and excreted. Nine amino acids are strictly essential for humans: lysine, isoleucine, leucine, phenylalanine, threonine, tryptophan, valine, histidine, and methionine.[2] Moreover, plants produce numerous structurally diverse nonproteinogenic amino acids. Several of these amino acids are critically involved in N assimilation and the translocation of N from source to sink tissues, whereas others function as precursors for phytohormones, including indole-3-acetic acid and ethylene, as well as for numerous secondary metabolites that mediate plant responses to abiotic and biotic signals.

Amino acids are grouped into four biosynthetic families based on their metabolic precursors: the glutamate family, the pyruvate/serine family, the aspartate family, and the phosphoenolpyruvate family. The biosynthetic pathways for the four types of amino acids in plants are primarily inferred from microbial pathways (Fig. 2.5). The C skeletons used by plants for amino acid biosynthesis come from a few intermediates of glycolysis and the Krebs cycle, as well as the PPP (for histidine). The N-containing amino group is derived from transamination reactions involving glutamine or glutamate, the primary products of ammonium assimilation. Amino acid biosynthesis constitutes a highly integrated network that connects primary C metabolism with N assimilation, supplying the essential building blocks for proteins, hormones, and a diverse range of specialized metabolites.

[2] Cysteine is classified as a conditionally essential amino acid, as it can be synthesized in the body from methionine under normal physiological conditions.

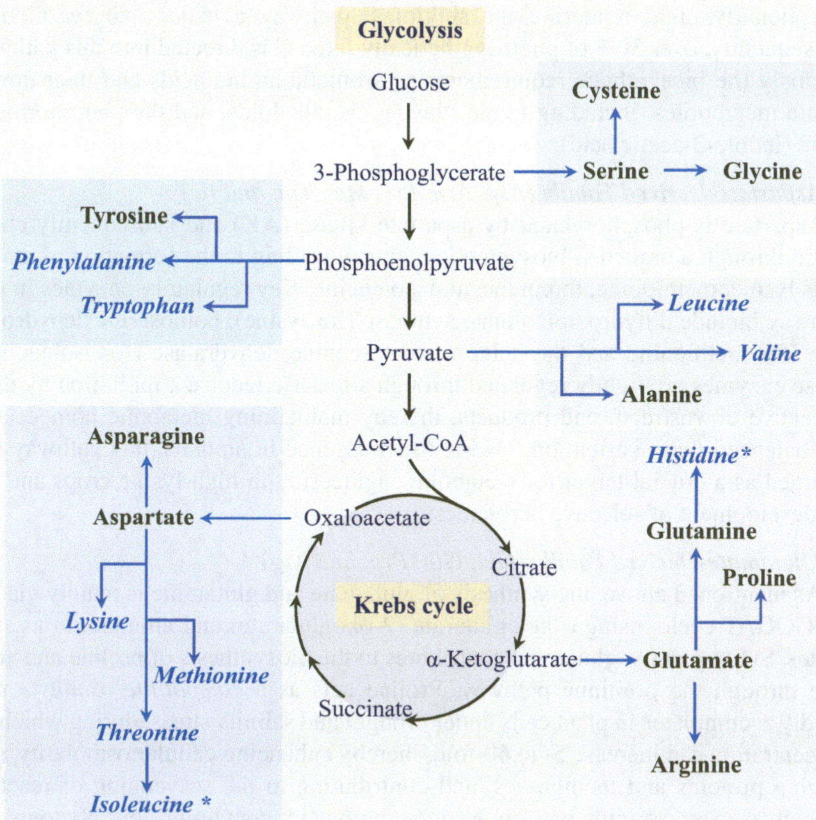

Fig. 2.5 Overview of the biosynthesis of the 20 proteinogenic amino acids in plants. The C skeletons of amino acids originate from various intermediates of central C metabolism. Based on their respective precursors, amino acids are classified into four families: those derived from glutamate, pyruvate/serine, aspartate, and phosphoenolpyruvate. It should be noted that isoleucine is involved in both the aspartate and pyruvate/serine metabolic pathways. During histidine biosynthesis, the C skeleton is derived from the pentose phosphate pathway, and the N is directly donated by glutamine. The nine essential amino acids, which animals are unable to synthesize, are presented in blue italics. (Redrawn after Buchanan et al. (2015))

1. *Phosphoenolpyruvate-Derived Family (Phe, Tyr, and Trp)*

The shikimate pathway mediates the enzymatic conversion of phosphoenolpyruvate (PEP) and erythrose-4-phosphate (E4P) into chorismate, which functions as the central precursor for the biosynthesis of the aromatic amino acids. Chorismate is subsequently directed toward phenylalanine biosynthesis via prephenate dehydratase, towards tyrosine through prephenate dehydrogenase, and towards tryptophan via anthranilate synthase. Crucial regulatory enzymes involved in this metabolic pathway encompass 3-deoxy-D-arabino-heptulosonate-7-phosphate (DAHP) synthase, 5-enolpyruvylshikimate-3-phosphate (EPSP) synthase, and chorismate mutase. In vascular plants, the demand for aromatic polymers such as lignin is exceptionally high, rendering the shikimate pathway a major hub for C flux. Consequently, over 30% of photosynthetically fixed C is directed into this pathway to satisfy the biosynthetic requirements of aromatic amino acids and their downstream metabolites, including lignin, flavonoids, alkaloids, and the plant hormone auxin (indole-3-acetic acid).

2. *Aspartate-Derived Family (Asp, Asn, Lys, Met, Thr, and Ile)*
   Aspartate is phosphorylated by aspartate kinase (AK) and subsequently channelled through a branched biosynthetic pathway leading to the formation of amino acids lysine, methionine, threonine, and isoleucine. Key regulatory enzymes in this pathway include dihydrodipicolinate synthase (for lysine), homoserine dehydrogenase (for methionine and threonine), and threonine dehydratase (for isoleucine). These enzymes are tightly regulated through allosteric feedback inhibition by their respective downstream end products, thereby maintaining metabolic homeostasis and balanced flux distribution. Owing to its absence in animals, this pathway has emerged as a crucial target for metabolic engineering in high-lysine crops and for the development of selective herbicides.

3. *Glutamate-Derived Family (Glu, Gln, Pro, and Arg)*
   As mentioned above, the synthesis of glutamate and glutamine is mainly via the GS/GOGAT cycle, using α-ketoglutarate (2-oxoglutarate) and ammonium as substrates. Subsequently, glutamine contributes to the biosynthesis of proline and arginine through the ornithine pathway. Proline acts as a *compatible osmolyte* that rapidly accumulates in plant cells under drought and salinity stress, during which its concentration can increase 5- to 60-fold, thereby enhancing cellular osmolarity, stabilizing proteins and membranes, and contributing to the scavenging of reactive oxygen species. Arginine is a key N-rich amino acid, containing four N atoms per molecule (approximately 32% by mass), which renders it an efficient compound for N storage. In seeds, tubers, and other storage tissues, arginine accumulates to high concentrations, serving as a readily mobilizable N reserve for germination and subsequent growth. Upon catabolism, arginine is converted to ornithine, which is then decarboxylated to putrescine via ornithine decarboxylase; further enzymatic reactions produce spermidine and spermine, the major polyamines that play essential roles in cell division, membrane stability, and stress responses.

4. *Pyruvate/Serine Families (Ala, Val, Leu, Ile, Ser, Gly, and Cys)*
   The serine and pyruvate amino acid biosynthetic pathways share key enzymes and are regulated by feedback inhibition mediated by the common end product, isoleucine. As a result, these two families are often treated as a single functional unit in integrated analyses. Pyruvate is transaminated to form alanine, while branched-chain amino acids (valine, leucine, and isoleucine) are synthesized from pyruvate via acetolactate synthase and subsequent dehydrogenation steps. Serine is derived from 3-phosphoglycerate through the action of phosphoglycerate dehydrogenase. Glycine and cysteine are then generated from serine by the serine hydroxymethyltransferase (SHMT) and cysteine synthase, respectively. SHMT links serine/glycine

interconversion to the folate cycle, providing one-C units for nucleotide biosynthesis. The cysteine synthase route is part of the sulphate assimilation pathway; the resulting cysteine supplies the sulphur required for glutathione synthesis, the major cellular redox buffer. Therefore, these pathways intersect with one-C metabolism and sulphur assimilation, thus connecting amino acid biosynthesis to the formation of folate derivatives and glutathione.

Across all the four families, feedback inhibition of the first committed enzyme, such as AK, DAHP synthase, and GS, serves as a primary regulatory mechanism. In addition, transcription factors that respond to C–N balance, light conditions, and environmental stresses simultaneously regulate the expression of genes encoding key metabolic enzymes. A comprehensive understanding of these regulatory pathways enables three major applications: (i) improving the nutritional quality of staple crops by overexpressing or deregulating rate-limiting enzymes; (ii) producing high-value secondary metabolites, such as phenylpropanoids and alkaloids, through metabolic engineering; and (iii) developing herbicides that target essential enzymes, including EPSP synthase (e.g. glyphosate) and acetolactate synthase.

3.3 Storage Protein Biosynthesis

The nutritional quality of *seed storage proteins* (SSPs) is of great significance for formulating feed for monogastric livestock and poultry, and its importance will increase as plant proteins substitute for animal proteins in processed foods (Shewry 2024). Legume seeds are particularly rich in protein, with contents typically ranging from 20 to 40% of the seed dry weight. In contrast, cereal grains contain lower levels of protein, averaging approximately 10% to 12% of dry weight. Despite this, cereals provide approximately three times more dietary protein than legume seeds due to their significantly higher production volume.

3.3.1 Classification of Seed Storage Proteins

SSPs generally contain high levels of the amino acids glutamine and asparagine compared to other plant proteins. They also possess high concentrations of amide groups and arginine, featuring a high N to C ratio that aligns with their role in N storage. Although SSPs are enzymatically inactive, many are evolutionarily derived from families of active enzymes, such as trypsin inhibitors or other hydrolases. In the seeds of many eudicots, storage proteins are predominantly accumulated in the embryonic cotyledons, whereas in most cereals, they are stored mainly in the endosperm tissue.

Based on their solubility, SSPs are categorized into albumins, globulins, prolamins, and glutelins, collectively known as Osborne fractions (Shewry 2024). Albumins are water-soluble, whereas globulins dissolve in dilute saline solutions (typically 0.5 M or 1 M NaCl). These two fractions primarily consist of metabolic and storage proteins. Globulins display structural domains similar to those of germin-like ancestors, which mediate cellular adaptations to water loss, desiccation, and osmotic imbalances. Globulins with sedimentation coefficients ($S_{20,w}$)[3] of 7–8 and 11–12 serve as major storage proteins in most crop plants, especially legumes. 2S albumin storage proteins are stored in the dicotyledonous seeds of *Brassica* species.

In cereal endosperm, the major SSPs are prolamins, which can only be effectively extracted by aqueous alcohols, typically 60%–70% (v/v) ethanol, although 50% (v/v) propanol is now more commonly employed. Prolamins are distinguished by their high proline and glutamine content and are solely found in cereals and other grass species. These proteins have been named after their species of origin: gliadin (wheat), hordein (barley), secalin (rye), avenin (oats), zein (maize), kafirin (sorghum), and oryzin (rice).

Glutelins can only be extracted using acidic or alkaline solvents. This fraction also encompasses insoluble structural and metabolic proteins. Upon reduction of the disulfide bonds that stabilize the glutenin polymers, the protein monomers become readily soluble in aqueous alcohols. These monomers exhibit amino acid sequences and properties that are closely related to those of classical prolamins. Glutelins are the most abundant group of SSPs in rice, accounting for 70% to 80% of the total SSPs. These proteins exhibit sequence homology with the 11–12S globulins found in legumes. The rice glutelins have a molecular weight (MW) ranging from 45 to 150 kDa and consist of two disulfide-linked subunits: an acidic α-polypeptide with a MW of 34–39 kDa and a basic β-polypeptide of 21–23 kDa.

3.3.2 Critical Role of Proteins in Cereal Grain Quality

Zeins, the predominant endosperm proteins in maize, account for approximately 50% of total seed proteins and comprise several alcohol-soluble polypeptides encoded by at least four gene families. The deficiency of two essential amino acids, lysine and tryptophan, in zeins restricts the application of maize proteins in the food and feed industries. Opaque 2 (O2) is an endosperm-specific transcription factor belonging to the basic leucine zipper (bZIP) family that regulates the expression of most zein genes. In the *o2* mutant, the synthesis of 22-kDa α- and β-zein proteins is markedly reduced. As a result of proteome rebalancing, the levels of non-zein proteins are elevated, compensating for the reduction in zein proteins and maintaining total protein levels in the *o2* endosperm. This shift increases the relative abundance of essential amino acids, particularly lysine and tryptophan, thereby significantly enhancing the nutritional value. The discovery of this mutation generated significant enthusiasm regarding its potential to enhance the nutritional value of the maize grain. However, the soft, floury endosperm of the mutant led to lower kernel density and increased vulnerability to insects, pathogens, and mechanical damage. These issues, coupled with reduced yield and protein content, hindered the widespread agronomic adoption of *o2* (Geetha et al. 1991).

Beyond their nutritional significance as a primary source of essential amino acids, the quantity and quality of cereal seed proteins also play a crucial role in food processing, notably influencing bread-making quality and suitability for brewing.

[3] Sedimentation coefficient in an aqueous solution at 20 °C.

The prolamin and glutelin fractions of wheat are known as gliadin and glutenin, respectively. These proteins are present in approximately equal proportions and constitute 70% or more of the total grain protein content. Glutenin in wheat is a complex polymer composed of high-molecular-weight glutenin subunits (HMW-GSs, with MW ranging from 67 to 90 kDa) and low-molecular-weight glutenin subunits (LMW-GSs, with MW ranging from 30 to 45 kDa), linked by disulfide bonds. The HMW-glutenins significantly affect the sensory properties of the bread. Together with gliadin, glutenin proteins constitute the gluten fraction, which imparts unique functional properties to wheat flour and dough. This fraction provides the cohesiveness and visco-elasticity that enable wheat dough to expand during fermentation, resulting in leavened foods. While the prolamins of barley and rye are closely related to those of wheat, they do not form a comparable visco-elastic network, highlighting the uniqueness of wheat in this context.

Across indica and japonica rice cultivars, grains with lower protein content (approximately 6%–7% of dry weight) generally yield softer and more cohesive kernels after cooking, whereas those with higher protein levels (>8%) tend to be firmer, less sticky, and are associated with lower sensory evaluation scores. The inverse correlation between protein content and palatability is primarily attributed to the presence of protein bodies occupying interstitial spaces between starch granules. During cooking, these protein bodies interact with starch through hydrogen bonding, disulfide linkages, and hydrophobic interactions, functioning as a physical constraint that limits starch swelling during gelatinization. However, specific protein fractions, such as glutelin, prolamin, albumin, and globulin, exert distinct influences on quality traits. Prolamin contributes to increased hardness and reduced stickiness by forming compact protein bodies that restrict starch granule swelling. Glutelin affects visual appearance, though its impact on textural properties is less pronounced compared to prolamin. Albumin enhances pasting viscosity and promotes the formation of a softer gel structure when present in higher concentrations. Globulin improves water-binding capacity, thereby modestly enhancing gel cohesiveness. Consequently, the overall sensory profile of the cooked rice is determined by the relative proportions and interactions among these protein fractions.

3.3.3 Biosynthesis of Seed Storage Proteins

In plant cells, protein biosynthesis takes place across three distinct intracellular locations: the cytoplasm, plastids, and mitochondria. The majority, approximately 75%, of all cellular proteins are produced in the cytoplasm through the translation of nuclear-encoded mRNAs. In photosynthetic cells, such as those in developing leaves, around 20% of the proteome is synthesized within chloroplasts, using transcripts originating from the chloroplast genome. Meanwhile, mitochondrial protein synthesis contributes a small yet crucial proportion, accounting for approximately 2%–5% of the total cellular protein output.

The biosynthesis of SSPs predominantly occurs on ribosomes (Fig. 2.6). It begins with the transcription of genes in the nucleus, yielding primary pre-mRNAs that undergo post-transcriptional modifications within the spliceosome. Subsequently, mature mRNAs are transported from the nucleus to the cytoplasm. In

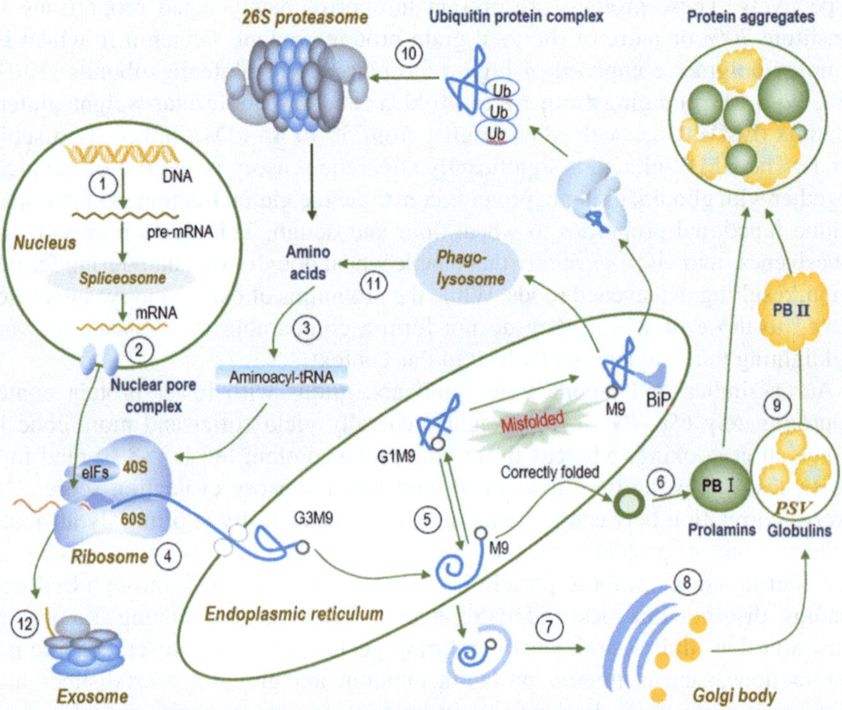

Fig. 2.6 A putative pathway of protein body formation processes. ① transcription, ② mRNA processing and transport, ③ aminoacyl-tRNA biosynthesis, ④ ribosome biosynthesis and translation, ⑤ protein processing in the ER, ⑥ ER-derived PB formation, ⑦ vesicle-mediated protein transport, ⑧ protein classification and packaging in the Golgi, ⑨ protein storage vacuole (PSV) formation, ⑩ ubiquitin-mediated proteolysis, ⑪ other damaged protein degradation, and ⑫ RNA degradation in the exosome. Membrane-bound and soluble proteins of the secretory pathway by the ER are frequently glycosylated. The core *N*-linked glycan exhibits a branched structure, such as G3M9 ($Glc_3Man_8GlcNAc_2$), G1M9 ($Glc_1Man_8GlcNAc_2$), and M9 ($Man_9GlcNAc_2$), which serve as structural configurations that precisely guide the folding of protein substrates and instruct quality control systems. BiP, binding protein; eIFs, eukaryotic initiation factors; ER, endoplasmic reticulum; PB, protein body; Ub, ubiquitin.

the cytoplasm, ribosomes recruit mRNA, amino acids, and various protein cofactors to facilitate translation. Through a series of sequential steps, including amino acid activation, initiation, elongation, termination, and release, a nascent polypeptide chain is synthesized. Once translation is complete, the mRNA is degraded by the exosome.

During transport from their site of synthesis to the protein body, the peptides undergo significant post-translational modifications. Initially, the peptide chain undergoes additional folding and sorting processes within the lumen of the endoplasmic reticulum (ER). This process involves quality control mediated by *N*-linked glycans such as G3M9, G1M9, and M9, as well as assistance from ER chaperones, including *binding proteins* (*BiPs*), *protein disulphide isomerases* (*PDIs*), and other *co-chaperones*. The uncorrected misfolded proteins are targeted for degradation via ubiquitin-mediated proteolysis in the 26S proteasome or are degraded within the phagolysosome, thereby releasing free amino acids for recycling. If the protein is correctly folded, the 7S and 11S globulins and 2S albumins are transported from the ER via the Golgi to *protein storage vacuoles* (*PSV*), forming the protein body II (PB II). Meanwhile, the prolamins also bud directly from the ER to form PB I. Finally, PB I and PB II are combined to form large protein aggregates.

4 Interactions Between C and N Metabolism at the Cellular Level

The cellular-level interaction between C and N can be illustrated by the process of photorespiration, in which the recycling of the released C is tightly coupled with secondary N assimilation via the GS/GOGAT cycle. This interaction also exhibits a diurnal pattern, wherein the availability of C skeletons influences N allocation between glutamine and asparagine. In addition, signalling molecules such as sugars and nitrate, along with sensor proteins including sucrose non-fermenting 1-related protein kinase 1 (SnRK1) and target of rapamycin kinase (TOR),[4] play pivotal roles in regulating C–N balance.

4.1 Photorespiration Integrates Carbon and Nitrogen Metabolism

Throughout geological history, oxygenic photosynthesis has significantly modified the atmospheric composition. Currently, the concentrations of O_2 are approximately 500 times higher than those of CO_2. In the CBC, C fixation is catalysed by ribulose-1,5-bisphosphate carboxylase/oxygenase (Rubisco), an enzyme that inherently binds CO_2 more readily than O_2. Yet, given the contemporary atmospheric composition, Rubisco frequently incorporates O_2 instead of CO_2, a reaction known as *photorespiration*, which yields 3-phosphoglycerate (3-PG) and 2-phosphoglycolate (2-PG). Crucially, 2-PG functions as a potent inhibitor of several essential CBC enzymes, including triose phosphate isomerase and SBPase (Smith et al. 2023).

The two C atoms of 2-PG cannot undergo further metabolism within the CBC. Instead, 2-PG is converted to 3-PG via photorespiration (also known as the C2 cycle), during which one C atom derived from the two molecules of 2-PG is released as CO_2; this results in a 25% loss of previously assimilated C. Moreover,

[4] SnRK1 and TOR regulate plant energy and nutrient status in response to sucrose signals. SnRK1 acts as an energy sensor, promoting catabolism and suppressing anabolism under low sucrose. TOR promotes growth by stimulating protein synthesis and cell proliferation when nutrients are abundant. The antagonistic interaction between SnRK1 and TOR maintains metabolic balance: SnRK1 inhibits TOR during nutrient scarcity, while TOR suppresses SnRK1 under resource-rich conditions.

this process leads to the release of ammonia and the consumption of ATP and reducing power (e.g. NADPH). Overall, photorespiration substantially reduces the efficiency of C assimilation in C3 plants, causing yield losses of approximately 30% or more.

The C2 cycle necessitates the coordination of three organelles: chloroplasts, mitochondria, and peroxisomes. During the C2 cycle, glycine decarboxylation releases ammonia within the mitochondrion. To prevent the accumulation of toxic ammonia and avoid N loss through volatilization, photo-respiring cells must efficiently recycle this ammonia within the chloroplast via the GS/GOGAT cycle. The reassimilation of ammonia involves the glutamate produced by peroxisomal transamination, which generates amino acids, such as serine and glycine, for incorporation into other metabolic pathways.

Although photorespiration diminishes the overall efficiency of photosynthesis, it is closely associated with the C and N metabolic pathways within leaf cells and may offer benefits under specific environmental circumstances. The cycle involves the metabolism of the amino acid glycine in a reaction that releases ammonia, which has the potential to result in significant losses of assimilated N. It is noteworthy that, under illuminated conditions, the rate of ammonium reassimilation generated by photorespiration is considerably higher than the rate of primary ammonium assimilation originating from nitrate uptake in the majority of plants. Consequently, photorespiration may play a vital role in optimizing metabolic functions in C3 plants, especially by improving N metabolism. The mechanisms are still under investigation, but they may be associated with the ability of photorespiration to alleviate over-reduction in the chloroplasts, thereby preventing photoinhibition. Furthermore, photorespiration may enhance stress tolerance in C3 plants by generating *reactive oxygen species* (*ROS*), which can activate the expression of stress-responsive signalling genes. Consequently, crop improvement strategies should be aimed at moderately reducing photorespiratory flux, typically by introducing alternative glycolate metabolic pathways, while maintaining its protective functions. In this way, yields can be improved without sacrificing stress resilience.

4.2 Carbon Availability Regulates the Assimilation of Nitrogen into Amino Acids

The assimilation of inorganic N into amino acids used for biosynthesis is a highly regulated process, modulated by both environmental cues, particularly light, and the intracellular availability of C and N intermediates. As early as the mid-nineteenth century, researchers observed that light exposure markedly alters asparagine levels across diverse plant species. Later work showed that asparagine accumulation in phloem sap and the enzymatic activity of AS are suppressed under illuminated conditions or in the presence of sucrose, whereas both increase substantially during dark adaptation. By contrast, glutamine displays an opposing regulatory pattern.

Genes accountable for ammonium assimilation are regulated by both light and cellular metabolism. For example, the genes encoding nitrate reductase and nitrite reductase are upregulated by both light and sucrose. Similarly, exposure to light stimulates the expression of genes responsible for chloroplast GS2 and Fd-GOGAT. Conversely, light downregulates the expression of genes for AS. Moreover, sugars also influence the expression of genes involved in N assimilation. Sucrose or glucose can, at a minimum, partially replace light in triggering the expression of genes for GS2 and Fd-GOGAT, as evidenced in tobacco and *Arabidopsis*. In contrast, sucrose represses the expression of AS genes when provided to dark-adapted maize explants or to whole *Arabidopsis* plants. The impacts of sucrose on GS and AS gene expression can be counteracted by amino acids. Amino acids repress the sucrose-induced GS expression, while they mitigate the sucrose-mediated inhibition of AS expression.

Building upon the bidirectional regulatory relationship between C availability and organic N levels in the control of GS and AS expression, Buchanan et al. (2015) developed a metabolic regulatory framework. In this framework, GS transcription is upregulated under light conditions or when sufficient C skeletons are available for ammonium incorporation. As a result, during illumination, N is primarily assimilated into glutamine, a metabolically versatile amino acid that serves as a key N donor in multiple biosynthetic pathways. In contrast, AS expression is induced in the dark, when photosynthetic C fixation ceases, or when the organic N pool greatly exceeds C supply. Under such C-limited or N-rich conditions, plants preferentially direct fixed N into asparagine, a N-rich, metabolically stable compound with an N:C ratio significantly higher than that of glutamine. This shift enables more efficient N storage and long-distance transport when C backbones are scarce (Fig. 2.4).

4.3 Signalling Coordination in the Interaction of Carbon and Nitrogen Metabolism

In the plant cell, the intricate *signalling* pathways that regulate cellular C and N metabolic synergy involve small molecules (e.g. nitrate, sucrose, and amino acids), signalling proteins (e.g. kinases and phosphatases), and transcription factors. These components control cellular metabolism through fine and/or coarse mechanisms.

To synchronize C skeleton production essential for amino acid biosynthesis, nitrate enhances the expression of genes involved in C catabolism. Upon nitrate exposure, cellular C flux shifts away from starch accumulation and is redirected toward the synthesis of amino acids and organic acids (e.g. malate) through coordinated modulation of both the expression levels and catalytic activities of key metabolic enzymes. For instance, high nitrate levels induce the expression of phosphoenolpyruvate carboxylase (PEPC) while reducing that of ADP-glucose pyrophosphorylase, which is the essential enzyme for starch biosynthesis. As a result, nitrate functions as a signalling molecule that guides plants to reallocate C from C-based biosynthesis to the synthesis of N-containing compounds.

On the other hand, C positively regulates N assimilation and metabolism at transcriptional and post-translational levels. Sugars upregulate the expression of genes responsible for nitrate uptake and assimilation. G6P plays a critical role in the signalling cascade of the sucrose-induced expression of nitrate uptake transporters NRT1.1 and NRT2.1. Trehalose 6-phosphate (T6P) activates NR and PEPC at the post-translational level, thereby guiding C allocation towards N metabolism (Fernie et al. 2020). SnRK1 and TOR function as key sensors of cellular C/N ratios. A high C status activates TOR to enhance N assimilation and growth, whereas a low C or high N level induces SnRK1 to strengthen stress adaptation and C remobilization (Morales-Herrera et al. 2024). This balance is further refined by metabolites such as 2-OG, which serves as a signal indicating the relationship between C supply and NR activity.

The transcription factor GROWTH-REGULATING FACTOR 4 (GRF4) and the growth inhibitor DELLA interact to maintain a balance between plant growth and development in rice. GRF4 promotes and integrates N assimilation, C fixation, and growth, whereas DELLA inhibits these physiological processes. As a result, the accumulation of DELLA in the Green Revolution cultivars not only leads to yield-increasing dwarfism but also decreases NUE. By fine-tuning the balance between GRF4 and DELLA to moderately promote GRF4 accumulation, Li et al. (2018) demonstrated that both NUE and grain yield can be enhanced simultaneously in modern rice cultivars. This strategy presents a promising approach for the next green revolution, aiming to achieve substantial yield increases while mitigating the negative environmental impacts associated with N fertilizer overuse.

The exogenous application of ABA suppresses C and N assimilation and reduces photosynthetic activity. This inhibitory effect depends on ABI5, a key component of the ABA signalling pathway (Long et al. 2025). ABI5 acts as a transcriptional repressor that directly binds to the promoters of C assimilation genes *RCA*, *RBCS2B*, *RBCS3B*, and the N assimilation gene *GLU1* (encoding chloroplast-localized Fd-GOGAT), thereby repressing their transcription. Reducing *ABI5* expression enhances photosynthetic performance, biomass accumulation, and grain yield in both *Arabidopsis* and rice. These findings reveal that ABI5 plays a crucial role in regulating photosynthesis via coordinated C and N metabolism, underscoring its potential as a genetic target for enhancing crop productivity.

5 The Relevance of C and N Interactions

The interaction between C and N metabolism during the entire life cycle of crop plants has a profound impact on grain yield, quality formation, and resource use efficiency. Specifically, starch (and sometimes lipids) and protein, the end products of C and N metabolism, respectively, play crucial roles in determining crop quality. Nevertheless, achieving an optimal balance in the chemical composition of harvested organs remains highly challenging due to complex C–N interactions across multiple biological levels.

5.1 Balancing the Accumulation of Starch and Protein in Cereal Grains

Optimizing both seed storage proteins (SSPs) and starch accumulation is essential for achieving high yield and superior quality in cereal crops. Starch, which makes up 75%–85% of the grain's dry weight, is the primary energy source for humans and directly influences grain yield and overall quality. Meanwhile, SSPs provide most of the essential amino acids, greatly enhancing the nutritional value and palatability of the grain. The processing characteristics and taste of cereal products are largely determined by the interaction between starch and SSPs.

Genes associated with SSPs are undergoing domestication in cereal crops. However, this process often unfolds via complex and, at times, unforeseen genetic and phenotypic pathways. Historically, artificial selection for high-yielding rice cultivars has favoured non-shattering panicles and larger grains with enhanced starch accumulation. However, this breeding strategy has inadvertently compromised both the quantity and functional quality of grain proteins. Furthermore, strategies designed to enhance EAA profiles and total protein levels often lead to undesirable phenotypic trade-offs, making it difficult to simultaneously improve nutritional value and maintain field performance (Tiozon Jr et al. 2026).

In wheat, grain filling, predominantly the accumulation of starch, commences shortly after anthesis and persists for the subsequent 20–30 days. Under optimal growth conditions, factors primarily acting within the grain itself, such as endosperm cellularization and differentiation, govern both the rate and duration of starch deposition. This process is largely genetically controlled and functions within the asymptotic phase of the rate-versus-supply curve. This intrinsic regulation guarantees a stable accumulation of starch, which accounts for the majority of grain weight and is essential for yield stability. In contrast, protein deposition exhibits a high degree of responsiveness to the external N supply. Although the inherent capacity for protein synthesis in wheat grain is genetically determined, the actual deposition rate is regulated by the availability and redistribution of N from the leaves and roots. This positions protein accumulation within the linear phase of the rate-versus-supply curve, where the deposition rate is directly proportional to N availability. As a result, protein content demonstrates significant variability, reflecting agronomic practices (e.g. N fertilization) and environmental conditions that influence N uptake and transport.

Consequently, the rates and durations of starch and protein accumulation in wheat grains are independent processes, each governed by distinct regulatory mechanisms (Impa et al. 2020). From a physiological standpoint, there is no inherent conflict in selecting cultivars or managing crops to achieve high grain or starch yields while maintaining acceptable, and even elevated, levels of grain protein content (Jenner et al. 1991). Empirically, the starch content in grains frequently increases at the expense of other constituents, a phenomenon known as the "dilution effect", which underlies the well-established inverse relationship between grain yield and protein concentration. Nevertheless, certain wheat genotypes consistently deviate from this trend: some maintain high protein concentrations despite high yields, whereas others exhibit unexpectedly low protein levels even at moderate yields. These deviations are termed positive and negative grain protein deviation (GPD), respectively. Evidence suggests that GPD is governed by a limited number of genes located on chromosomes 3A and 5A, with certain genes functioning independently of grain protein content. Consequently, starch (grain) yield and protein yield should be regarded as inherently unrelated and distinct traits in crop breeding programs.

Numerous studies have been conducted to gain a deeper understanding of the genetic architecture underlying the synthesis of SSPs or starch, which involves a range of TFs. The *synteny* of cereal genomes shows that O2-like TFs are conserved throughout cereal evolution (Cao et al. 2024). O2 and prolamin-box binding factor 1 (PBF1) are major activators of SSP and starch synthesis in maize and rice. Both TFs interact physically to form a complex for synergistic regulation of the storage reservoir in grains. Transient expression assays validated that the maize O2 and PBF1 proteins function as transcriptional activators of rice SSP-encoding genes, suggesting their functional conservation in SSP accumulation. Transient expression assays of *TaSPA* and *TaPBF*, the orthologs of *O2* and *PBF1*, respectively, also confirmed that they promote SSP accumulation in wheat. The conserved module formed by O2 and PBF1 plays a central role in SSP and starch synthesis in major cereal crops, but has undergone functional differentiation during evolution. Future research should focus on identifying the regulatory factors that integrate SSP and starch synthesis to optimize yield and quality improvement.

5.2 Breaking the Inverse Correlation Between Protein and Oil Content in Legumes

In soybean seeds, storage proteins make up approximately 35%–40% of the dry weight, while lipids (specifically oil) account for around 18%–20%, primarily in the form of *triacylglycerol* (*TAG*). Carbohydrates, including *raffinose* family oligosaccharides (RFOs) and cell wall polysaccharides (CWPs), constitute roughly 35% or more of the final seed composition and play a crucial role in diverting C away from oil and protein production. The RFOs, such as raffinose and stachyose, cannot be digested by monogastric animals for energy production and are therefore regarded as anti-nutritional factors in livestock feed. Breeders have encountered significant challenges in simultaneously enhancing soybean seed yield and protein content. Indeed, it has long been recognized that while yield and oil content are typically positively correlated, both traits exhibit a negative correlation with protein content. Overcoming these inverse correlations to enhance the overall value of seeds without compromising yields remains an unachieved goal for most breeding and biotechnological endeavours.

The negative association between protein and oil production in soybean seed primarily stems from the composition of mature seed, which reflects the cumulative outcome of events throughout soybean seed development. Therefore, it does not provide specific information regarding metabolic fluctuations at each developmental stage. In developing soybean seeds, C is allocated among oil, protein, and carbohydrates. Kambhampati et al. (2021) assessed maternal nutrient supply via measurement of seed coat exudates and metabolite levels within the cotyledon throughout development to identify trends in the accumulation of central C and N metabolic intermediates. Active metabolic activity during late seed development was probed through transient labelling with ^{13}C substrates. Both lipid and protein levels peak during the early maturation phase and subsequently decline as residuals, predominantly in the form of CWPs, begin to accumulate. The results indicate a decrease in lipid and protein contents during seed maturation, accompanied by a corresponding increase in carbohydrates.

Recent endeavours to enhance seed quality have concentrated on directing lipases to minimize lipid degradation during the late developmental stage, along with regulating the biosynthesis pathways of RFOs, thus improving seed nutritional quality without influencing maturation or germination. Aznar-Moreno et al. (2022) demonstrated that suppressing lipase-mediated turnover of TAGs during late embryogenesis increases fatty acid content and reduces RFOs accumulation. Three soybean genes orthologous to the *Arabidopsis* lipase gene *SUGAR-DEPENDENT 1* (*SDP1*) are upregulated during this developmental phase. Suppression of these genes leads to increased oil accumulation, with lipid content in the most effective lines exceeding 24% of seed weight. Furthermore, transgenic lines with suppressed lipase activity exhibit larger seed size compared to WT plants, resulting in a greater than 20% increase in total lipid. Raffinose and stachyose levels are reduced in the transgenic lines, with an average decrease of 15% in total RFOs. Notably, despite the elevated oil content, protein levels remain unaffected and show a tendency to increase. These findings support a role for *SDP1* in TAG turnover to provide C skeletons for other metabolic processes, including RFOs biosynthesis, and present novel avenues for enhancing soybean seed composition. Therefore, targeted manipulation of key regulatory nodes in C partitioning may effectively decouple the traditionally observed inverse relationship between oil and protein content.

6 Concluding Remarks and Perspectives

C and N assimilation and metabolism are dominant biochemical processes in crop physiology. Their products, e.g. starch, sucrose, proteins, and lipids, comprise the major components of the harvest organs. The quantity of these products directly determines crop yield, while their composition fundamentally influences crop quality. C and N metabolism are intricately interconnected across multiple organizational levels, and strategies for adjusting the C–N balance must account for this interdependence. At the cellular level, C and N metabolism are intricately intertwined. Due to their close biochemical coordination and reciprocal regulation, it is impossible to meaningfully modify either process in isolation by genetic approaches. At the whole-plant level, prioritizing photosynthetic C fixation without a corresponding improvement in N assimilation results in metabolic imbalances that limit growth and development. Conversely, an excessive supply of N frequently downregulates photosynthesis through feedback inhibition, thus reducing C assimilation. Achieving precise spatiotemporal coordination between C and N metabolism remains a central challenge for simultaneously enhancing crop yield and quality.

In the future, with continuous advancements in genome editing technology and synthetic biology, the deep integration of molecular design and artificial intelligence will enable the precise tailoring of C and N metabolism in crop plants. More importantly, crop physiology, the discipline aimed at understanding interactions across various biological levels, should have a critical role by providing the key link between genetic modification and crop performance. There is an urgent need to construct a systematic platform that integrates genomics, biochemistry, physiology, phenomics, and ecology to fully evaluate the impact of crop biotechnology under field conditions. By integrating fundamental plant biology with crop science, we can usher in a novel agricultural paradigm that balances the quantity and quality of crop products.

Summary

1. Assimilation refers to the process by which organisms convert exterior nutrients into their own building blocks and store energy.
2. Metabolism is the sum of intracellular chemical reactions that occur in a living organism, serving as energy sources for essential processes and providing substrates for the synthesis of new organic compounds.
3. The process of crop yield formation can be regarded as the outcome of interactions between C and N. Fundamentally, it involves applying N fertilizer to the soil, which enables plants to assimilate C into carbohydrates via enzymes, such as Rubisco, that are synthesized using the added nitrogen.
4. Starch synthesis in cereal grains begins with ADP-glucose in the plastid stroma. It mainly involves the synthesis of amylose by GBSS embedded within the growing starch granule, and the formation of branched amylopectin through the coordinated actions of four SSs, two BEs, and a specific DBE, all of which act on the granule surface.
5. The C skeletons used by plants for amino acid biosynthesis are derived from only a handful of intermediates of glycolysis, the PPP, and the Krebs cycle.
6. The GS/GOGAT cycle is the principal N assimilation pathway in plants. Glu, Gln, Asp, and Asn are the major forms for N translocation.
7. The intricate C–N interactions during the growth stage of crop plants have a profound impact on crop physiology, posing a significant challenge to achieving an optimal balance between grain yield and quality.

Important Terms

Anabolism; AS/AspAT/GDH/GOGAT/GS; Assimilation; Catabolism; Compartmentalization; Interaction; Metabolism; N primary assimilation; N secondary assimilation; Plastids

Review Questions

Multiple Choice

1. (   ) are found in companion cells of the leaf vein, promoting phloem loading for N translocation to sinks.
   (a) GS1/Fd-GOGAT
   (b) GS2/Fd-GOGAT
   (c) GS1/NADH-GOGAT
   (d) GS2/NADH-GOGAT
2. In amyloplasts, carbon for starch synthesis is derived from the metabolism of (   ) translocated from source tissues.
   (a) Glucose
   (b) Fructose
   (c) Sucrose
   (d) G-1-P
3. In the chloroplast, carbon for starch synthesis is derived from (   ).
   (a) Calvin-Benson cycle
   (b) Sucrose catabolism
   (c) Glycolysis
   (d) The pentose phosphate pathway
4. (   ) is a colourless plastid that is involved in the synthesis of monoterpenes and other volatile compounds found in essential oils.
   (a) Leucoplast
   (b) Elaioplast
   (c) Etioplast
   (d) Amyloplast
5. Developmentally programmed metabolic changes and responses to significant and sustained environmental shifts often involve (   ) mechanisms of enzymes. For instance, the levels of all enzymes involved in the Calvin-Benson cycle increase during leaf development and acquisition of photosynthetic capacity, whereas these enzyme levels decrease during leaf senescence.
   (a) Fine control
   (b) Covalent modification

(c) Non-covalent modification
(d) Coarse control

6. ( ) is the non-covalent modification mechanism that operates on a shorter timescale to modulate enzyme activities as compared to coarse control.

(a) Phosphorylation/dephosphorylation of serine residues
(b) Oxidation/reduction of cysteine residues' sulfhydryl (–SH)
(c) A nonbinding site of an enzyme molecule reacting with an allosteric effector
(d) Modulating the rates of enzyme protein synthesis and turnover

7. The hexose phosphate pool is composed of three interconvertible hexose phosphates, EXCEPT ( ).

(a) G1P
(b) G6P
(c) F1P
(d) F6P

8. The reaction of sucrose degradation catalysed by ( ) is irreversible.

(a) SPS
(b) SPP
(c) Sucrose synthase
(d) Invertase

9. The process of protein biosynthesis occurs on ( ), which are complex assemblies composed of proteins and RNA that securely anchor tRNA and mRNA in position while facilitating the catalysis of peptide bond formation between amino acid residues.

(a) Ribosome
(b) Endoplasmic reticulum
(c) Nucleus
(d) Golgi

10. ( ) can stimulate seed germination by inducing de novo synthesis of α-amylase and other hydrolytic enzymes in the aleurone layer of cereal grains.

(a) GAs
(b) CKs
(c) Auxin
(d) SLs

11. The assimilation of secondary ammonium during photorespiration in leaf chloroplasts is primarily mediated by ( ).

(a) GS1/Fd-GOGAT
(b) GS2/Fd-GOGAT
(c) GS1/NADH-GOGAT
(d) GS2/NADH-GOGAT

True or False

1. GS, AS, GOGAT, and AspAT all participate in N assimilation (   ).
2. It is generally accepted that the primary role for GDH in vivo is in glutamate anabolism (   ).
3. Amino acid biosynthesis pathways in plants have been inferred largely from microbial pathways (   ).
4. For the 20 "proteinogenic" amino acids, adult humans cannot synthesize nine of them, such as lysine, methionine, and aspartate (   ).
5. Growth rings of starch granule in cereal endosperm are probably a structural feature rather than representing periodic phases of growth (   ).
6. GBSS differs from other starch synthases as it is exclusively bound to starch and becomes buried within the granule as it grows (   ).
7. The cellular abundance of enzyme proteins is regulated by coarse control, while fine control operates on a shorter timescale to adjust enzyme activities within the established framework by coarse control (   ).

Blank Filling

1. The production of branched amylopectin requires the concerted actions of four isoenzymes of soluble starch synthases, two of branching enzymes, and a specialized ________________, all acting at the starch granule surface.
2. The process of assimilating ammonium from the soil is commonly referred to as ________________ nitrogen assimilation.
3. Plant cells avoid ammonium toxicity by rapidly converting this type of ammonium into amino acids via the GS/GOGAT cycle, which is known as ________________ nitrogen assimilation.

Short Answer

1. Diagrammatic illustration of the processes of sucrose biosynthesis and degradation.
2. Diagram how NH_4^+ is incorporated into the four important amino acids including Glu, Gln, Asp, and Asn.
3. Interpret how light and C metabolism regulate N assimilation into amino acids.
4. Exemplify the interaction of C and N metabolism at the cellular level.
5. Contrast the anabolism of starch in leaves with that in germinating cereal grains.

Concept Map

Create a concept map using the terms below: *DNA; Nucleus; Spliceosome; mRNA; tRNA; Phagolysosome; Endoplasmic reticulum; Exosome; 26S proteasome; Ubiquitin ligase complex; Protein body II; Protein body I; Prolamins; Globulins; Golgi apparatus; Ribosome; Amino acids.*

Essay Question

1. Evaluate the challenges associated with the simultaneous enhancement of grain yield and quality, using wheat, rice, or soybean as an example.
2. C and N metabolism are intricately intertwined at various levels, posing a huge challenge when attempting to modulate physiological processes in order to achieve a good balance between the chemical components of crop seeds (better quality). In your field, whether it is breeding or management, what are the possible ways to breed or grow crops with higher quality?

References

Aznar-Moreno JA, Mukherjee T, Morley SA et al (2022) Suppression of *SDP1* improves soybean seed composition by increasing oil and reducing undigestible oligosaccharides. Front Plant Sci 13:863254. https://doi.org/10.3389/fpls.2022.863254

Buchanan BB, Gruissem W, Jones RL (2015) Biochemistry & molecular biology of plants, 2nd edn. Wiley Blackwell, Hoboken

Cao SH, Liu BY, Wang DW et al (2024) Orchestrating seed storage protein and starch accumulation toward overcoming yield-quality trade-off in cereal crops. J Integr Plant Biol 66(3):468–483. https://doi.org/10.1111/jipb.13633

Cheng WH, Taliercio EW, Chourey PS (1996) The *MINIATURE1* seed locus of maize encodes a cell wall invertase required for normal development of endosperm and maternal cells in the pedicel. Plant Cell 8:971–983. https://doi.org/10.1105/tpc.8.6.971

Du L, Xu F, al Fang J e (2018) Endosperm sugar accumulation caused by mutation of *PHS8/ISA1* leads to pre-harvest sprouting in rice. Plant J 95(3):545–556. https://doi.org/10.1111/tpj.13970

Fernie AR, Bachem CWB, Helariutta Y et al (2020) Synchronization of developmental, molecular and metabolic aspects of source–sink interactions. Nat Plants 6(2):55–66. https://doi.org/10.1038/s41477-020-0590-x

Geetha KB, Lending CR, al Lopes MA e (1991) Opaque-2 modifiers increase γ-zein synthesis and alter its spatial distribution in maize endosperm. Plant Cell 3(11):1207–1219. https://doi.org/10.1105/tpc.3.11.1207

Hakata M, Kuroda M, Miyashita T et al (2012) Suppression of α-amylase genes improves quality of rice grain ripened under high temperature. Plant Biotechnol J 10(9):1110–1117. https://doi.org/10.1111/j.1467-7652.2012.00741.x

Impa SM, Vennapusa AR, Bheemanahalli R et al (2020) High night temperature induced changes in grain starch metabolism alters starch, protein, and lipid accumulation in winter wheat. Plant Cell Environ 43(2):431–447. https://doi.org/10.1111/pce.13671

Jenner CF, Ugalde TD, Aspinall D (1991) The physiology of starch and protein deposition in the endosperm of wheat. Funct Plant Biol 18(3):211–226. https://www.researchgate.net/publication/262994876

Kambhampati S, Aznar-Moreno JA, Bailey SR et al (2021) Temporal changes in metabolism late in seed development affect biomass composition. Plant Physiol 186(2):874–890. https://doi.org/10.1093/plphys/kiab116

Krapp A (2015) Plant nitrogen assimilation and its regulation: a complex puzzle with missing pieces. Curr Opin Plant Biol 25:115–122. https://doi.org/10.1016/j.pbi.2015.05.010

Li S, Tian YH, Wu K et al (2018) Modulating plant growth–metabolism coordination for sustainable agriculture. Nature 560(7720):595–600. https://doi.org/10.1038/s41586-018-0415-5

Liu XJ, Hu B, Chu CC (2022) Nitrogen assimilation in plants: current status and future prospects. J Genet Genomics 49(5):394–404. https://doi.org/10.1016/j.jgg.2021.12.006

Liu H, Gao XH, Fan WS et al (2025) Optimizing carbon and nitrogen metabolism in plants: from fundamental principles to practical applications. J Integr Plant Biol 67(6):1447–1466. https://doi.org/10.1111/jipb.13919

Long Q, Huang Y, Xie H et al (2025) ABA insensitive 5 mediates abscisic acid regulation of plant photosynthesis by coordinating carbon and nitrogen assimilation. New Phytol 248:1903–1919. https://doi.org/10.1111/nph.70409

Morales-Herrera S, Paul MJ, al Dijck PV e (2024) SnRK1/TOR/T6P: three musketeers guarding energy for root growth. Trends Plant Sci 29(10):1066–1076. https://doi.org/10.1016/j.tplants.2024.03.006

Shewry PR (2024) Can we increase the use of wheat and other cereals as sources of protein? J Cereal Sci 117:103899. https://doi.org/10.1016/j.jcs.2024.103899

Smith AM, Couplant G, Dolan L et al (2009) Plant biology. Taylor & Francis Group, LLC, New York

Smith EN, van Aalst M, Tosens T et al (2023) Improving photosynthetic efficiency toward food security: strategies, advances, and perspectives. Mol Plant 16(10):1547–1563. https://doi.org/10.1016/j.molp.2023.08.017

Sreenivasulu N, Wobus U (2013) Seed-development programs: a systems biology–based comparison between dicots and monocots. Annu Rev Plant Biol 64:189–217. https://doi.org/10.1146/annurev-arplant-050312-120215

Sreenivasulu N, Borisjuk L, Junker BH et al (2010) Barley grain development: toward an integrative view. Int Rev Int Rev Cell Mol Biol 281:49–89. https://doi.org/10.1016/S1937-6448(10)81002-0

Tegeder M, Masclaux-Daubresse C (2018) Source and sink mechanisms of nitrogen transport and use. New Phytol 217(1):35–53. https://doi.org/10.1111/nph.14876

Tiozon R Jr, Zhan J, De Guzman CD et al (2026) Cereal protein biofortification at the interface of nutrition, yield and sustainability. Nat Plants. https://doi.org/10.1038/s41477-026-02252-5

Wang ET, Wang JJ, Zhu XD et al (2008) Control of rice grain-filling and yield by a gene with a potential signature of domestication. Nat Genet 40(11):1370–1374. https://doi.org/10.1038/ng.220

Embryo–Endosperm Interaction

3

A Model System for Unlocking the Secrets of Crop Quality

Contents

Z. Liu, *Crop Physiology*, https://doi.org/10.1007/978-981-92-2960-4_3

Abstract

This chapter investigates organ-level physiological processes by examining the critical interaction between the embryo and the endosperm, highlighting its essential role in dominating cereal grain filling and quality. Although advancements in functional genomics have revealed critical genetic programs for seed development, the mechanisms of embryo–endosperm communication and its impact on grain filling are still not fully understood. This chapter first presents the classification and physicochemical basis of quality traits, followed by an overview of the morphological and developmental characteristics of cereal grains, establishing a spatiotemporal framework for analysis. It then focuses on the bidirectional signalling in embryo–endosperm interactions and employs rice grain chalkiness as a representative example to elucidate the regulatory mechanisms behind complex quality traits. Finally, it identifies four pivotal scientific and agronomic questions that warrant further investigation.

Keywords

Embryo · Endosperm · Development · Interaction · Crop quality · Cereal grain

Did You Know

- The embryo makes up a relatively small portion of the whole grain in rice (2%) and wheat (3%), but it is considerably larger in maize, representing 10%–15% of the grain weight, with an average of about 12%.
- The endosperm constitutes the primary component of cereal grains, accounting for approximately 91%–92% in rice, 81%–84% in wheat, and 80%–84% in maize.
- It is widely recognized that the endosperm of most grasses becomes starchy and dry upon maturity, although notable exceptions exist. For example, in a survey of 169 grass genera, nine were found to have semisolid endosperms, and 30 contained species with liquid or soft endosperms at maturity, a condition that can persist for several decades.
- The rice endosperm may have 12–22 cells in the dorsal radius, 10–18 cells in the ventral radius, 10–17 cells in the lateral radius, and 103–256 cells along the longitudinal axis. It is thus estimated that the endosperm of one rice grain contains between 10 and 20 thousand cells.

Learning Objectives

After reading this chapter, students will be able to:

- Comprehend the complexity of crop quality formation
- Understand the morphological foundation of cereal grain quality
- Describe the developmental processes of the embryo and endosperm
- Summarize how the embryo and endosperm interact at their interface
- Make use of the embryo–endosperm interaction to dissect the complex quality trait

Introductory Remarks

The production of *seeds*, which ensures the protection, nourishment, and dispersal of the embryo through maternal tissues, undoubtedly contributes significantly to the remarkable evolutionary success of seed plants that currently dominate terrestrial ecosystems. The ecological dominance of angiosperms can be attributed to their distinctive capacity for *double fertilization*, which leads to the simultaneous formation of both an *embryo* and *endosperm*, enclosed within the maternal seed coat. In *eudicots*, the mature seed primarily comprises the embryo, while the endosperm tissues are transient and eventually absorbed by the embryo during later developmental phases. Conversely, *monocots* such as wheat, maize, and rice have a persistent endosperm in their grains, serving as the principal storage tissue of nutrients (Wang et al. 2022).

The endosperm serves as a critical source of nourishment for the embryo, facilitating its growth and germination. Consequently, successful embryo development is highly dependent on proper endosperm formation. Failure in endosperm development will significantly hinder embryo growth. This intimate relationship requires efficient signal exchanges between the endosperm and the embryo to ensure coordinated growth and development. However, attaining an in-depth understanding of the intricate interactions between the embryo and the endosperm requires a solid foundation in the evolutionary origins of these processes, a task that presents considerable difficulty. Furthermore, these interactions are influenced by multiple internal and external factors, including environmental signals, parental conflicts, and, in the case of crop species, artificial selection for target traits such as large ears or high lysine content (Doll and Ingram 2022).

The fundamental science of functional genomics has disclosed the key mechanisms governing seed development. For the endosperm, research has partially elucidated the cellular processes involved in its development across various crop species and characterized the genes responsible for endosperm differentiation. Additionally, the regulatory networks governing the accumulation of storage products and the epigenetic mechanisms underlying *imprinted gene* expression are gradually being revealed (Liu et al. 2022).

(continued)

Meanwhile, functional genomic studies in monocots, such as rice, maize, and wheat, have deepened our understanding of the molecular aspects of grain embryogenesis (Armenta-Medina et al. 2020). However, the communication mechanisms between the endosperm and embryo are still poorly understood, and their functional roles in grain development and quality formation remain to be fully elucidated.

The order of genes has presumably undergone alterations during evolution because of chromosomal rearrangements, deletions, and insertions. Nevertheless, the gene arrangement remains largely conserved among related species. For example, cereal crops, such as wheat, barley, maize, rice, sorghum, millet, and sugarcane, display different chromosome numbers. However, when the chromosomes of each species are fragmented and rearranged, a conserved gene order becomes evident, presumably reflecting the genetic composition of their shared grass ancestor. The conservation of gene order, known as *synteny*, aids in the identification of genes accountable for valuable phenotypes in crop species. As more plant genome sequences become publicly available, *comparative genomics* will accelerate the discovery of functionally conserved genes across species. Consequently, knowledge regarding gene function can be efficiently transferred from *model plants* to crops and even to wild relatives.

Rice holds significant global importance as a vital staple food globally, particularly in Asia, Africa, and South America. Moreover, it has emerged as a pivotal model system for plant science due to several key factors: (a) possessing the smallest genome among major crop plants[1]; (b) exhibiting substantial synteny with grass family members, including vital cereal crops; (c) availability of high-precision genome sequences; and (d) access to efficient transformation technology, among other advantages. Rice genomic studies have provided fundamental insights into the complex and dynamic processes governing crop productivity and resilience.

Recently, improving milling, appearance, and sensory properties has become a key objective in rice breeding and production. However, our understanding of the molecular and physiological mechanisms underlying rice quality is still limited. Using the model plant rice as a paradigm, this chapter focuses on the dialogue between the embryo and endosperm, elucidating its pivotal role in regulating cereal grain quality. Notably, it presents the current knowledge and prospects of this emerging research topic on embryo–endosperm interaction from a crop physiological perspective, aiming to bridge the gap between fundamental plant science (molecular study) and practical technology (crop breeding and management).

[1] Rice was the first crop to undergo genome sequencing due to its chromosome number of $2n = 24$ and smallest estimated genome size (400–430 Mb) among major cereal crops, which made sequencing economically feasible. In contrast, wheat is a polyploid with a much larger genome size of approximately 16,000 Mb, containing $2n = 6x$ chromosomes with three sets of related genomes (A, B, and D), as well as significant amounts of duplicated (25%–30%) and repeated DNA (over 80%) (Wijerathna-Yapa et al. 2024).

Concept Map

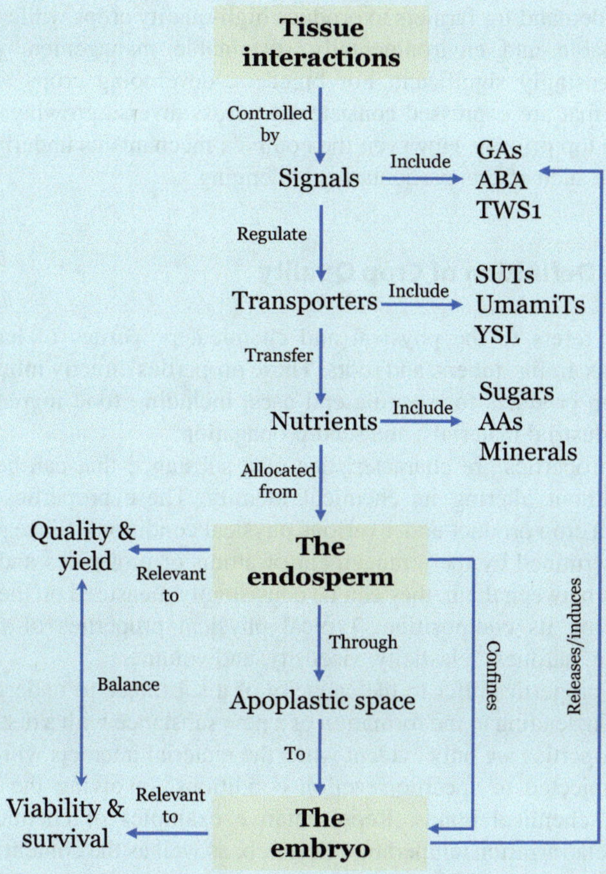

Annotation: This map illustrates the key physiological processes at the organ level, the interaction between the two major seed tissues, the embryo and the endosperm, in cereal grains. Briefly, to obtain sufficient nutrients, the embryo releases signalling molecules such as hormones to regulate the allocation of nutrients towards itself via apoplastic pathways. Transporters, such as UmamiTs and SUTs, located at the interface between the two tissues, mediate this transmembrane movement of nutrients. The volume and composition of the endosperm ultimately influence grain weight and quality, whereas the viability and vigour of the embryo determine the germination potential of the seed. Therefore, it is essential to maintain a precise balance between embryo and endosperm development.

1 Basic Concepts of Crop Quality

The growing demand for farmers to produce high-quality crops while adopting economically viable and environmentally sustainable management practices has become increasingly significant. For breeders, developing crops with desirable quality traits that are expressed consistently across diverse growing environments has become a top priority. However, the complex mechanisms underlying the quality traits make such efforts particularly challenging.

1.1 The Definition of Crop Quality

Crop quality refers to the physical and chemical properties of harvested plant organs, such as grains, tubers, and roots. These properties directly influence the suitability of crop products for specific end uses, including food ingredients, animal feeds, raw industrial materials, and seed propagation.

Physical properties are characteristics of a substance that can be observed or measured without altering its chemical identity. These properties describe the behaviour of a crop product under various physical conditions. Since physical properties are determined by the arrangement of atoms or molecules and the intermolecular forces between them, they can be consistently measured on the same sample without altering its composition. Typical physical properties of crop products include colour, hardness, elasticity, viscosity, and volume.

Chemical properties refer to the capacity of a substance to undergo a chemical transformation, leading to the formation of a new substance with a distinct composition. Such properties are only evident when the material interacts with other chemicals or is subjected to specific reaction conditions, involving the breaking and formation of chemical bonds. Representative examples of chemical properties include the gelatinization temperature of starch, as well as the content and composition of proteins.

The definition of crop quality varies with the specific stage within the value chain, from breeding and field cultivation to food or feed processing, industrial manufacturing, and ultimately consumer acceptance. For example, the quality of wheat at harvest is determined by key metrics like test weight and chemical composition. When it is sold to grain dealers, its market value relies on four crucial attributes: seed purity, test weight, moisture content, and protein concentration. In the milling process, a high flour yield is essential for both efficiency and profitability. In baking, dough strength, which is largely governed by gluten and protein quantity, directly influences loaf volume and texture.

1.2 The Classification of Quality Traits

Quality traits can generally be classified into two primary categories: physical and chemical traits. In practical applications, based on their performance at harvest and during human consumption or industrial processing, quality traits can be further categorized into four distinct types as follows. The consistent manifestation of these traits over multiple years and various locations significantly contributes to a cultivar's reputation (Fitzgerald et al. 2009).

1. The Appearance Quality: the size, shape, colour, and aesthetic value of the intact harvested organs, such as the test weight of wheat grains, the shape of rice kernels (long, medium, and short), and the grain chalkiness (opaque part). These traits are basically physical and immediately apparent to consumers. They vary according to local cuisine and cultural preferences, and significantly influence the determination of market value.
2. The Sensory Quality: the hardness, roughness, stickiness, elasticity, and palatability or fragrance when consumed by humans. These characteristics can be either chemical or physical in nature, such as the gelatinization temperature for rice cooking, and the hardness and stickiness of cooked rice. They are gaining increasing importance as the living standard increases in developing nations.
3. The Processing Quality: the physical and chemical properties expressed during industrial processes, such as the oil content of soybeans and canola, the strength of cotton fibres, and the milling rate of rice. These traits are critical for industrial crops utilized as raw materials to produce oil, fibre, sugar, rubber, and pharmaceuticals. For instance, a 1% increase in the oil content of soybeans would result in an additional ten thousand tonnes of oil produced using the same extraction technology for an oil factory with a processing capacity of one million tonnes.
4. The Nutritional Quality: mainly the nutrient value of the harvested organs in relation to both humans and animals, such as the composition of starch and storage proteins, essential amino acids like lysine and methionine, and micronutrients, including minerals and vitamins (Table 3.1). In humans, starch is primarily digested in the small intestine. However, a portion of starch, estimated to range from 3% to 30%, may resist digestion in the small intestine and pass into the large intestine. This type of starch, known as *resistant starch* (RS), is considered by some researchers to exhibit beneficial effects like certain forms of dietary fibre. Another crucial indicator of nutritional quality is the micronutrient content. Hidden hunger, micronutrient deficiency, is a global health crisis largely driven by poverty. Over 3 billion people, mainly in Africa, Asia, and Latin America, cannot afford diets rich in essential vitamins and minerals. Biofortifying staple crops with enhanced micronutrients is a practical, proven solution to this crisis.

1.3 The Physiochemical Foundation of Quality Traits

Quality traits are commonly complex, as governed by multiple genes and influenced by environments and management practices. The strategies and tools required to produce grains with specific quality characteristics are not as well-established as those for achieving high yields due to the complex mechanisms underlying quality

Table 3.1 Fifty-one essential nutrients for sustaining human life

Classification		Nutrients
Macronutrients	1. Air, water, & energy (3)	Oxygen, water, carbohydrates
	2. Amino acids (9)	Histidine, isoleucine, leucine, lysine, methionine, phenylalanine, threonine, tryptophan, valine
	3. Fatty acids (2)	Linoleic acid, linolenic acid
Micronutrients	4. Macrominerals (7)	Na, K, Ca, Mg, S, P, Cl
	5. Microminerals (17)	Fe, Zn, Cu, Mn, I, F, Se, Mo, Co (in B_{12}), B, Ni, Cr, V, Si, As, Li, Sn
	6. Vitamins (13)	A (retinol), *D* (calciferol), E (α-tocopherol), K (phylloquinone), C (ascorbic acid), B_1 (thiamin), B_2 (riboflavin), B_3 (niacin), B_5 (pantothenic acid), B_6 (pyridoxine), B_7/H (biotin), B_9/M (folic acid), B_{12} (cobalamin)

Redrawn from Graham et al. (2007)

Note: Numerous other beneficial substances found in food are increasingly recognized for their positive contributions to overall health and well-being, including phytochemicals, antioxidants, and dietary fibre. In addition to water and oxygen, humans need to consume at least 49 known nutrients in adequate amounts regularly to live healthy, productive lives. Plants can supply 47 out of the 49 essential nutrients, apart from vitamins D and B_{12}

formation. A more comprehensive understanding of the physiological basis of grain quality should lay a foundation for the development of novel breeding and agronomic strategies to achieve a balance between quality and yield.

Each quality trait is underpinned by a distinct chemical basis. For cereals, starch, the major component, significantly influences the physical, sensory, and nutritional properties of grain products, either directly or through interactions with other components such as proteins, lipids, and water. Proteins also play a crucial role in the functional properties of cereal foods, especially in the bread-making performance of wheat or the eating quality of rice. Although lipids are present to a lesser extent, the polar lipids found in wheat flour positively impact dough properties and contribute to producing high-quality bread with good texture. Therefore, the initial step of dissecting a quality trait is to identify the chemical foundation of the target quality trait.

Some quality traits reflect the physical properties, such as processing and appearance quality. To clarify the mechanisms underlying these traits, it is necessary to first "translate" these physical properties into chemical ones. For instance, kernel hardness in maize is determined by the balance between vitreous and floury endosperm portions, which is influenced by the packing patterns of starch and protein (Otegui et al. 2021). Starch granules are embedded within a protein matrix that stretches as the starch granules grow. During kernel desiccation, the protein matrix tears in sections where it is thin and labile, leading to the formation of air-filled spaces and resulting in the floury endosperm. Conversely, if starch granules develop within thicker and more robust regions of the protein matrix, no void spaces will form during kernel desiccation, resulting in a dense, vitreous appearance and a hard texture. Therefore, if we aim to dissect the physiological mechanism for maize grain hardness, it is crucial to begin with its chemical properties, the histochemistry of starch and proteins.

2 Morphological Foundation of Cereal Grain Quality

Cereal crops are primarily grasses that are cultivated for their edible seeds, technically a type of fruit known as a caryopsis. The caryopsis originates from a monocarpellary ovary with a single ovule and contains the main storage tissue, the endosperm. For most cereals, the value of the crop is largely determined by the endosperm, both in quantitative and qualitative terms.

As illustrated in Fig. 3.1 with rice as an example, the caryopsis of cereal crops shares a similar structural organization, comprising tissues derived from two generations: (1) the outermost layer of the pericarp, originating from the carpel of the parent plant, and (2) the internal seed, a product of double fertilization, composed of seed coat, the endosperm, and the embryo. The main characteristic of the caryopsis is the fusion of the pericarp with the residual seed integuments. The pericarp performs multiple functions, including protection, photosynthesis, mineral deposition, and transient starch storage (Cerri and Reale 2020). In wheat and rice, chloroplasts and chlorophyll are present in the pericarp, where they perform photosynthetic assimilation, supplying essential sugars during the early grain-filling phase.

The chemical composition of cereal grains exhibits spatial variation across different tissues. The majority of lipids and essential micronutrients, including vitamins, iron, and zinc, are concentrated in the embryo and aleurone layer, while starch and proteins are mainly present in the endosperm. This spatial distribution significantly influences the end-use quality of cereal grains. During the milling process, cereal grains are broken down to extract the pure starchy endosperm by removing the embryo, maternal tissues, aleurone layer, and even a portion of the starchy endosperm. The chemical composition of resulting products, such as wheat flour and milled rice, differs significantly from that of their original forms.

Apparently, the end-use quality of cereal crops is primarily determined by their endosperm, which constitutes the most important edible component of the grain. In rice, the physicochemical properties of the starchy endosperm largely determine its overall quality (Fig. 3.1). For instance, eating quality is influenced by amylose content, amylopectin structure, protein content, and lipid composition, whereas *grain chalkiness* arises from loosely packed starch granules and protein bodies in the endosperm (Xi et al. 2014). However, the composition, structure, and size of the endosperm are influenced by its interactions with other tissues, particularly the embryo, throughout seed development (Lafon-Placette and Köhler 2014; Ingram 2020). The simultaneous development of the embryo and endosperm within maternal tissues (the seed coat and pericarp) necessitates precise, dynamic coordination between these two compartments.

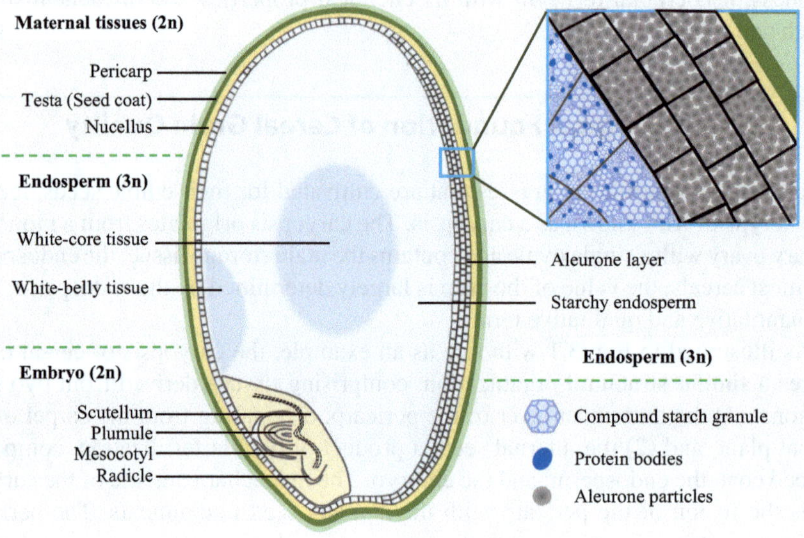

Fig. 3.1 The morphological anatomy of mature rice caryopsis. The caryopsis of rice consists of three genetically distinct tissues: the diploid embryo (2*n*), the triploid endosperm (3*n*), and the diploid maternal tissue (2*n*). The outer layer of the rice grain consists of supportive and protective maternal tissues, including the pericarp (light green), testa (green), and nucellus (yellow). The endosperm is composed of the aleurone layer (white) and starchy endosperm (light grey), which are densely packed with starch granules and protein bodies. Opaque tissues (light blue), referred to as chalkiness, commonly occur in the ventral (white-belly) and central (white-core) regions of the endosperm. Typically, the aleurone layer exhibits two to three layers on the dorsal side, while only a single layer is present on the ventral side of a rice grain. Aleurone particles are fundamentally subcellular inclusions, mainly composed of phytate (phytic acid) crystals encased by a proteinaceous matrix. Their structure features dense, electron-dense granules, which store essential minerals such as phosphorus, calcium, magnesium, and potassium in the form of phytate salts. Functionally, they act as crucial nutrient reservoirs, supplying phosphorus and other vital minerals to the seedling during germination. However, phytate is an antinutritional factor for monogastric animals such as humans, poultry, and pigs. The embryo consists of two main structures: the scutellum and the embryonic axis, which includes the plumule, mesocotyl, and radicle. Note that the diagrams are not drawn to scale. (Redrawn from An et al. (2020))

3 Morphogenesis of Cereal Grains

3.1 General Process of Seed Development

Angiosperm seeds are composed of a combination of distinct genotypes and tissue types. The formation of seeds results from a process known as double fertilization: one sperm cell fuses with the diploid central cells to form the triploid endosperm, and the second sperm cell fuses with the egg cell to produce a diploid zygote, the embryo. The developing embryo and endosperm are enclosed within the seed coat, which is derived from the maternal tissues of the ovule.

The seed life cycle can be broadly divided into three distinct phases: embryogenesis, maturation, and germination (Sami et al. 2025). Embryogenesis involves a series of coordinated cell divisions that establish the basic body plan of the plant. Subsequently, during the maturation phase, seeds accumulate nutrient reserves and develop crucial physiological characteristics, such as the ability to germinate, enter dormancy, tolerate dehydration, and preserve viability over prolonged periods. This enables them to withstand a diverse array of environmental stresses. After maturation, seeds enter a quiescent, desiccated state, enabling them to maintain viability until favourable conditions for germination arise. Overall, seed maturation serves as a crucial bridge between embryo development and germination, representing a key phase in the seed life cycle during which essential survival traits are acquired.

Cereal grain development exhibits fundamental differences from that of *Arabidopsis*,[2] the canonical eudicot model species. In *Arabidopsis*, the embryo occupies the dominant position within the mature seed, and nutrient reserves are predominantly stored in the cotyledons. The endosperm, formed transiently during early embryogenesis, is completely absorbed by the developing embryo and disappears entirely upon seed maturity. In contrast, the cereal endosperm experiences continuous growth and gradual nutrient accumulation during the entire process of caryopsis development. Eventually, it constitutes over 80% of the dry weight of the mature grain and serves as the main storage sink for starch and proteins (Fig. 3.2).

3.2 Embryonic Development

Embryonic development is a highly orchestrated process in which cells precisely coordinate proliferation, migration, differentiation, and tissue assembly, each occurring timely and appropriately to build functional organs and body structures. These developmental processes are genetically encoded and are influenced by two crucial factors: cellular lineage, which denotes the developmental history of each cell, and the signalling pathways that coordinate intercellular interactions during organogenesis.

Rice embryogenesis is a complex process through which a zygote is transformed into a fully developed embryo capable of germinating to give rise to the next generation. As a model system, it aids in uncovering the developmental mechanisms that are shared among monocots, such as maize and wheat. This embryonic development typically progresses through five phases: zygotic, globular, coleoptile, juvenile vegetative, and maturation (Fig. 3.3). Each phase shapes the future plant, from a single cell to a mature, dormant embryo.

[2] *Arabidopsis* is the short form for *Arabidopsis thaliana*. It is extensively utilized as a model organism owing to its small physical size, rapid generation time, simple genetic manipulation, high fecundity, and compact genome. Moreover, its complete genomic sequence has been fully elucidated, and a vast array of mutants have been documented.

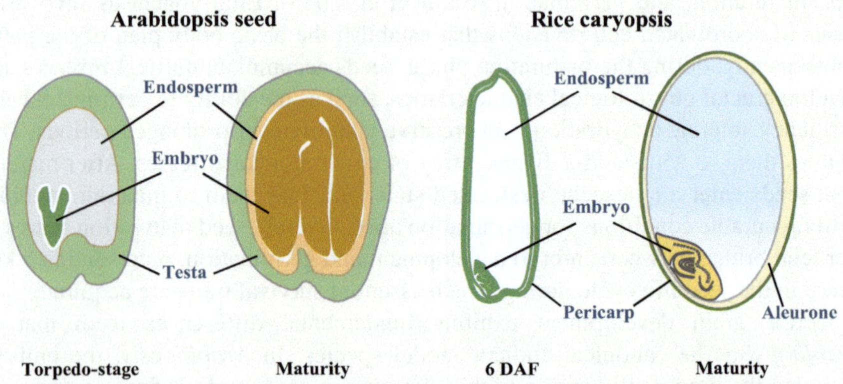

Fig. 3.2 Developmental differences between *Arabidopsis* seed and rice caryopsis. The structure of the rice caryopsis and the *Arabidopsis* seed is highly similar during early developmental stages, with the endosperm occupying the bulk of the seed volume. As development progresses, the endosperm in *Arabidopsis* is gradually and almost completely absorbed by the growing embryo. In contrast, at maturity, the rice endosperm persists and differentiates into specialized storage tissues, the starchy endosperm and the aleurone layer. In addition, in *Arabidopsis*, the outermost protective layer is the seed coat (testa), derived solely from the ovular integuments. Conversely, the rice caryopsis is encased by a pericarp, the fruit wall derived from the ovary, which forms the outermost protective layer. Beneath the pericarp lies a relatively thin testa, which is often fused with the pericarp and lacks the distinct multilayered integument structure observed in *Arabidopsis*. DAF, days after fertilization. (Redrawn from Zhou et al. (2013))

1. *Zygotic Phase (0 DAF).* Pollination in rice occurs immediately upon flower opening and lasts for approximately 15 min around noon in summer. Fertilization takes place several hours after pollination. The fusion of egg and sperm forms the zygote, a totipotent cell that gives rise to the major cell lineages and tissues of the mature plant through coordinated cell division and cell fate specification (Armenta-Medina et al. 2020).
2. *Globular Phase (1–4 DAF).* At 1 DAF, the zygotic nucleus migrates toward the apical pole and subsequently undergoes asymmetric division, resulting in two cells. The basal cell differentiates into a large vacuolated cell. In contrast, the apical cell undergoes further divisions, leading to the formation of the entire embryo. Subsequent rounds of variable cell divisions result in the formation of a multilayered, globular structure. At 4 DAF, the first morphogenetic event occurs: formation of the shoot apical meristem (SAM), scutellum, coleoptile primordium, and radicle primordium. Because these structures are at fixed positions, the embryonic axis and compartmentalization are inferred to be established.
3. *Coleoptile Phase (5 DAF).* The first leaf primordium is recognized at 5 DAF, and the adaxial–abaxial axis of the embryo becomes distinctly established. The coleoptile primordium starts to emerge from the adaxial side of the embryo, while the scutellum develops from the abaxial region. Importantly, both the SAM and

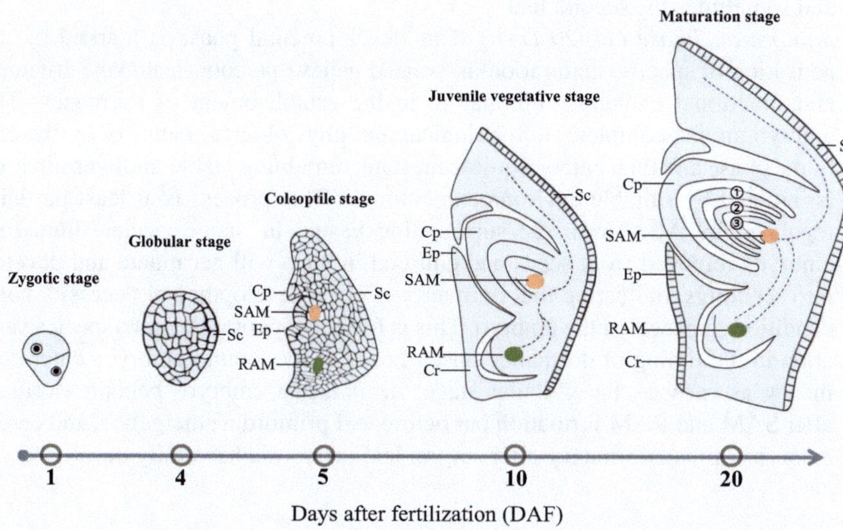

Fig. 3.3 The developmental phases of the rice embryo. (1) Zygotic phase (0 DAF): Fertilization occurs several hours after pollination, resulting in the formation of the zygote via the fusion of the egg and the sperm. (2) Globular phase (1–4 DAF): The fertilized zygote undergoes rapid cell divisions and initiates the process of differentiation. The embryo assumes a predominantly spherical morphology, and the fundamental axial polarity begins to be established. (3) Coleoptile phase (5 DAF). The coleoptile primordium emerges from the adaxial region of the embryo, whereas the scutellum originates from the abaxial region. Notably, both the SAM and RAM are established during this developmental stage. (4) Juvenile vegetative phase (6–10 DAF). The embryo now exhibits clearly defined regions, including the scutellum, coleoptile, and embryonic root primordium. Tiller primordia differentiate at the axils of the coleoptile, the first leaf, and occasionally the second leaf. (5) Maturation phase (11–20 DAF). The coleoptile undergoes elongation and emerges through the seed coat, while the radicle and shoot primordia are fully differentiated, positioning the seed for successful germination. Cp coleoptile, Cr coleorhiza, DAF days after fertilization, Ep epiblast, RAM root apical meristem, SAM shoot apical meristem, Sc scutellum. ①–③, the first to third leaf primordia. The timing of DAF is not depicted to scale. (Redrawn from Armenta-Medina et al. (2020))

the root apical meristem (RAM) are formed. These meristematic tissues serve as the driving forces for subsequent germination and seedling development, thereby establishing the foundation for the organogenesis of the next generation.

4. *Juvenile Vegetative Phase (6–10 DAF)*. The embryo reaches its mature morphology at 7–8 DAF, when organ differentiation is complete. A primary characteristic of this developmental phase is the initiation of leaf differentiation by the SAM. The early vegetative stages of the embryonic seedling are established within the embryo before the onset of dormancy. The plumule contains the first, second, and occasionally the third leaf primordia, all of which have already undergone differentiation and encompass a central growing point (Box 3.1).

Tiller primordia are differentiated at the leaf axils of the coleoptile, the first leaf, and sometimes the second leaf.

5. *Maturation Phase (11–20 DAF)*. This developmental phase is marked by the activation of specific maturation-associated genes that coordinate key structural and functional changes, culminating in the establishment of dormancy. The embryo attains complete morphological and physiological maturity by the end of the phase and then enters a quiescent state, remaining viable until germination is initiated by suitable environmental stimuli. This process is at least partially regulated by ABA from the surrounding tissue. In many species, immature embryos removed from seeds and cultured in vitro will germinate and develop into seedlings, indicating that dormancy is not a developmental necessity but a condition imposed on the embryo. This is further supported by interspecies variation in the timing of dormancy onset. For instance, orchid embryos enter dormancy as early as the globular stage; *Arabidopsis* embryos become dormant after SAM and RAM formation but before leaf primordia emergence; and cereal embryos initiate dormancy after several leaf primordia have fully developed.

Box 3.1: The Number of Leaf Primordia in Cereal Grains

Leaf primordia are the initial leaf buds that emerge from the shoot apical meristem prior to seed germination. The number of these primordia present in a mature seed differs among the three major cereal crops and establishes the developmental "starting point" for each seedling. In rice, a mature embryo contains two to three differentiated foliage-leaf primordia at the time of sowing. Similarly, wheat embryos also possess two to three leaf primordia. In contrast, maize embryos are more developmentally advanced, with five to six leaf primordia already visible in the mature seed.

The counts of leaf primordia serve as functional determinants of early seedling development, rather than merely anatomical details. In rice and wheat, the presence of two to three pre-emergent leaf primordia supports a compact growth habit and generates sufficient leaf area to facilitate the transition to photosynthetic independence. In contrast, maize initiates four to six primordia, enabling faster leaf blade emergence and sustaining its characteristically rapid vegetative growth as a C4 crop. Because the number of primordia directly governs the timing of first true leaf emergence and thereby influences early vigour, leaf expansion rate, and photosynthetic capacity, crop growth models treat it as a fundamental parameter for simulating emergence dynamics and early biomass accumulation. Accurately quantifying this trait empowers breeders to design climate-resilient cultivars and enables modelers to refine predictions under diverse agronomic and environmental scenarios.

3.3 Endosperm Development

The primary component of cereal grains is the endosperm, which consists of a mass of starchy endosperm (SE) cells surrounded by a layer of aleurone (AL) cells. At the base are transfer cells (TC), and adjacent to the embryo are cells of the embryo-surrounding region (ESR). Grain endosperm development in grasses exhibits a modular pattern, as outlined by Olsen (2020). Initially, the endosperm is composed predominantly of SE cells. Elevated sucrose concentration induces the differentiation of the outer layer of AL cells at the endosperm surface, a process that can be recapitulated in vitro under sucrose-enriched culture conditions. Subsequently, the basal coenocytic region of the triploid endosperm responds to glucose-mediated signalling, leading to the initiation of TC formation. External stimuli, including hormonal and sugar signals, further activate cells in the ESR. Beyond these environmentally influenced developmental events, the overall progression of endosperm development proceeds largely autonomously, governed by intrinsic gene regulatory networks and metabolic cues. Notably, low-sucrose culture media fail to trigger SE and AL differentiation, highlighting the dual role of sucrose as both an essential metabolite and a critical developmental signal.

Among cereals, endosperm development has been most extensively studied in maize, owing to a combination of historical, economic, anatomical, and genetic factors (Sabelli and Larkins 2009). Anatomically, the maize kernel has the largest caryopsis within the cereal family. It features distinct and readily identifiable cell types, including the SE, AL, and basal endosperm transfer layer (BETL), which are highly suitable for histological analysis. As classified by Sabelli and Larkins (2009), the development of the maize endosperm proceeds through several distinct yet overlapping phases: (i) Early development, including double fertilization, *coenocyte* formation, and *cellularization*; (ii) differentiation, involving the formation of major cell types, periods of *mitosis* and *endoreduplication*, and the accumulation of storage reserves; and (iii) maturation, comprising *programmed cell death (PCD)*, dormancy induction, and desiccation.

In rice, Wu et al. (2016) classified endosperm development into four distinct phases: the coenocytic phase, the cellularization phase, the storage product accumulation phase, and the maturation phase (Fig. 3.4). This staging system provides a more accurate temporal delineation of developmental events, and the cytomorphological changes characteristic of each phase are described below.

1. *Coenocyte (0–2 DAF)*. Following fertilization, the primary endosperm nucleus undergoes a series of divisions without the immediate formation of cell walls, resulting in the generation of multiple free nuclei. This coenocytic phase is characterized by the rapid proliferation of nuclei in the absence of concurrent cellular differentiation or cell wall formation. All nuclei, along with their surrounding cytoplasm, become localized to the periphery of the embryo sac, whereas the central region is occupied by a large central vacuole.
2. *Cellularization (3–5 DAF)*. The process of cellularization begins with the formation of cell walls within the endosperm. These cell walls first appear at the

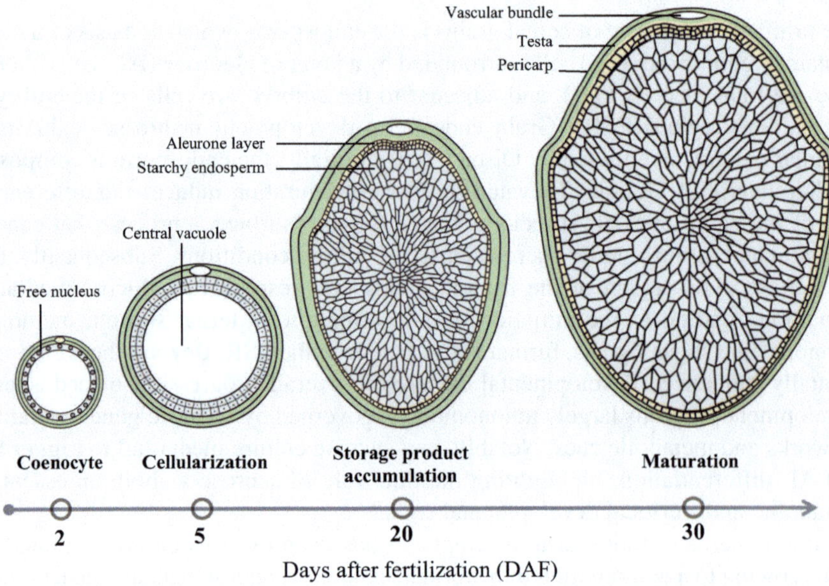

Fig. 3.4 The developmental process of rice endosperm. (1) Coenocytic phase (0–2 DAF): The central cell undergoes rapid and synchronous nuclear divisions in the absence of cytokinesis, resulting in a multinucleate coenocyte. Absence of cell walls characterizes this structure, with the cytoplasm containing numerous free nuclei distributed around a large central vacuole. (2) Cellularization phase (3–5 DAF): The nuclei become enclosed by primary cell walls that originate at the periphery and extend inward, resulting in the formation of 5–7 layers of thin-walled peripheral cells. This process transforms the coenocyte into a fully multicellular tissue. (3) Storage-product accumulation phase (6–20 DAF): The endosperm differentiates into an outer aleurone layer, composed of living cells that store proteins, lipids, and vitamins, and an inner starchy endosperm, consisting of dead cells densely packed with starch granules. During this process, starch and storage protein biosynthesis are significantly enhanced, while PCD is initiated in the starchy endosperm cells. (4) Maturation phase (21–30 DAF): The endosperm attains its maximum volume, with starch granules densely packed in the cytoplasm, while the aleurone layer remains the sole metabolically active tissue. As the grain undergoes desiccation, water content progressively decreases, and cellular tissues harden, thereby preparing the seed for dormancy and subsequent germination. The timing of DAF is not depicted to scale. DAF days after fertilization. (Redrawn from Wu et al. (2016))

periphery of the central vacuole, effectively compartmentalizing adjacent coenocytic nuclei. Cellularization progresses gradually from the periphery towards the centre of the endosperm. By 5 DAF, the embryo sac cavity (the large vacuole) is filled with thin-walled endosperm cells, and small starch grains surrounding the endosperm nuclei become distinguishable. This phase is marked by the initiation of cell differentiation, during which the endosperm begins to develop into distinct cellular structures, such as the outer aleurone layer and the inner starchy endosperm. The peripheral cells that eventually form the aleurone layer are typically smaller than those in the central region and contain little to no starch grains.

3. *Storage Product Accumulation (6–20 DAF).* From 5 DAF onwards, starch grains begin to accumulate in the cellularized endosperm, starting in the central region and gradually extending towards the periphery. Within the central endosperm cells, starch grains are distributed along the cell periphery, while the central region is occupied by large vacuoles. By 9 DAF, the central cells are densely filled with starch grains, whereas cells in the peripheral region contain fewer starch grains. Starch grains in the dorsal cells are generally larger than those in the lateral cells. Aleurone cells are distinguished from starchy endosperm cells starting at 6 DAF, and this distinction becomes more evident between 7 and 9 DAF due to the differential accumulation of storage products. PCD initiates in the central region of the starchy endosperm at 8 DAF and subsequently progresses centrifugally toward the periphery, extending over the entire starchy endosperm by 18 DAF. In contrast, aleurone cells remained viable at the end of this phase.
4. *Maturation (21–30 DAF).* Approximately 20 DAF, the endosperm enters the maturation stage. The volume and density of endosperm cells reach their maximum levels, and the accumulation of starch and storage proteins peaks. The deposition and crystallization of these compounds contribute to the formation of a semi-transparent starchy endosperm. During this phase, endosperm cell walls begin to undergo degradation, and PCD is progressively completed. Aleurone cells remain metabolically active, whereas the cells of the starchy endosperm gradually lose their defined cellular boundaries and achieve full maturation. Additionally, subaleurone cells accumulate a significant amount of protein bodies.

3.4 Integrative Staging System for Seed Development

In cereal grains, the coordinated development of the embryo and endosperm is a highly intricate and finely regulated process, which is critical in determining yield and quality. As outlined above, precise classification of developmental stages has been established for the embryo and endosperm, offering a clear morphological basis for genomic or molecular studies on seed development. However, in crop science, there remains a lack of a practical staging system that can be effectively used to assess or predict the timing of grain filling and maturation. In particular, the integration of developmental features of both the embryo and endosperm has rarely been considered. To address this gap, Tao et al. (2021) proposed a staging system for rice based on molecular, physiological, and anatomical evidence and provided brief descriptions of cellular features of each phase (Fig. 3.5).

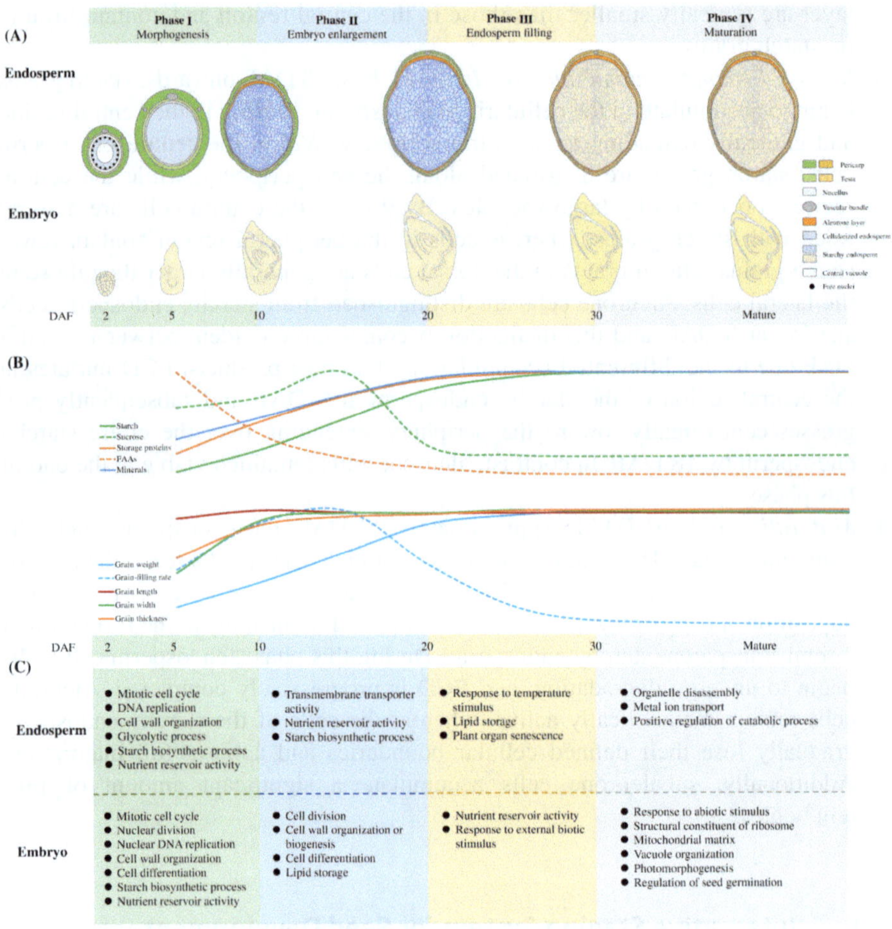

Fig. 3.5 A comprehensive overview of rice grain development: integrating embryo and endosperm dynamics. (**A**) Morphological changes in the embryo (longitudinal section) and endosperm (transverse section). The varying colours of the pericarp and testa indicate the progression of degradation in maternal tissues, while the colour changes in the starchy endosperm correspond to the grain-filling process. (**B**) Dynamics of storage product accumulation and volume changes during caryopsis development. (**C**) Molecular signatures associated with the development of the embryo and endosperm. DAF days after fertilization. (Reprinted from Tao et al. (2021))

Phase I: Morphogenesis (0–10 DAF)

After *double fertilization*, *patterning* and *differentiation* occur concurrently in both the embryo and endosperm. This phase is predominantly characterized by embryo and endosperm morphogenesis. The embryo progresses through the globular stage until 3 DAF. By 5 DAF, the first leaf *primordium* becomes distinguishable. Subsequently, there is a significant enlargement of embryonic organs, particularly the scutellum, which continues until 10 DAF when most morphogenetic processes are completed.

During this phase, the endosperm undergoes sequential developmental stages: *nuclear endosperm* (0–2 DAF), *cellular endosperm* (3–5 DAF), and aleurone/starchy endosperm differentiation (6–9 DAF). These processes result in the formation of two distinct subregions, the aleurone layer and the starchy endosperm, and initiate starch and protein accumulation.

At 10 DAF, the embryo completes its morphological differentiation and development, acquiring the capacity for germination. Meanwhile, cell division in the endosperm ceases around 10–11 DAF, after which the cells undergo only thickening. Thus, the embryo and endosperm show synchronized development in terms of cell division and differentiation.

Phase II: Embryo Enlargement (11–20 DAF)

The prominent feature of this period is the increase in embryo enlargement. The embryo reaches its maximum volume at 20 DAF, at which point the SAM and RAM are completely enclosed by the coleoptile and coleorhiza, respectively. The early vegetative development of the embryonic seedling has already commenced, as evidenced by the formation of two or three leaf primordia. During this period, the starchy endosperm reaches its peak rate of storage accumulation, while aleurone cells are densely packed with aleurone particles and *spherosomes*. This phase is characterized by intense interactions between the embryo and endosperm, playing a crucial role in grain filling and quality (Tao et al. 2021).

Phase III: Endosperm Filling (21–30 DAF)

The embryo enters a dormant state, while the endosperm continues to accumulate starch and proteins, achieving its maximum weight at 30 DAF. By the end of this phase, the majority of the starchy endosperm and maternal tissues undergo PCD, thereby losing their biological activity. In contrast, the aleurone cells and the embryo tissues remain viable. This phase is critical for grain yield (weight) and quality formation. However, there is limited information available regarding the cellular processes occurring during this period, particularly the regulatory role of the mature embryo in endosperm filling.

Phase IV: Maturation (31 DAF–maturity)

Upon the completion of starch and protein accumulation, the embryo acquires desiccation tolerance and undergoes a developmentally programmed dehydration process that triggers dormancy, a state characterized by metabolic *quiescence*. During seed maturation and *desiccation*, the starchy endosperm cells undergo complete PCD. This phase lasts approximately 20–40 days, with no significant increase in grain dry matter. Seeds are prone to germination under high-temperature and humid conditions, increasing their susceptibility to preharvest sprouting.

As shown in Fig. 3.5, this staging system from a physiological perspective reflects the distinctions and connections among key developmental events, metabolic activities, and molecular signatures during rice grain development. It provides a more precise distinction among morphogenesis, embryo enlargement, and endosperm filling, as compared to the previously oversimplified classification of

Table 3.2 Master process for rice quality from the perspective of embryo–endosperm interaction

Phase	Duration (DAF)		Embryo	Endosperm	Interaction	Traits	Environ.
I Morphogenesis		0–10	Embryogenesis is essentially complete	Cellularization is complete; Reserve accumulation starts	Strong	Chalkiness	HT
II Embryo enlargement		10–20	Embryo development is finished	Cell differentiation ends; Endosperm filling peaks	Strong	Chalkiness Sensory	HT
III Endosperm filling		20–30	Desiccation and dormancy	Endosperm filling is complete; Desiccation starts	Weak	Sensory	HT Drought
IV Maturation		30–	Maturation	Desiccation and maturation	Weak	Milling	LT Drought

Note: Interaction refers to the communication between the embryo and the endosperm, as defined by Tao et al. (2021)
DAF days after fertilization, *Environ.* environment, *HT* high temperature, *LT* low temperature

these processes under a single phase, the grain filling. More importantly, this staging system identifies the critical timing of grain filling and quality formation, making it valuable for developing agronomic practices aimed at enhancing reserve accumulation or promoting early maturity. Accordingly, the relevance of each phase to the formation of quality traits and their influencing factors is outlined in Table 3.2, which is helpful for the improvement of individual quality traits.

4 Interactions Between Embryo and Endosperm During Grain Filling

The term "*interaction*" denotes a bilateral or reciprocal process, specifically the effect that arises when two entities influence each other. On one hand, the endosperm serves as a critical support system, nurturing and regulating the development of the embryo. On the other hand, the embryo acts as the development centre, prioritizing nutrient allocation since it represents the next generation. It is reasonable to hypothesize that when nutrients are insufficient for its development, the embryo may release signals to remobilize the storage carbohydrates or proteins in the endosperm. The development of both the embryo and endosperm must be tightly linked through physical and chemical interactions to ensure viable seed formation. Elucidating these interactions represents an intriguing yet challenging task, especially regarding the mechanism of nutrient allocation between the embryo and endosperm. Accumulating evidence from studies in *Arabidopsis*, as well as rice and maize, supports the existence of bidirectional communication between these two compartments (Fig. 3.6).

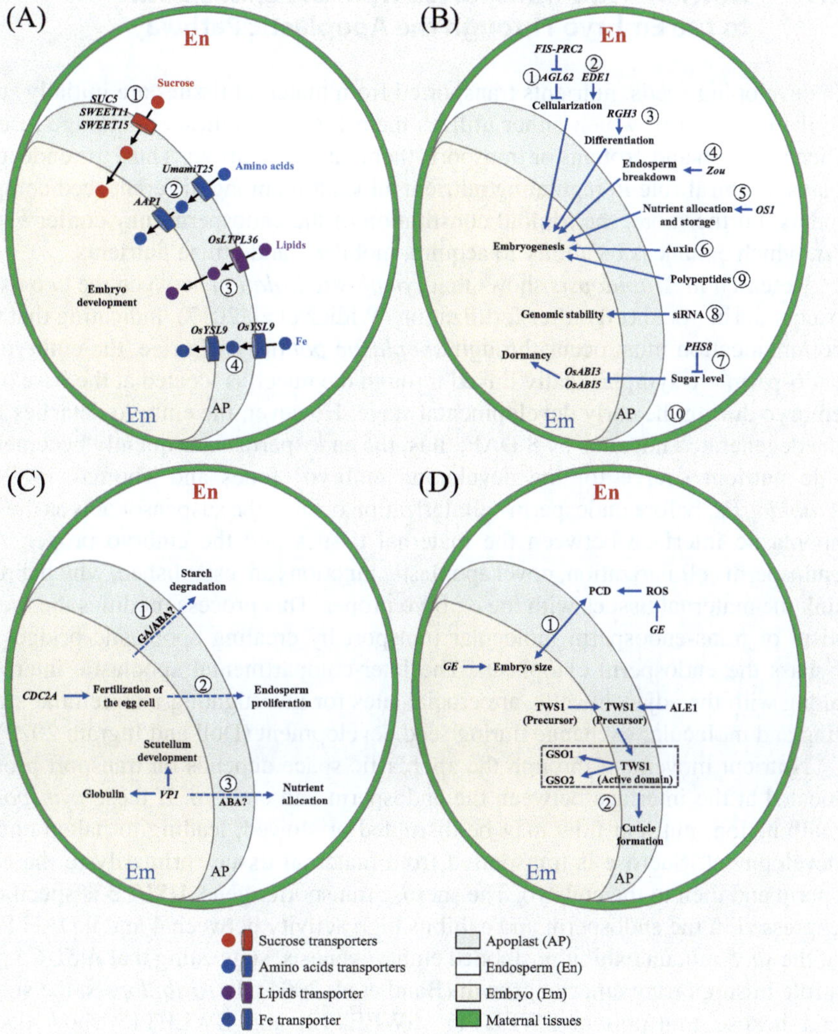

Fig. 3.6 Schematic illustration of embryo–endosperm interactions: rice as a model. The simultaneous development of the embryo and endosperm within the confined space of the seed coat necessitates complex bidirectional dialogue between these two compartments. This interaction is reflected in the following four aspects. (**A**) Nutrient transport from the endosperm to the developing embryo through apoplastic routes. Sucrose, hexose, and amino acid transporters are discussed in the main text. For further details on lipid transporter OsLTPL36 and iron transporter OsYSL9, please refer to Wang et al. (2015) and Senoura et al. (2017), respectively. (**B**) The dependence of embryo development on the endosperm. Genes involved in the regulation of endosperm development and nutrient allocation are discussed in the main text. Note that AGL62, a type I MADS-box transcription factor, is epigenetically regulated by the FIS-PRC2 complex (Hehenberger et al. 2012). For further details on other regulatory factors, the following references are recommended: auxin (Chen et al. 2014), polypeptides (Widiez et al. 2017), *PHS8* (Du et al. 2018), *siRNA* (Lafon-Placette and Köhler 2014), and T6P (Martínez-Barajas et al. 2011). (**C**) Modification of endosperm development by the embryo through hormonal signalling. (**D**) Bidirectional signalling pathways between embryo and endosperm, as mediated by *GE* and TWS1. Dotted lines represent putative or unverified signalling routes. (Redrawn from An et al. (2020))

4.1 Nutrients Are Transported from the Endosperm to the Embryo Through the Apoplastic Pathway

In developing seeds, nutrients transported from maternal tissues are initially stored in the endosperm, which either utilizes them for the synthesis of storage reserves such as starch and proteins or transports them into the embryo. Thus, the endosperm plays a central role in regulating nutrient allocation among different seed compartments. Furthermore, the triploid constitution of the endosperm may confer *heterosis*, which enhances its ability to acquire, mobilize, and utilize nutrients.

Research in *Arabidopsis* shows that *symplastic isolation* between the two tissues occurs during or shortly after fertilization (Widiez et al. 2017), indicating that their communication must occur through *apoplastic pathways*. In rice, the embryo and endosperm are symplastically linked through the nucellus located at the base of the embryo during the early developmental stage. However, the embryo detaches from the degenerated nucellus by 8 DAF; thus, the endosperm subsequently becomes the sole nutrient source for the developing embryo (Jones and Thomas 1989). In *Arabidopsis*, before endosperm cellularization occurs, the suspensor acts as the only apoplastic interface between the maternal tissues and the embryo proper. After endosperm cellularization, novel apoplastic junctions are established, which directly link the maternal tissues with the embryo proper. This process modifies the mechanism of trans-endosperm molecular transport by creating apoplastic bridges that bypass the endosperm cytoplasm. The inter-compartmental apoplastic interfaces, along with the adjacent cells, are crucial sites for investigating intercellular signalling and molecular exchange during seed development (Doll and Ingram 2022).

Nutrient movement through the apoplastic space depends on transport proteins located at the interface between the endosperm and embryo. If these *transporters* malfunction, nutrient flow may be disrupted or slowed, leading to halted embryo development. Sucrose is transported from maternal tissues primarily to the endosperm and then to the embryo. The sucrose transporter gene *AtSUC5* is specifically expressed in the endosperm and exhibits high activity between 4 and 9 DAF. Seeds of the *suc5* mutant exhibit postponed embryogenesis, suggesting that *AtSUC5* plays a role in supporting embryo growth (Baud et al. 2005). In *Arabidopsis*, the sucrose and hexose transporters SWEET11, SWEET12, and SWEET15 show distinct spatio-temporal expression patterns (Chen et al. 2015). SWEET11 is mainly expressed in the endosperm and seed coat during the linear cotyledon and maturation green stages. SWEET12 is primarily found in the seed coat during these stages, but is also present in the suspensor and micropylar regions during the globular stage. SWEET15 is expressed in the endosperm during both the globular and maturation green stages. A *sweet11;12;15* triple mutant produces wrinkled seeds with retarded embryogenesis, highlighting the crucial role of the coordination of these transporters in delivering sugar to the developing embryo via the endosperm.

Members of the UmamiT family (usually multiple amino acids move in and out transporter) play a role in the process of nitrogen loading into the embryo. In *Arabidopsis*, *UMAMIT25* is specifically expressed in the endosperm. Knockout lines of *umamit25* lowered amino acid content during embryo development, suggesting that *UMAMIT25* is likely critical for importing amino acids from the endosperm into the embryo (Besnard et al. 2018). AAP1 (amino acid permease 1) functions as an amino acid transporter localized in the abaxial epidermal cells and parenchyma of the *Arabidopsis* embryo (Sanders et al. 2009). In mature *aap1* seeds, the total nitrogen and carbon content is significantly reduced in the embryo, while the level of free amino acids in the seed coat and endosperm is elevated. This indicates that the impaired uptake of amino acids by the *aap1* embryo disrupts nitrogen homeostasis in the seed coat and endosperm.

4.2 Embryo Development Is Critically Reliant on the Endosperm

Since endosperm formation is strictly reliant on successful fertilization, plants avoid investing resources in unfertilized ovules, thus avoiding unnecessary depletion of maternal nutritional resources. The resulting endosperm functions as a nutritive tissue, providing the embryo with sucrose, amino acids, minerals, and growth regulators, thereby tightly linking embryonic growth with nutrient availability. In addition to providing essential nutrients for the embryo, the endosperm may also exert physical and metabolic constraints to inhibit or even arrest embryogenesis. Therefore, embryo development is critically reliant on the endosperm, as evidenced by *hetero-fertilization* experiments in maize, which demonstrated that the genetic relationship between the embryo and endosperm influences nutrient allocation to the embryo (Wu et al. 2013).

The role of endosperm cellularization in maintaining embryo viability has been elucidated through studies on mutants lacking the endosperm-specific *FERTILIZATION INDEPENDENT SEED POLYCOMB REPRESSOR COMPLEX 2* (*FIS-PRC2*). Dysfunction of *FIS-PRC2* results in abnormal endosperm cellularization, thereby arresting the embryo development at the heart stage (Köhler et al. 2003). Furthermore, this developmental retardation can be alleviated through in vitro culture of isolated embryos, indicating that defective endosperm cellularization directly impedes embryogenesis (Hehenberger et al. 2012). Similarly, loss-of-function of *ENDOSPERM DEFECTIVE1* (*EDE1*) results in impaired endosperm cellularization and delayed embryo development at the late heart stage (Pignocchi et al. 2009).

Endosperm differentiation is regulated by parental genomes (Fouquet et al. 2011). The embryo-surrounding region (ESR) of maize kernels is a specialized layer of endosperm that forms shortly after fertilization. Molecular markers such as *ZmEsr* are detectable as early as 4 DAF and remain active through at least 28 DAF, indicating sustained ESR functionality throughout seed development. The ESR performs three essential functions. First, it serves as a nutrient conduit, facilitating the transport of sucrose from the endosperm to the embryo. Second, it acts as a physical barrier, isolating the embryo from the bulk endosperm to protect it and maintain a distinct microenvironment. Third, the ESR likely mediates intercompartmental biochemical signalling; evidence suggests that ESR cells regulate excess auxin produced in the endosperm, thereby modulating the timing of stem cell specification in the embryo (Dresselhaus et al. 2025). Collectively, these roles are vital for proper embryogenesis and seed maturation. *Rough Endosperm 3* (*RGH3*) encodes a putative RNA splicing factor. In the *rgh3* mutant, the differentiation of both the ESR and the basal endosperm transfer layer (BETL) is impaired. Consequently, mature *rgh3* seeds display a rough and pitted endosperm surface, accompanied by varying degrees of embryonic abnormalities. These phenotypic alterations in the endosperm and embryo indicate that the *rgh3* mutation exerts a negative regulatory effect on embryogenesis.

Endosperm breakdown creates the necessary physical space and promotes the mobilization of stored reserves to support embryo development. ZHOUPI (ZOU), a basic helix–loop–helix (bHLH) transcription factor specifically localized around the ESR, plays a crucial role in mediating endosperm degradation and nutrient remobilization in *Arabidopsis* (Ingram 2010). The *zou* mutant retains a significantly greater number of endosperm cells at maturity compared to the WT, and its embryo exhibits defects in cuticle formation and epidermal cell adhesion (Yang et al. 2008). Likewise, in maize, *ZmZOU* performs a similar active function in the degradation of the ESR and suspensor during embryo development (Grimault et al. 2015).

The basal endosperm transfer layer (BETL) in maize kernels arises from the basal region of the endosperm following fertilization. As the coenocytic endosperm undergoes cellularization, the basal cells adjacent to the maternal vascular tissue in the pedicel differentiate into specialized transfer cells. These cells develop intricate wall ingrowths, significantly increasing their plasma membrane surface area to form the BETL, which serves as the primary conduit for transporting nutrients from the maternal plant to the developing embryo and endosperm. *OS1* (*OPAQUE ENDOSPERM AND SMALL GERM 1*) is predominantly expressed in the BETL and central endosperm, encoding a putative transcription factor with an RWP-RK domain (Song et al. 2019). The *os1* mutation leads to an opaque endosperm phenotype with a characteristic crown-like appearance, along with a smaller embryo. RNA-seq analysis shows that many genes associated with nutrient transport and storage are significantly downregulated in the *os1* mutant, indicating that *OS1* plays a critical role in regulating the transport of nutrients from maternal to filial tissues during seed development.

4.3 Endosperm Development Is Regulated by the Embryo

The endosperm is commonly described as altruistic, as it supports the development of the embryo, the next generation, at the cost of its own growth (Wu et al. 2013; Ingram 2020). This phenomenon raises an intriguing hypothesis: the embryo may exert a regulatory influence on the endosperm, a notion increasingly supported by empirical findings. Direct evidence comes from scanning electron microscopy (SEM) images, which reveal the degradation of starch granules in endosperm cells adjacent to the scutellum (Fig. 3.7). The presence of micro-pores or pits on the starch granule surfaces clearly indicates α-amylase-mediated starch hydrolysis (Lin

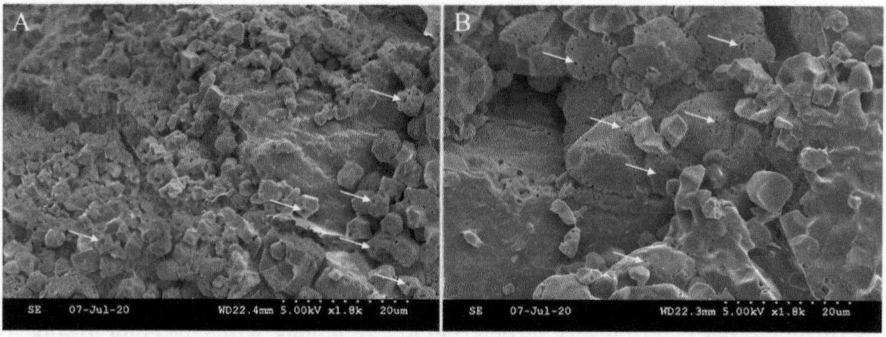

Fig. 3.7 Starch decomposition in the endosperm cells adjacent to the scutellum of the rice embryo. Micro-pores or pits (indicated by the white arrow) on the surface of starch granules suggest that α-amylase mediates the hydrolysis of starch reserves. Location A is closer to the embryo (scutellum) than location B. Notably, starch granules at location A are smaller than those at location B. This difference may result from the nutrient depletion experienced by starch granules at location A, which is likely caused by the metabolic demands of the developing embryo. (Reprinted from An et al. (2020))

et al. 2016; Nakata et al. 2017). These observations suggest that starch degradation is likely driven by the embryo's high demand for sugars during development. However, it remains unclear whether GA- or ABA-mediated signalling pathways are involved in this process.

Nowack et al. (2006) identified a mutation in the *Arabidopsis Cdc2* homolog *CDC2A*, in which mutant pollen fertilizes only one cell in the embryo sac instead of both, enabling the genetic analysis of double fertilization. Their findings showed that fertilization of the egg cell alone can trigger endosperm development, indicating early signalling from the embryo to the central cell. A novel interface tissue, termed the endosperm adjacent to the scutellum (EAS), was identified through genome-wide expression profiling of tissues located at the boundary between the embryo and endosperm in maize seeds (Doll et al. 2020b). The EAS develops around 9 DAF and remains active for approximately 11 days, during which it displays a distinct transcriptomic profile enriched in genes encoding transport proteins. In an embryo-specific mutant lacking a differentiated embryo, the expression patterns of EAS marker genes are altered, suggesting that the embryo plays a regulatory role in the development or function of this specialized endosperm tissue.

The embryo plays a pivotal role in regulating nutrient allocation between itself and the endosperm. VIVIPAROUS 1 (VP1), a member of the B3 family of transcription factors, is involved in ABA-mediated transcriptional regulation (McCarty et al. 1991). *VP1* begins to be specifically expressed in the scutellum starting at 5 DAF, and loss of its function results in preharvest sprouting, a phenomenon known as the viviparous phenotype in maize. Recent studies indicate that *VP1* monitors the nutritional status of the endosperm through communication between the endosperm and the embryo, potentially via the ABA signalling pathway. This regulatory mechanism not only promotes scutellum development but also facilitates the remobilization of nutrients from the endosperm, thereby enhancing the embryo nutrient storage (Zheng et al. 2019).

In most flowering plants, the developing embryo actively invades the surrounding endosperm, triggering progressive remodelling and eventual elimination of endosperm tissue to accommodate embryonic expansion. In *Arabidopsis*, this process occurs through a lytic mechanism, leading to the degradation of most of the endosperm, with only a thin, living peripheral layer remaining at maturity. In species with persistent or semi-persistent endosperm, such as cereals, embryo invasion is spatially restricted. Although the cereal endosperm is largely retained until seed maturation, viability is limited to the outer aleurone layer. The central starchy endosperm undergoes PCD, preserving dead cellular remnants, while the embryo appears to play a minimal role in initiating this process. In maize, two distinct zones of endosperm removal have been identified adjacent to the embryo. The embryo-surrounding region (ESR) is a small, dense cytoplasmic domain likely involved in nutrient transfer and developmental support. It degenerates by approximately 14 DAF, coinciding with the loss of the embryonic suspensor and dependent on *ZmZOU*, the maize ortholog of *ZHOUPI*. Following embryo emergence from the ESR, cell death becomes apparent around the expanding scutellum, particularly within the EAS, resulting in the deposition of cell wall remnants at the interface. Notably, *ZmZOU* does not regulate EAS degradation or PCD in the starchy endosperm, and it remains unclear whether the elimination of the ESR or EAS is directly driven by embryo growth (Doll and Ingram 2022).

Although certain evidence indicates that endosperm development may be influenced by embryo-related signals under specific conditions, alternative findings suggest that it can proceed independently. Xiong et al. (2021) demonstrated that in the absence of a *zygote* and embryo, cytological processes such as endosperm initiation, coenocyte formation, free nuclear cellularization, and endosperm degeneration proceed similarly to those observed in the WT. Furthermore, while rapid embryo expansion accelerates endosperm breakdown, endosperm development appears to be an autonomously organized process, independent of egg cell fertilization and embryo–endosperm interactions. This study thus underscores the dual nature of the endosperm: it plays an altruistic role in nourishing the embryo, yet retains a self-organized developmental capacity during seed development.

4.4 Bidirectional Dialogues Facilitate the Coordinated Development of the Embryo and the Endosperm

The simultaneous development within the limited space of the seed coat requires bidirectional communication between the embryo and endosperm to ensure proper seed formation and a viable *sporophytic* generation. Belmonte et al. (2013) created gene expression profiles of *Arabidopsis* seeds from fertilization to maturity and found significant overlap between embryo and endosperm gene programs, suggesting strong coordination in their biological processes. This is also supported by studies of mutants with altered embryo-to-endosperm size ratios. The rice *GIANT*

EMBRYO (*GE*) gene, encoding CYP78A13, regulates the size balance between the embryo and endosperm (Nagasawa et al. 2013). It is expressed mainly in tissues at the embryo-endosperm interface, functioning in both the embryo and endosperm to maintain this size balance. This balance is associated with ROS homeostasis and PCD regulation in the endosperm. Given the physical constraints imposed by the rice bracts (palea and lemma), PCD may contribute to resolving the spatial competition between the embryo and the endosperm (Nagasawa et al. 2013).

The bidirectional interaction between the embryo and the endosperm, exemplified by the sulphated peptide TWISTED SEED 1 (TWS1), plays a key role in seed development, especially in forming the embryonic cuticle (Doll et al. 2020a). TWS1 is initially synthesized within the embryo and subsequently transported into the endosperm. Within the endosperm, a specific protease known as ALE1 cleaves the precursor protein, thereby converting it into an active peptide form. The activated mature TWS1 peptide is then translocated back into the embryo, where it binds to two receptor-like kinases located on the cell surface, GSO1 and GSO2. Upon activation of these receptors, a downstream signalling cascade is initiated, directing the embryo to form and maintain its cuticle, a thin waxy layer that functions as a waterproof barrier. The cuticle is essential for preventing desiccation of the developing plant during seed germination. The spatial separation of TWS1 and ALE1 production in distinct tissues ensures tight regulation of this pathway: the embryo acts solely as the signal sender, while the endosperm serves exclusively as the activator. This compartmentalization restricts TWS1 activity to the appropriate spatiotemporal context, ensuring that the peptide functions specifically as the ligand for GSO1 and GSO2 to promote cuticle reinforcement without unintended activation in other tissues. In summary, the communication between the embryo and endosperm operates as a bidirectional signalling system: the embryo secretes a precursor signal, the endosperm processes it into its active form, and the processed peptide returns to instruct the embryo to complete the formation of its protective cuticle, thus safeguarding the seed against dehydration during germination.

5 Relevance to Cereal Grain Quality: Rice Grain Chalkiness as an Example

Physically and chemically, rice quality is defined by the characteristics of the starchy endosperm, which is densely packed with starch granules and protein bodies. Most studies on quality traits have focused on the endosperm, concluding that carbon and nitrogen metabolisms, along with their interaction, form the biochemical basis for rice quality. However, the role of the embryo merits further investigation, as it plays a crucial regulatory role in endosperm formation and storage compound accumulation, as discussed previously.

Chalkiness is an opaque tissue characteristic that commonly appears in the ventral (white-belly) and central (white-core) regions of the rice endosperm (Fig. 3.8). Inadequate packing of either starch or protein may result in the formation of chalky tissue. This characteristic is highly unfavourable, as it exerts a negative impact not

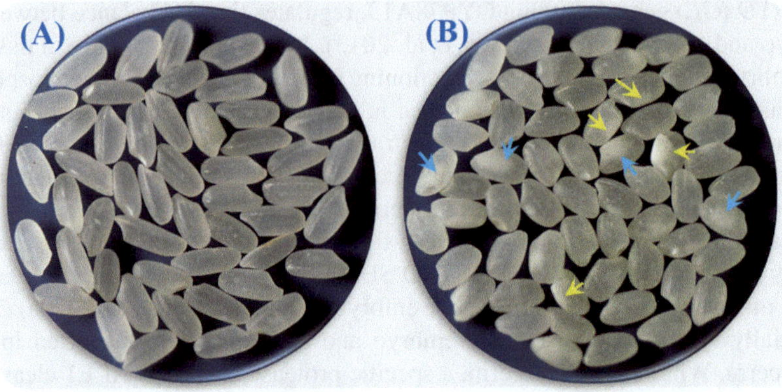

Fig. 3.8 Appearance quality of polished japonica rice from Northeast (**A**) and Southeast (**B**) China. In Northeast China, a region characterized by its cold climate, japonica rice exhibits an attractive appearance due to its low grain chalkiness. In contrast, in Southeast China, where temperatures are higher during the grain-filling phase, japonica rice displays a less desirable appearance because of higher grain chalkiness. Blue arrows indicate the white-core endosperm tissue, whereas yellow arrows denote the white-belly tissue

only on the appearance quality but also on the milling, cooking, and sensory properties of rice. The frequent appearance of chalky grains presents a substantial challenge for rice producers, especially in the context of global warming. Owing to the intricate interactions between regulatory genes and environmental conditions, the physiological and molecular mechanisms underlying grain chalkiness remain poorly understood. For example, using doubled haploid (DH), recombinant inbred line (RIL), or MAGIC populations,[3] hundreds of QTLs have been identified across all 12 rice chromosomes, with only a few genes, such as *Chalk5*, *WCR1*, and *Chalk9*, having been characterized. Therefore, grain chalkiness can serve as a valuable model system for studying complex quality traits in crops.

Lin et al. (2014) identified a notched-belly mutant that exhibits a high incidence of chalky tissue specifically at the basal endosperm near the embryo, whereas the upper part of the grain remains translucent. Utilizing this mutant, they established a novel comparative system to investigate the physiological factors contributing to grain chalkiness. Importantly, the comparison occurs within individual grains, ensuring a uniform genetic background and minimizing confounding effects from genetic and environmental factors. Using this system, Lin et al. (2017) conducted integrated proteome and transcriptome analyses and discovered several key pathways linked to chalky tissue formation, especially the interaction between C and N metabolism, reduced ribosomal proteins, and lower levels of the 13 kDa prolamin subunit. Their study also showed that the embryo significantly affects endosperm composition, lowering the contents of proteins, amino acids, and minerals in the chalky endosperm (Lin et al. 2016).

[3] Multi-parent advanced generation intercross.

Another example of embryo involvement in chalkiness formation is the heat-induced expression of α-amylase in the endosperm. Hakata et al. (2012) found that heat stress increased the expression of α-amylase genes (*Amy1A*, *Amy1C*, *Amy3A*, *Amy3D*, and *Amy3E*) and enzyme activity, while decreasing ABA levels, a hormone that suppresses α-amylase. Moreover, RNAi suppression of these genes reduced the incidence of chalky grains under heat stress, demonstrating that α-amylase activation is critical for chalky tissue formation. It is well established that during germination, the embryo releases GA, which stimulates the production of starch hydrolases in the aleurone layer and scutellum. Nevertheless, it remains unclear whether a comparable mechanism operates in developing cereal grains.

The role of the embryo in regulating the formation of grain chalkiness has also been demonstrated by the investigation of the physiological response of grain development to high night temperature (HNT). Chang et al. (2021) utilized picolitre pressure-probe electrospray ionization mass spectrometry (picoPPESI-MS) to conduct single-cell metabolomic analyses in specific regions of HNT-exposed grains, namely, the scutellum, the basal endosperm with chalkiness, and the transparent middle endosperm. Microscopic observations indicated that HNT increases the cell expansion rate in the scutellum. At the peak of embryo expansion, marked spatial metabolic differences emerged, particularly in the ascorbate–glutathione pathway and purine metabolism. Moreover, accumulations of sugars and amino acids were higher in embryonic scutellum cells, which may be associated with reduced seed dormancy and accelerated embryo growth under HNT conditions. In addition to accelerating embryo enlargement, HNT concurrently reduced endosperm cell size and promoted chalkiness formation in the basal endosperm closer to the embryo. Microscopic observations revealed that preserved cytosolic space, primarily composed of numerous large vacuoles, plays a crucial role in the formation of the chalky tissue, presumably by inhibiting starch and protein accumulation. These findings corroborate the previously proposed hypothesis of a close relationship between embryo growth and chalky endosperm formation under heat stress (Wada et al. 2019).

6 Concluding Remarks and Perspectives

Recent studies have revealed a robust bidirectional dialogue between the embryo and endosperm during seed development, a feature that may be evolutionarily conserved across diverse plant species. The mechanistic insight primarily derives from research on the model plant *Arabidopsis*. Nevertheless, the relatively small size of its seeds poses an impediment to high-resolution analysis of this crosstalk, as the technical difficulty of isolating adjacent tissues without contamination or damage remains unresolved. Moreover, structural and developmental disparities limit the direct applicability of findings from *Arabidopsis* to major cereals such as rice, maize, and wheat. Consequently, several crucial questions regarding the mechanisms and implications of embryo–endosperm interactions in major crops need to be addressed, as detailed below.

1. Signalling Pathways Governing the Developmental Transition of the Endosperm

The developmental shifts in the endosperm, particularly the cellularization of the coenocyte, are notable and abrupt, involving comprehensive and rapid reprogramming of gene expression (Sabelli and Larkins 2009). The duration of the coenocytic phase has attracted substantial research attention, as it is positively linked with grain size and yield (Ishimaru et al. 2013; Olsen 2020). Under heat stress, endosperm cells experience accelerated proliferation and expansion, as evidenced by Morita et al. (2005) and Shi et al. (2017). This leads to incomplete starch accumulation and the formation of chalky endosperm because of the insufficient filling of enlarged cells. In contrast, Tao et al. (2021) suggest that postponed endosperm development also induces the formation of grain chalkiness. Therefore, producing perfect rice grains free of chalky tissue requires a delicate balance between endosperm development and filling. The exact timing of the transition from cellular differentiation to storage accumulation is thus of great significance for reducing grain chalkiness. However, the signals that initiate these critical developmental changes in the endosperm remain poorly understood. Could these signals originate from the embryo during its morphogenetic process?

2. Hormonal Regulation of Starch Degradation During Grain Filling

The degradation of starch during the germination of cereal grains, specifically in barley and wheat, has been extensively studied using biochemical analytical methods. Research suggests that GAs from the embryo trigger the synthesis and secretion of starch hydrolases by the aleurone cells and scutellum. Nevertheless, it is still uncertain whether a similar mechanism regulates starch hydrolysis in the endosperm during grain filling. Is there a functional link between starch degradation during germination and grain filling? What is the role of GAs in regulating starch hydrolysis in the developing endosperm? Is the dynamic equilibrium between starch biosynthesis and degradation controlled by the ratio of GAs to ABA? Furthermore, how do the aleurone cells facilitate the communication between the embryo and the endosperm?

3. Cellular Events During the Maturation Phase

Accumulating evidence demonstrates that the durations of embryonic development and endosperm filling are highly conserved across rice cultivars, with the former completing around 10 DAF, whereas the latter ends by approximately 30 DAF (Itoh et al. 2005; Zhu et al. 2011; Fu et al. 2013; Wu et al. 2016). Subsequently, seeds enter the maturation phase lasting 20–40 days, during which the grain weight remains largely unchanged. Maturation serves as a critical transitional phase linking embryogenesis and germination. During this phase, essential adaptive traits, such as dormancy, desiccation tolerance, and extended viability, are established. Moreover, maturation exerts a substantial influence on seed developmental competence, phenotypic plasticity, and long-term ecological performance, as gene expression is highly responsive to external environmental signals during this period (Sami et al. 2025). In cereal crops, the cellular and molecular events that occur during the maturation phase are still not well understood. If the developmental trajectory during this phase is evolutionarily conserved across cereals, what implications might embryo–endosperm interactions have for preharvest sprouting susceptibility, seed viability, and end-use quality?

4. The Biophysical Characteristics of the Embryo–Endosperm Interface
The boundary between the embryo and endosperm forms a highly intricate and continually evolving apoplastic structure, functioning as a crucial route for nutrient transport during embryogenesis (Doll and Ingram 2022). This structure is shaped by reciprocal biochemical and biophysical interactions between the embryo and the endosperm, but some key questions regarding its architectural and functional aspects remain unresolved. What is the precise chemical composition and physical structure of this apoplastic interface? How do developmental and environmental cues trigger its dynamic remodelling? Can we quantitatively assess the biomechanical properties of this interface?

All questions ultimately centre on the physiological and molecular mechanisms that coordinate communication between the embryo and the endosperm, a crucial yet inadequately understood aspect of seed development. Currently, there exists a significant gap between basic plant science and applied technologies in crop breeding and management. This gap is evident in the relatively slow progress of high-yield, quality-oriented breeding compared to the rapid advancements in functional genomics. A key factor contributing to this situation is the challenge of translating mechanistic insights from the cellular level to the organ and whole-plant levels, where yield and quality traits are developed and integrated. It is essential to strengthen collaborative efforts among crop geneticists, breeders, physiologists, and agronomists to more effectively elucidate the key mechanisms underlying embryo–endosperm interactions. Emphasis should be placed on re-evaluating the role of embryo development in grain filling and on refining strategies that simultaneously improve grain yield and quality.

Summary

1. Quality traits can be classified into two main categories: physical and chemical properties. In practical applications, they are further divided into four types: appearance, sensory, processing, and nutritional quality.
2. Each quality trait is rooted in specific chemical properties; therefore, the first step in modifying a quality trait is to identify and understand its underlying chemical basis.
3. The quality of cereal grain for end-use applications is predominantly determined by the composition, structure, and size characteristics of its endosperm. However, endosperm characteristics are influenced by interactions with other seed tissues, especially the embryo, throughout seed development.

4. The development of cereal endosperm involves synchronous coenocytic nuclear division, cellularization, tissue differentiation, and the accumulation of storage compounds, giving rise to two primary tissues: the outer aleurone layer and the inner starchy endosperm.
5. Mounting evidence from studies on *Arabidopsis*, rice, and maize indicates that there is bidirectional molecular crosstalk between the embryo and endosperm, which is essential for the coordinated development of seeds.
6. The embryo and endosperm become symplastically isolated either at the time of fertilization or shortly thereafter, necessitating the use of apoplastic pathways for nutrient and signal exchange between these two compartments.
7. The endosperm functions as the primary nutrient source to fuel embryo growth; however, it can also exert substantial physical and metabolic constraints on the embryo, potentially hindering or even completely preventing its development.

Important Terms

Apoplastic transportation; Cellularization; Coenocyte; Communication; Double fertilization; Embryogenesis; Interaction; Interface; Morphogenesis; PCD; Quality

Review Questions

Multiple Choice

1. Concerning the relation between endosperm and embryo, which of the statements below is NOT right (   )

   (a) During seed development, the embryo and endosperm are physically separated from each other by the apoplast at or soon after fertilization. This suggests that there must be communication between these two compartments through the symplastic continuum.
   (b) The endosperm serves as a vital nutrient source for supporting embryo growth, but it can also pose significant physical and metabolic challenges to the embryo, potentially hindering or even preventing its development entirely.
   (c) It is widely acknowledged that the endosperm plays an altruistic role, willingly providing nourishment to support the growth of the next generation—the embryo.
   (d) The simultaneous growth of the embryo and endosperm inside the limited space of the seed coat necessitates effective communication in both directions to ensure successful seed development and establish a new capable sporophytic generation.

2. (   ) is predominantly expressed in tissues at the embryo-endosperm interface and functions in both the embryo and endosperm to maintain size balance between the two.

   (a) *Zou*
   (b) *VP1*
   (c) *GE*
   (d) *UMAMIT25*

3. (   ) is specifically localized to the embryo-surrounding region, where it plays a crucial role in mediating endosperm degradation and nutrient remobilization in *Arabidopsis*.

   (a) *Zou*
   (b) *VP1*
   (c) *GE*
   (d) *FIS-PRC2*

4. Dysfunction of (   ) results in abnormal endosperm cellularization, thereby arresting embryo development at the heart stage.

   (a) *Zou*
   (b) SWEET15
   (c) *GE*
   (d) *FIS-PRC2*

5. Accumulation of (   ) during seed development has been associated with maturation of the seed, development of desiccation tolerance, and suppression of vivipary.

   (a) GAs
   (b) CKs
   (c) Auxin
   (d) ABA

True or False

1. The concept of quality for each specific end use and for each stage of the commercial chain in every crop has its own proper criteria (   )
2. The strategies and tools needed to produce grains with specific quality characteristics aren't as well-established as those for achieving high yields because the complex mechanisms that determine quality are still not fully understood (   )

Blank Filling

1. The development of rice seed can be divided into four stages: morphogenesis (0–10 DAF), embryo ___________ (10–20 DAF), endosperm ___________ (20–30 DAF), and maturation (30 DAF–).
2. Crop quality refers to the ___________ and ___________ attributes of the harvested or sink organs, such as grains, tubers, and roots, which determine their suitability for specific end uses.

Short Answer

1. Describe the interactions between embryo and endosperm during grain development of cereal crops.
2. Draw a picture of the grain of a cereal crop, such as rice, wheat, or maize, and explain how its morphological characteristics influence its end-use quality.

Essay Question

1. Select one quality trait of cereals, and interpret the significance of embryo–endosperm interaction for clarifying the mechanism of it.
2. Justify the necessity of interdisciplinary coordination in formulating strategies to enhance cereal crop quality.
3. Demonstrate the staging system for seed development of a crop you are familiar with or interested in, and provide brief descriptions of the agronomic relevance of each stage.

Authentic Performance Task

Analyze a specific biological system, such as a seed, an individual plant, or a complex crop stand, by deconstructing it into its constituent components to identify the critical interactions that influence its functionality.

References

An L, Tao Y, Chen H et al (2020) Embryo-endosperm interaction and its agronomic relevance to rice quality. Front Plant Sci 11:587641. https://doi.org/10.3389/fpls.2020.587641

Armenta-Medina A, Gillmor CS, Gao P et al (2020) Developmental and genomic architecture of plant embryogenesis: from model plant to crops. Plant Commun 2(1):100136. https://doi.org/10.1016/j.xplc.2020.100136

Baud S, Wuillème S, Lemoine R et al (2005) The *ATSUC5* sucrose transporter specifically expressed in the endosperm is involved in early seed development in Arabidopsis. Plant J 43(6):824–836. https://doi.org/10.1111/j.1365-313X.2005.02496.x

Belmonte MF, Kirkbride RC, Stone SL et al (2013) Comprehensive developmental profiles of gene activity in regions and subregions of the *Arabidopsis* seed. Proc Natl Acad Sci USA 110(5):E435–E444. https://doi.org/10.1073/pnas.1222061110

Besnard J, Zhao C, Avice JC et al (2018) Arabidopsis UMAMIT24 and 25 are amino acid exporters involved in seed loading. J Exp Bot 69(21):5221–5232. https://doi.org/10.1093/jxb/ery302

Cerri M, Reale L (2020) Anatomical traits of the principal fruits: an overview. Sci Hortic 270:109390. https://doi.org/10.1016/j.scienta.2020.109390

Chang FY, Hatakeyma Y, Nonami H et al (2021) Metabolic coordination of rice seed development to nighttime warming: in-situ determination of cellular redox states using picolitre pressure-probe electrospray-ionization mass spectrometry. Environ Exp Bot 188:104515. https://doi.org/10.1016/j.envexpbot.2021.104515

Chen J, Lausser A, Dresselhaus T (2014) Hormonal responses during early embryogenesis in maize. Biochem Soc Trans 42(2):325–331. https://doi.org/10.1042/bst20130260

Chen LQ, Lin IW, Qu XQ et al (2015) A cascade of sequentially expressed sucrose transporters in the seed coat and endosperm provides nutrition for the Arabidopsis embryo. Plant Cell 27(3):607–619. https://doi.org/10.1105/tpc.114.134585

Doll NM, Ingram GC (2022) Embryo-endosperm interactions. Annu Rev Plant Biol 73:293–321. https://doi.org/10.1146/annurev-arplant-102820-091838

Doll NM, Just J, Brunaud V et al (2020a) Transcriptomics at maize embryo/endosperm interfaces identifies a transcriptionally distinct endosperm subdomain adjacent to the embryo scutellum. Plant Cell 32(4):833–852. https://doi.org/10.1105/tpc.19.00756

Doll NM, Royek S, Fujita S et al (2020b) A two-way molecular dialogue between embryo and endosperm is required for seed development. Science 367(6476):431–435. https://doi.org/10.1126/science.aaz4131

Dresselhaus T, Balboni M, Berg L et al (2025) How meristems shape plant architecture in cereals-Cereal Stem Cell Systems (CSCS) Consortium. Plant Cell 37(7):koaf150. https://doi.org/10.1093/plcell/koaf150

Du L, Xu F, Fang J et al (2018) Endosperm sugar accumulation caused by mutation of PHS8/ISA1 leads to pre-harvest sprouting in rice. Plant J 95:545–556. https://doi.org/10.1111/tpj.13970

Fitzgerald MA, McCouch SR, Hall RD (2009) Not just a grain of rice: the quest for quality. Trends Plant Sci 14(3):133–139. https://doi.org/10.1016/j.tplants.2008.12.004

Fouquet R, Martin F, Fajardo DS et al (2011) Maize *rough endosperm3* encodes an RNA splicing factor required for endosperm cell differentiation and has a nonautonomous effect on embryo development. Plant Cell 23(12):4280–4297. https://doi.org/10.1105/tpc.111.092163

Fu J, Xu YJ, Chen L et al (2013) Changes in enzyme activities involved in starch synthesis and hormone concentrations in superior and inferior spikelets and their association with grain filling of super rice. Rice Sci 20(2):120–128. https://doi.org/10.1016/S1672-6308(13)60116-X

Graham RD, Welch RM, Saunders DA (2007) Advances in agronomy. In: Sparks DL (ed) Nutritious subsistence food systems, 1st edn. Academic Press, New York, pp 1–74

Grimault A, Gendrot G, Chamot S et al (2015) *ZmZHOUPI*, an endosperm-specific basic helix-loop-helix transcription factor involved in maize seed development. Plant J 84(3):574–586. https://doi.org/10.1111/tpj.13024

Hakata M, Kuroda M, Miyashita T et al (2012) Suppression of α-amylase genes improves quality of rice grain ripened under high temperature. Plant Biotechnol J 10:1110–1117. https://doi.org/10.1111/j.1467-7652.2012.00741.x

Hehenberger E, Kradolfer D, Köhler C (2012) Endosperm cellularization defines an important developmental transition for embryo development. Development 139(11):2031–2039. https://doi.org/10.1242/dev.077057

Ingram GC (2010) Family life at close quarters: communication and constraint in angiosperm seed development. Protoplasma 247(3–4):195–214. https://doi.org/10.1007/s00709-010-0184-y

Ingram GC (2020) Family plot: the impact of the endosperm and other extra-embryonic seed tissues on angiosperm zygotic embryogenesis. F1000Res 9:F1000. https://doi.org/10.12688/f1000research.21527.1

Ishimaru K, Hirotsu N, Madoka Y et al (2013) Loss of function of the IAA-glucose hydrolase gene *TGW6* enhances rice grain weight and increases yield. Nat Genet 45(6):707–711. https://doi.org/10.1038/ng.2612

Itoh J, Nonomura K, Ikeda K et al (2005) Rice plant development: From zygote to spikelet. Plant Cell Physiol 46(1):23–47. https://doi.org/10.1093/pcp/pci501

Jones TJ, Thomas L (1989) Histochemistry and ultrastructure of rice (*Oryza Sativa*) zygotic embryogenesis. Am J Bot 76(4):504–520. https://doi.org/10.2307/2444345

Köhler C, Hennig L, Bouveret R et al (2003) *Arabidopsis* MSI1 is a component of the MEA/FIE *Polycomb* group complex and required for seed development. EMBO J 22(18):4804–4814. https://doi.org/10.1093/emboj/cdg444

Lafon-Placette C, Köhler C (2014) Embryo and endosperm, partners in seed development. Curr Opin Plant Biol 17:64–69. https://doi.org/10.1016/j.pbi.2013.11.008

Lin ZM, Zhang XC, Yang XY et al (2014) Proteomic analysis of proteins related to rice grain chalkiness using *ITRAQ* and a novel comparison system based on a notched-belly mutant with white-belly. BMC Plant Biol 14:163. https://doi.org/10.1186/1471-2229-14-163

Lin ZM, Zheng DY, Zhang XC et al (2016) Chalky part differs in chemical composition from translucent part of japonica rice grains as revealed by a notched-belly mutant with white-belly. J Sci Food Agric 96(11):3937–3943. https://doi.org/10.1002/jsfa.7793

Lin ZM, Wang ZX, Zhang XC et al (2017) Complementary proteome and transcriptome profiling in developing grains of a notched-belly rice mutant reveals key pathways involved in chalkiness formation. Plant Cell Physiol 58(3):560–573. https://doi.org/10.1093/pcp/pcx001

Liu JX, Wu MW, Liu CM (2022) Cereal endosperms: development and storage product accumulation. Annu Rev Plant Biol 73:255–291. https://doi.org/10.1146/annurev-arplant-070221-024405

Martínez-Barajas E, Delatte T, Schluepmann H et al (2011) Wheat grain development is characterized by remarkable trehalose 6-phosphate accumulation pregrain filling: tissue distribution and relationship to *SNF1*-related protein kinase1 activity. Plant Physiol 156(1):373–381. https://doi.org/10.1104/pp.111.174524

McCarty DR, Hattori T, Carson CB et al (1991) The *Viviparous-1* developmental gene of maize encodes a novel transcriptional activator. Cell 66(5):895–905. https://doi.org/10.1016/0092-8674(91)90436-3

Morita S, Yonemaru J, Takanashi J (2005) Grain growth and endosperm cell size under high night temperatures in rice (*Oryza sativa* L.). Ann Bot 95(4):695–701. https://doi.org/10.1093/aob/mci071

Nagasawa N, Hibara KI, Heppard EP et al (2013) *GIANT EMBRYO* encodes *CYP78A13*, required for proper size balance between embryo and endosperm in rice. Plant J 75(4):592–605. https://doi.org/10.1111/tpj.12223

Nakata M, Fukamatsu Y, Miyashita T et al (2017) High temperature-induced expression of rice α-amylases in developing endosperm produces chalky grains. Front Plant Sci 8:2089. https://doi.org/10.3389/fpls.2017.02089

Nowack MK, Grini PE, Jakoby MJ et al (2006) A positive signal from the fertilization of the egg cell sets off endosperm proliferation in angiosperm embryogenesis. Nat Genet 38(1):63–67. https://doi.org/10.1038/ng1694

Olsen OA (2020) The modular control of cereal endosperm development. Trends Plant Sci 25(3):279–290. https://doi.org/10.1016/j.tplants.2019.12.003

Otegui ME, Cirilo AG, Uhart SA et al (2021) Maize. In: Victor OS, Daniel FC (eds) Crop physiology case histories for major crops. Academic Press, New York, pp 2–43

Pignocchi C, Minns GE, Nesi N et al (2009) Endosperm defective1 is a novel microtubule-associated protein essential for seed development in *Arabidopsis*. Plant Cell 21(1):90–105. https://doi.org/10.1105/tpc.108.061812

Sabelli PA, Larkins BA (2009) The development of endosperm in grasses. Plant Physiol 149(1):14–26. https://doi.org/10.1104/pp.108.129437

Sami AA, Bentsink L, Artur MAS (2025) The phylotranscriptomic profile of angiosperm seed development follows a reverse hourglass pattern. Plant Cell 37(12):koaf266. https://doi.org/10.1093/plcell/koaf266

Sanders A, Collier R, Trethewy A et al (2009) *AAP1* regulates import of amino acids into developing Arabidopsis embryos. Plant J 59(4):540–552. https://doi.org/10.1111/j.1365-313x.2009.03890.x

Senoura T, Sakashita E, Kobayashi T et al (2017) The iron-chelate transporter *Osysl9* plays a role in iron distribution in developing rice grains. Plant Mol Biol 95(4–5):375–387. https://doi.org/10.1007/s11103-017-0656-y

Shi WJ, Yin XY, Struik PC et al (2017) High day- and night-time temperatures affect grain growth dynamics in contrasting rice genotypes. J Exp Bot 68(18):5233–5245. https://doi.org/10.1093/jxb/erx344

Song WB, Zhu JJ, Zhao HM et al (2019) *OS1* functions in the allocation of nutrients between the endosperm and embryo in maize seeds. J Integr Plant Biol 61(6):706–727. https://doi.org/10.1111/jipb.12755

Tao Y, An L, Xiao F et al (2021) Integration of embryo–endosperm interaction into a holistic and dynamic picture of seed development using a rice mutant with notched-belly kernels. Crop J 10(3):729–742. https://doi.org/10.1016/j.cj.2021.10.007

Wada H, Hatakeyama Y, Onda Y et al (2019) Multiple strategies for heat adaptation to prevent chalkiness in the rice endosperm. J Exp Bot 70:1299–1311. https://doi.org/10.1093/jxb/ery427

Wang X, Zhou W, Lu ZH et al (2015) A lipid transfer protein, *OsLTPL36*, is essential for seed development and seed quality in rice. Plant Sci 239:200–208. https://doi.org/10.1016/j.plantsci.2015.07.016

Wang W, Xiong H, Sun KT et al (2022) New insights into cell-cell communications during seed development in flowering plants. J Integr Plant Biol 64(2):215–229. https://doi.org/10.1111/jipb.13170

Widiez T, Ingram GC, Gutiérrez-Marcos JF (2017) Maize kernel development. In: Larkins BA (ed) Embryo-endosperm-sporophyte interactions in maize seeds, 1st edn. CAB International, Beijing, pp 95–107

Wijerathna-Yapa A, Bishnoi R, Ranawaka B et al (2024) Rice-wheat comparative genomics: gains and gaps. Crop J 12(3):656–669. https://doi.org/10.1101/gr.1113003

Wu CC, Diggle PK, Friedman WE (2013) Kin recognition within a seed and the effect of genetic relatedness of an endosperm to its compatriot embryo on maize seed development. Proc Natl Acad Sci USA 110(6):2217–2222. https://doi.org/10.1073/pnas.1220885110

Wu XB, Liu JX, Li DQ et al (2016) Rice caryopsis development II: dynamic changes in the endosperm. J Integr Plant Biol 58(9):786–798. https://doi.org/10.1111/jipb.12488

Xi M, Lin ZM, Zhang XC et al (2014) Endosperm structure of white-belly and white-core rice grains shown by scanning electron microscopy. Plant Prod 17(4):285–290. https://doi.org/10.1626/pps.17.285

Xiong HX, Wang W, Sun MX (2021) Endosperm development is an autonomously programmed process independent of embryogenesis. Plant Cell 33(4):1151–1160. https://doi.org/10.1093/plcell/koab007

Yang SX, Johnston N, Talideh E et al (2008) The endosperm-specific *ZHOUPI* gene of *Arabidopsis thaliana* regulates endosperm breakdown and embryonic epidermal development. Development 135(21):3501–3509. https://doi.org/10.1242/dev.026708

Zheng XX, Li Q, Li CS et al (2019) Intra-kernel reallocation of proteins in maize depends on *VP1*-mediated scutellum development and nutrient assimilation. Plant Cell 31(11):2613–2635. https://doi.org/10.1105/tpc.19.00444

Zhou SR, Yin LL, Xue HW (2013) Functional genomics based understanding of rice endosperm development. Curr Opin Plant Biol 16(2):236–246. https://doi.org/10.1016/j.pbi.2013.03.001

Zhu GH, Ye NH, Yang JC et al (2011) Regulation of expression of starch synthesis genes by ethylene and ABA in relation to the development of rice inferior and superior spikelets. J Exp Bot 62(11):3907–3916. https://doi.org/10.1016/j.pbi.2013.03.001

Plant Architecture

4

Optimizing Spatial Arrangement of Vegetative and Reproductive Organs for an Ideotype

Contents

Z. Liu, *Crop Physiology*, https://doi.org/10.1007/978-981-92-2960-4_4

Abstract

Plant architecture refers to the three-dimensional arrangement of a plant's organs. It serves as a traditional criterion for classifying plant species and significantly influences the adaptability and productivity of crop plants. This chapter commences by exploring the regulatory mechanisms that govern plant architecture, placing specific emphasis on the hormonal and genetic regulation of plant height, branching patterns, leaf and tiller angles, inflorescence structure, and growth habit, predominantly in cereal crops. Subsequently, it elucidates the concept and principles of ideotype along with some examples from wheat and rice. Finally, it explores the potential of cutting-edge genome editing technologies to revolutionize ideotype breeding and outlines future directions from the perspective of biotechnology–information technology (BIT) integration.

Keywords

Plant architecture · Phytohormones · Genetic regulation · Ideotype

Did You Know

- The discovery of auxin, the first plant hormone, can be traced back to De Candolle (1832), who conducted the initial observations on phototropism. He observed that indoor plants orient themselves towards windows in the process of seeking light rather than air.
- Christiane Nüsslein-Volhard identified Bicoid as a morphogen, which is a transcription factor existing in a concentration gradient within the syncytial blastoderm of the common fruit fly (*Drosophila melanogaster*) embryo. In recognition of this pioneering discovery, she was conferred the 1995 Nobel Prize in Physiology or Medicine.
- Gibberellins were first identified by Eichi Kurosawa in 1926 and isolated in the 1930s as natural products in the fungus *Gibberella fujikuroi* (now known as *Fusarium fujikuroi*); it is from this fungus that they derive their distinctive name. Rice plants infected with *F. fujikuroi* exhibit excessive internode elongation, leading to stem lodging and yield loss.
- In the nineteenth century, a significant number of households utilized coal gas, a complex gaseous mixture containing ethylene, for gas lighting. This usage created an inhospitable environment for most houseplants. Nevertheless, *Aspidistra elatior* demonstrated remarkable tolerance to ethylene, rendering it the most popular indoor plant of the period and earning it the nickname "cast-iron plant".
- In the early 1960s, researchers postulated that the rapid germination and growth of pollen grains were correlated with a growth-promoting substance. A crude extract of rape pollen induced significant stem elongation in common beans, a response distinct from GA-mediated stem elongation. This groundbreaking work led to the isolation and identification of brassinolide, the first steroidal plant growth regulator.

(continued)

- Norman Ernest Borlaug was awarded the Nobel Peace Prize in 1970 in recognition of his contributions to the Green Revolution. This initiative involved the development and widespread dissemination of high-yielding, disease-resistant dwarf wheat cultivars, which significantly increased cereal production in Mexico, India, Pakistan, and numerous other developing countries.

Learning Objectives

After reading this chapter, students will be able to:

- Understand the morphogenetic foundation of plant architecture
- Summarize how plant architecture is regulated by hormonal and genetic factors
- Describe the genetic and molecular foundation of the Green Revolution
- Explain the genetic basis for the repression of tillering in maize and rice
- Comprehend the core principle of the ideotype, the weak competition
- Outline strategies for breeding ideotypes in cereal crops

Introductory Remarks

The structure of a plant refers to the physical entity composed of carbohydrates, proteins, lipids, minerals, water, and other constituents. The construction of a plant body incurs costs in terms of both materials and energy, which plants allocate to vegetative structures such as stems, leaves, and roots. These resources could instead have been allocated to reproductive organs, thereby increasing the quantity and quality of offspring (i.e. seeds). According to evolutionary theory, plants must derive sufficient benefits from their supporting structures to offset these costs. This benefit lies in providing a vegetative framework that enables the acquisition and assimilation of valuable natural resources, including solar energy, water, and mineral nutrients. Therefore, optimizing the spatial arrangement of vegetative and reproductive organs is central to shaping an ideal plant form for higher productivity.

Over the past 10,000 years, humans have selectively domesticated and cultivated specific plants to provide food and materials. In this process, specific architectural and phenological characteristics that enhance the yield of harvestable organs, particularly seeds and fruits, have been prioritized. Repeated over generations, domestication has resulted in crops that display significant morphological differences from their wild relatives, sometimes in a remarkable way. One of the most notable morphological alterations brought about by domestication is the transformation of teosinte, a highly branched wild grass native to Central America, into modern maize (Fig. 4.1).

Research on the genetic mechanisms governing plant architecture has revealed a surprisingly simple genetic basis for its remarkable phenotypic diversity. Despite the complexity of plant forms, key hormonal pathways act as

(continued)

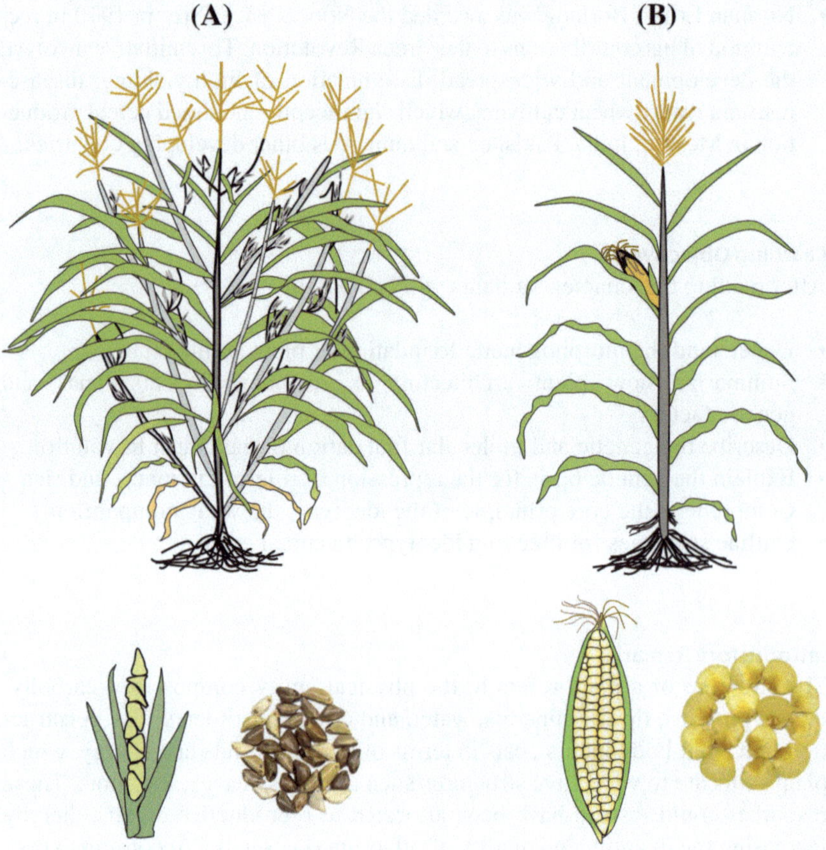

Fig. 4.1 Morphological comparison of teosinte (**A**) and maize (**B**). Modern maize was domesticated from teosinte. The two plants display remarkably distinct architectures, especially in terms of branching pattern and inflorescence morphology. Teosinte cobs carry only six to ten grains, which are covered with a very hard, inedible shell and do not have a husk formed by leaves. In contrast, modern maize cobs contain a substantially greater quantity of soft and edible grains, enveloped by a husk, which facilitates the harvesting and storage processes. These disparities are so substantial that it took scientists many years to uncover their close genetic relationship. The primary gene accountable for this fundamental alteration, diminished branching and augmented ear size, is *TEOSINTE BRANCHED 1* (*TB1*), and its orthologs assume crucial roles in numerous species, such as *Arabidopsis*, rice, and pea

central regulators, integrating internal and external signals to determine diverse architectural phenotypes. This chapter focuses on the regulatory mechanisms governing plant architecture, with emphasis on the hormonal and *genetic regulation* of plant height, branching patterns, leaf and tiller angles, inflorescence structure, and growth habit in major crops. These insights provide a scientific basis for designing both molecular interventions and agronomic practices to optimize plant architecture, thereby enhancing crop productivity and resilience.

Concept Map

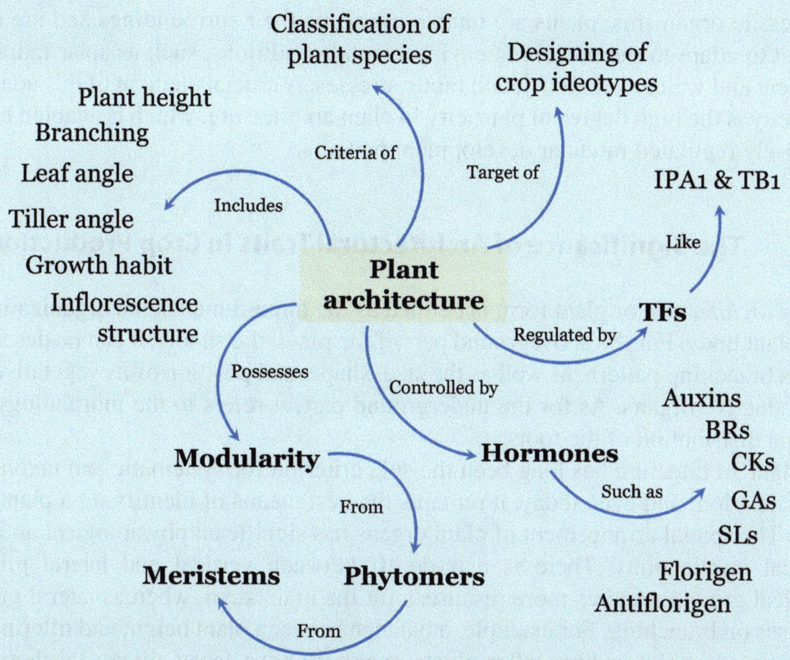

Annotation: Plant architecture has been, and continues to be, a key criterion for plant classification. More importantly, it represents a primary breeding target in the development of ideotypes designed to maximize the utilization of natural resources, solar energy in particular, for a target area. Architectural traits of agronomic importance encompass plant height, branching pattern, leaf and tiller angles, growth habit, and inflorescence structure. The formation of plant architecture follows a modular pattern, characterized by the repeated development of phytomers through the sustained activity of meristems, namely apical, axillary, intercalary, secondary, and inflorescence meristems, which continuously generate new structural units. Despite their phenotypic complexity, the formation of these architectural traits is governed by a relatively simple genetic foundation. Some traits are regulated by transcription factors (TFs) such as IPA1 and TB1, whereas the majority are controlled by phytohormones, including auxins (root branching), brassinosteroids (BRs; leaf angle), cytokinins (CKs; inflorescence development), gibberellins (GAs; plant height), strigolactones (SLs; shoot branching), and florigen/antiflorigen systems (growth habit).

1 Plant Architecture

As sessile organisms, plants are unable to escape their surroundings and are compelled to adapt to the prevailing environmental conditions, such as solar radiation, nutrient and water availability, and biotic stresses. A crucial element of this adaptive capacity is the high degree of plasticity in plant architecture, which is enabled by the precisely regulated modular development patterns.

1.1 The Significance of Architectural Traits in Crop Production

Plant architecture or plant form is defined as the three-dimensional organization of the plant body. For the aboveground part of the plant, the shoots, it can be described by its branching pattern, as well as the size, shape, and position of its vegetative and reproductive organs. As for the underground part, it refers to the morphology and spatial distribution of the roots.

Plant architecture has long been the sole criterion for systematic and taxonomic classification, and even today, it remains the best means of identifying a plant species. The spatial arrangement of plant organs has significant physiological and ecological implications. There is a trade-off between vertical and lateral growth. Vertical growth requires more resources for the main stem, whereas lateral growth focuses on branching. For example, a balance between plant height and tillering has been noted in rice, where taller plants generally have fewer tillers. Furthermore, plant architecture holds significant agronomic importance as it strongly influences the suitability for cultivation, yield, and resource use efficiency of a given species.

Mechanization is widely recognized as an essential pathway to sustainable crop production in developing countries. As rural labour shortages worsen and food demand rises with growing populations, agricultural machinery, such as tractors and combine harvesters, plays a critical role in increasing yields, reducing manual labour, and enhancing operational efficiency. To fully leverage mechanization, breeding programs are increasingly focused on developing mechanization-friendly crop cultivars. These cultivars are specifically designed for compatibility with planting, management, and harvesting machinery, enabling faster operations, lower labour dependency, and greater yield stability. In major crops such as rice, maize, wheat, soybean, and cotton, breeding efforts focus on several key structural traits that support efficient machine-based farming systems, as presented below.

First, a compact and uniform plant architecture, exemplified by semi-dwarf rice and wheat cultivars and lodging-resistant maize hybrids, reduces plant height, thereby preventing lodging and allowing combines to move smoothly through fields without damaging crops. Second, strong and thick stems provide essential structural support, enabling plants to bear the weight of heavy heads and withstand environmental stresses such as wind and mechanical contact, a critical trait in maize and soybean breeding. Third, synchronized flowering and subsequent uniform maturity ensure that all plants in a field reach physiological maturity simultaneously, making single-pass harvesting both feasible and efficient in crops such as rice, wheat, and cotton. Fourth, optimized placement of reproductive organs, such as mid-stalk ears in maize, centrally located cotton bolls, and well-positioned rice panicles, improves harvester access and reduces grain or fibre loss during harvest. Finally, reduced shattering in rice and wheat, minimized pod splitting in soybeans, and non-dehiscent cotton bolls are vital genetic improvements that prevent premature seed dispersal and significantly reduce losses during mechanical threshing.

1.2 Cellular Process for Morphogenesis

The primary parameters delineating plant architecture are plant height, branching pattern, and growth habit. These properties are ultimately governed by the coordinated activities of individual cells, which carry out three fundamental processes: division, expansion, and differentiation. Cellular aggregates form tissues, which subsequently integrate into functional units termed organs. The dimensions and relative arrangement of these organs endow plants with their distinctive architectural patterns. Therefore, to comprehensively comprehend morphogenesis, the generation of plant form, it is imperative to dissect the cellular events that regulate development and growth.

Alan Turing (1912–1954), who introduced the concept of *morphogens*, diffusible substances that can establish concentration gradients, was a pioneer in proposing a chemical basis for the *patterning* of organisms (Zsögön and Peres 2018). Turing demonstrated that these gradients could offer spatial information, guiding cells in ascertaining their relative positions. For a chemical compound to be classified as a morphogen, it must meet three criteria: (1) it must exert its effects in a concentration-dependent manner, (2) it must be produced at a specific location and diffuse from that site, and (3) it must act directly on target cells. The local concentration of the morphogen is detected by cells, initiating a signalling cascade that activates a specific subset of transcription factors. This activation leads to differential gene expression patterns, subsequently leading to distinct cellular outcomes such as division, expansion, or differentiation. Turing's theory was validated decades later through the discovery and analysis of morphogens in *Drosophila melanogaster* (the fruit fly).

In plants, functional equivalents of morphogens include small RNAs, *signal peptides*, transcription factors, and molecules with potent morphogenetic effects, such as phytohormones. Auxin acts as a morphogen, exerting a substantial influence on shoot elongation, branching, and the morphogenesis of leaves and roots. At the cellular level, auxin triggers cell expansion, a mechanism well elucidated by the "acid growth" hypothesis. This hypothesis posits that auxin activates plasma membrane H^+-ATPase (AHA), leading to apoplastic acidification that activates cell wall-loosening enzymes, thus promoting cell expansion (Gallei et al. 2020). This process is precisely regulated through both rapid non-transcriptional pathways and delayed transcriptional responses, ensuring coordinated cell elongation in response to developmental and environmental cues (Fig. 4.2).

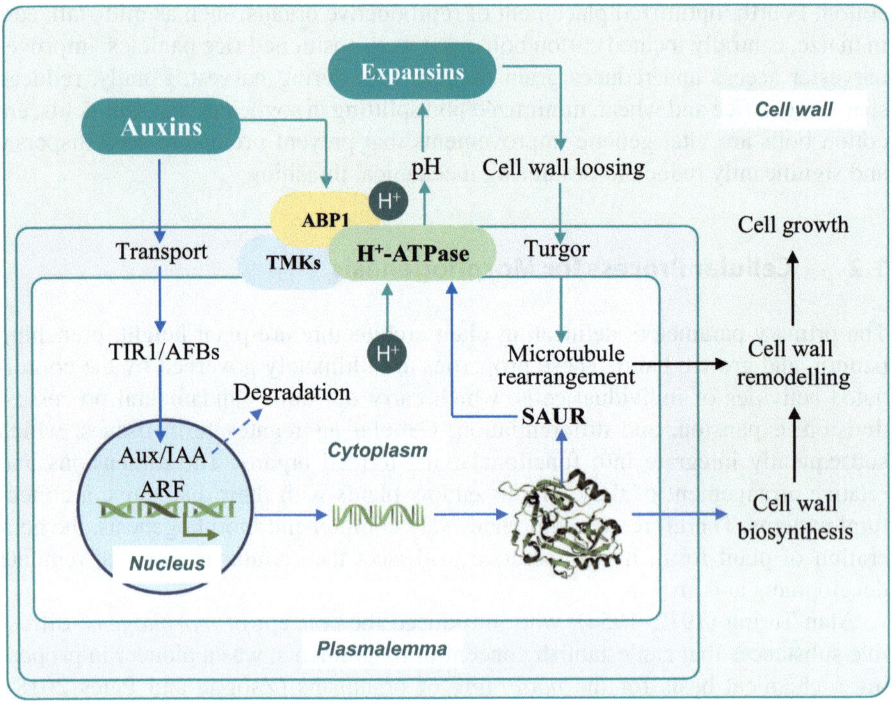

Fig. 4.2 Auxin functioning as a morphogen to induce cell expansion. Apoplastic, non-transcriptional pathways are indicated by dark cyan arrows, nuclear transcriptional pathways by blue arrows, and cell wall-related activities by black arrows. (1) In the apoplastic space, auxin binds to the auxin-binding protein (ABP1), which, in conjunction with the transmembrane kinases TMK1 and TMK4, forms a cell-surface receptor complex that triggers the activation of plasma membrane H^+-ATPases. (2) Upon entering the cell, intracellular auxin signal transduction is initiated through the binding of auxin to a small family of F-box proteins that function as intracellular auxin receptors. This family includes TRANSPORT INHIBITOR RESISTANT 1 (TIR1) and related AUXIN SIGNALING F-BOX PROTEINS AFB1, AFB2, and AFB3. Aux/IAA proteins function as transcriptional repressors that play a pivotal role in auxin signalling through their interaction with auxin response factors (ARFs), thereby suppressing the expression of early or primary auxin-responsive genes, including *SMALL AUXIN-UP RNA* (*SAUR*) genes. Elevated auxin concentrations trigger the degradation of the Aux/IAA complex, thereby initiating transcriptional activation. (Redrawn from Murray et al. (2012) and Zsögön and Peres (2018))

1. *Apoplastic, Non-Transcriptional Pathway*

In the extracellular apoplastic space, the phytohormone auxin initially binds to secreted auxin-binding protein 1 (ABP1). In an auxin-dependent manner, ABP1 subsequently interacts with the extracellular domains of plasma membrane-localized receptor-like kinases TMK1 and TMK4. This ternary complex, comprising ABP1, TMK1, and TMK4, functions as a cell-surface system that perceives extracellular auxin and triggers the activation of plasma membrane AHA. The activated proton pumps transport H^+ into the apoplast, lowering the pH from approximately 6.0 to

5.0. The resultant acidic environment activates pH-dependent cell wall-modifying proteins, including *expansins*, xyloglucan endotransglucosylase/hydrolases (XTHs), and other hydrolases, thereby breaking the hydrogen bonds between cellulose microfibrils and matrix polysaccharides. This process facilitates cell wall loosening and reduces mechanical resistance to turgor pressure. The consequent decrease in turgor pressure lowers water potential within the cell, driving water influx from the apoplast into the cytoplasm. This water uptake restores cellular turgor pressure, ultimately promoting cell expansion. The cytoskeleton, specifically the dynamic microtubule network, functions as a guiding mechanism for cellulose synthase complexes during cellulose deposition. Microtubule orientation is also sensitive to mechanical stress, establishing a direct connection between physical signals and morphogenetic patterning.

2. *Nuclear, Transcriptional Pathway*

Upon entering the cell, auxin is initially perceived by nuclear F-box receptors TIR1/AFB. The binding of auxin facilitates the ubiquitination and subsequent proteasomal degradation of Aux/IAA proteins, thereby releasing AUXIN-RESPONSIVE FACTOR (ARF) transcription factors to activate auxin-responsive genes. Among the primary transcriptional targets are the *SMALL AUXIN-UP RNA* (*SAUR*) and *GH3* gene families, whose protein products promote the phosphorylation of plasma membrane AHA. Phosphorylated AHAs actively transport protons out of the cell, resulting in a reduction of apoplastic pH (~6.0) to approximately 4.5–5.0. This acidic environment activates expansins and other cell wall-loosening enzymes, enabling slippage within the cellulose-microfibril network and facilitating cell expansion. Concurrently, this pathway regulates a suite of genes associated with cell wall biosynthesis and remodelling, ensuring the sustained progression of cell expansion. This transcriptional cascade, although relatively slow (occurring over minutes to hours), can be rapidly initiated due to the pre-existing poised state of many *SAUR* genes for swift induction.

In summary, the TMK-AHA pathway rapidly activates H^+-ATPase within seconds to several minutes, leading to transient acidification of the cell wall and an immediate increase in cell wall extensibility. Subsequently, the TIR1/AFB–ARF–SAUR signalling route initiates a transcriptional program that promotes the expression of downstream genes encoding proteins like SAUR, thereby sustaining and amplifying the cell wall acidification effect.

1.3 Modularity of Plant Morphogenesis

Forms and patterns in plants are emergent properties of cell collectives. Thus, development is not merely the process of creating a multicellular body from a single cell, as such a process would result in an unstructured cell mass or tumour. Instead, it is a series of coordinated events that lead to the formation and patterning of organized tissues, ultimately giving the adult organism a specific form. Although plant architecture shows significant variation across different crop species, it is formed through the repetition of small, modular units. This process, referred to as *modular construction*, is governed by common developmental processes and principles based on *meristems* and *phytomers*.

1.3.1 The Meristems

Plant morphology emerges from the coordinated expansion of modular units and dynamic shifts in meristem activity, enabling structural and functional adaptability (Dresselhaus et al. 2025). The meristem is a specialized group of undifferentiated cells capable of continuous division, expansion, and differentiation, thereby giving rise to new tissues and organs. Within the meristem, *stem cells* reside in defined microenvironments known as *stem cell niches* (SCNs), typically composed of slowly dividing organizing centre cells surrounded by actively dividing stem cells (Fig. 4.3A). *Organogenesis* at apical meristems occurs through a dynamic balance between the maintenance of undifferentiated cells and commitment to differentiate. Vegetative meristems establish the fundamental body plan by producing both aerial and underground structures. Subsequently, some of them transition into reproductive meristems, which initiate floral development and germ cell formation.

Shoot architecture is shaped by the organization and activities of apical, axillary, intercalary, lateral, or secondary, and inflorescence meristems (Fig. 4.3B). *The shoot apical meristem* (SAM) is the meristem at the shoot apex that forms all other shoot meristems and organs. *Intercalary meristem* (ICM) in the stems contributes to elongation growth. *Axillary meristem* (AM) located in the axils of leaves gives rise to lateral branches or inflorescences. *Inflorescence meristem* (IM) is responsible for generating panicles, spikes, and flowers. *Lateral meristem* (LM) in the lateral tissues of the stem drives secondary growth by giving rise to the vascular cambium and the cork cambium. The cambial layer produces new xylem inwardly and phloem outwardly, increasing the stem's girth and providing structural support as well as a protective bark. Thus, the lateral meristem enables plants to thicken and sustain continuous upright growth.

In flowering plants, AMs, located in the leaf axils, can develop into buds that elongate and form branches, giving rise to secondary growth axes. These branches grow similarly to the primary axis, with potential for further branching, thereby contributing to increasingly complex plant architectures. Additionally, AMs may differentiate into floral structures, integrating vegetative growth with reproductive development. Thus, AMs produce both vegetative and inflorescence *branches*. The number of branches is mainly determined by the number of leaves on the primary stem because there is normally only one AM per leaf.

Monocots, such as cereals, exhibit two distinct types of vegetative branching. The first type of branching may potentially arise from AMs situated in the upper nodes, namely the elongation nodes. For example, ratooning rice exhibits regrowth of new shoots from AMs following the harvest of the panicle, a process that disrupts apical dominance. However, this phenomenon is relatively uncommon, as these lateral branches are typically inhibited and remain in a bud-like state. The second type

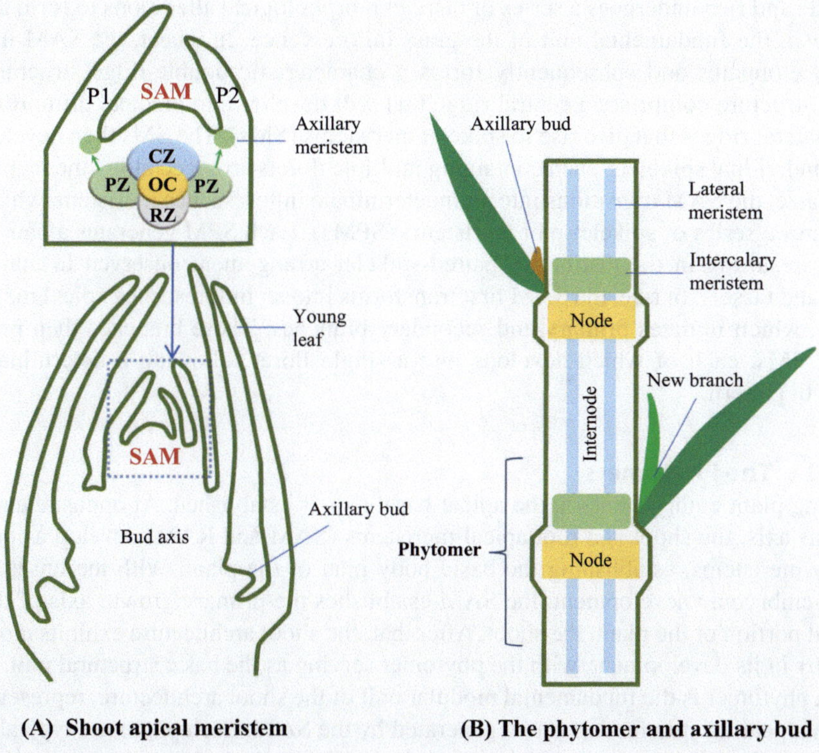

Fig. 4.3 Shoot apical meristems (**A**) and the phytomer (**B**) of cereal crops. The shoot apical meristem (SAM) is organized into four distinct functional zones: (1) Central zone (CZ): Located at the centre of the SAM, the CZ serves as a reservoir of undifferentiated and indeterminate stem cells. These cells are crucial for maintaining the meristem activity, and cell division occurs infrequently here. (2) Peripheral zone (PZ): Surrounding the SAM, this region exhibits more active cell division. Cells from the PZ are incorporated into leaf primordia and give rise to axillary meristems (AMs), which later develop into vegetative branches or inflorescences. (3) Rib zone (RZ): Situated in the proximal region of the SAM, the RZ receives displaced cells from the CZ and contributes to the formation of the stem body, generating major internal tissues such as the pith and vascular core. (4) Organizing centre (OC): Acting as a signalling hub, the OC regulates stem cell proliferation in the CZ and coordinates differentiation processes in both the PZ and RZ, thereby ensuring proper organization and sustained function of the SAM. The images are not drawn to scale. Abbreviations: CZ central zone, OC organizing centre, P1 and P2 leaf primordia, PZ peripheral zone, RZ rib zone. (Redrawn from Wang et al. (2018))

of branching originates from AMs located at lower, non-elongated *tillering* nodes near the ground level. These AMs develop into *tillers* that are morphologically indistinguishable from the main culm. Tillers generate their own adventitious roots, thereby attaining functional independence from the main culm. Tillering capacity is a trait displayed by numerous species in the Poaceae family, ensuring an adequate number of culms for a successful harvest.

During the transition from vegetative to reproductive growth, the SAM in wheat, maize, and rice undergoes a series of distinct morphological alterations to form the *spikelet*, the fundamental unit of the grass inflorescence. In wheat, the SAM initially elongates and subsequently forms a characteristic double-ridge structure. This structure comprises a central ridge that will develop into bracteal primordia, and lateral ridges that give rise to spikelet meristems (SMs). The SMs then develop into individual spikelets, each containing multiple florets arranged along the rachis. In maize, the SAM transitions into an indeterminate inflorescence meristem, which initiates a series of spikelet-pair meristems (SPMs). Each SPM generates a pair of SMs, resulting in the distinctive paired-spikelet arrangement observed in maize ears and tassels. In rice, the SAM first transforms into an inflorescence apical meristem, which initiates primary and secondary branches. These branches then produce SMs, each of which develops into a single floret following a determinate growth pattern.

1.3.2 The Phytomers

During plant embryogenesis, the apical-basal axis is established. At opposite ends of this axis, the shoot and root apical meristems (SAM and RAM) develop as primary meristems, establishing the basic body plan of the plant. With the onset of post-embryonic development, the SAM establishes the primary growth axis of the aerial portion of the plant, the shoot. After that, the shoot architecture exhibits *modularity* in its development, with the phytomer serving as the basic structural unit.

A phytomer is the fundamental modular unit of the shoot architecture, representing a repeating structural segment generated by the SAM. Each phytomer typically consists of a node, one or more leaves attached at that node, an internode connecting to the next node, and an axillary bud located in the leaf axil (Fig. 4.3B). As the fundamental unit of growth and development, the phytomer facilitates stem elongation through internode extension, enables photosynthesis via its leaves, and offers the potential for lateral branching through the axillary bud. Since phytomers are produced in a sequential manner, they collectively document the plant's developmental history and facilitate modular plasticity, enabling adaptation to environmental fluctuations.

Variations in the position, size, and shape of individual phytomers, together with the differential regulation of axillary bud outgrowth, form the morphological basis for the extensive diversity in shoot architecture among seed plants. Viewing the phytomer as a repeating structural and functional unit is central to functional–structural plant modelling (FSPM), where it serves as the fundamental building block for simulating whole-plant architecture, resource allocation, and growth dynamics. This modular framework facilitates cross-species comparisons, enhances breeding strategies by linking organ-level traits to overall plant performance, and provides an integrative platform for combining genetics, physiology, and biomechanics. Consequently, the phytomer concept bridges descriptive morphology with quantitative approaches, solidifying its role as a cornerstone of modern plant biology.

2 Overview of Plant Hormones

2.1 Phytohormones

The form and function of multicellular organisms rely on the chemical signals exchanged among cells, tissues, and organs. *Phytohormones*, which exist in extremely low concentrations, serve as chemical messengers to coordinate cellular functions in plants. They are synthesized in one cell and regulate cellular processes in another cell through interaction with specific receptor proteins that initiate intracellular signalling cascades. Plants have nine major hormones[1]: *auxins*, *cytokinins* (CKs), *abscisic acid* (ABA), *ethylene* (ETH), *gibberellins* (GAs), *brassinosteroids* (BRs), *salicylic acid* (SA), *jasmonates*[2] (JAs), and *strigolactones* (SLs). Additionally, several peptides, such as CLAVATA3,[3] along with proteins like *florigen* and *antiflorigen*, exert localized control over organogenesis. Consequently, these molecules can also be regarded as phytohormones.

Phytohormones interact in complex, context-dependent ways to regulate plant development, through synergy, antagonism, or functional complementarity. Synergistic interactions occur when two or more hormones amplify each other's effects. For example, auxin and CKs jointly promote the activity of the SAM, while ethylene enhances ABA-mediated drought responses. In contrast, antagonistic interactions occur when one hormone inhibits the action of another. A well-documented instance is the opposing roles of GAs and ABA in promoting and suppressing seed germination, respectively. Complementarity becomes evident when different hormones regulate distinct phases or aspects of the same developmental process. For example, auxin controls root cell patterning, whereas CKs regulate cell division within the same root region. These synergistic, antagonistic, and complementary interactions enable plants to precisely fine-tune their developmental programs and respond rapidly to fluctuating environments. A deeper understanding of these hormonal interactions empowers breeders to develop stress-resilient cultivars and improve crop productivity under challenging environments.

[1] Polyamines (PAs) found most frequently in plants and other organisms are putrescine, spermidine, and spermine. PAs induce a wide range of physiological responses in plants, including cell division, tuber formation, root initiation, embryogenesis, flower development, fruit ripening, and stress tolerance. Given that PAs are present in much higher concentrations in plants compared to other hormones, and considering that millimolar concentrations of PAs are typically required to induce a biological response, it is debatable whether PAs function as true hormones.

[2] Jasmonic acid (JA) and related compounds, often referred to as jasmonates (JAs) collectively, were initially identified as the inhibitors of plant growth.

[3] CLAVATA3 (CLV3) is a peptide signal initially identified in the analysis of *clv* mutants in the model plant *Arabidopsis*, as a regulator of meristem homeostasis and floral organ numbers.

2.2 Hormonal Signalling

Hormonal regulation of plant development demonstrates significant diversity in specific mechanisms; however, all core pathways possess common characteristics. Typically, the perception of an external signal or an internal developmental cue modifies the balance of hormonal homeostasis in favour of anabolism. Subsequently, the synthesized hormone is transported to its target site, where it binds to a specific receptor. This interaction initiates a series of molecular events, including modifications in gene expression, post-translational alterations such as phosphorylation, *protein turnover*, and ion flux across membranes. These biochemical responses ultimately result in a physiological or developmental outcome appropriate to the initial stimulus. Moreover, plants employ negative feedback mechanisms that attenuate signalling by inhibiting further hormone production. Hormone catabolism and sequestration mechanisms decrease active hormone levels after signal transduction, thereby restoring them to the baseline. By resetting hormonal states, plants preserve their capacity to sense and respond to new stimuli, ensuring continuous adaptability and precise control overgrowth and development.

One of the primary topics in plant hormone biology is the interrelated nature of hormones and their physiological effects, a phenomenon commonly termed *crosstalk*. Crosstalk refers to situations in which one or more components of a signal transduction pathway impact or regulate another distinct pathway. Hormone crosstalk specifically describes the interconnected signalling networks of plant hormones that interact through synergistic, antagonistic, or additive effects, thereby facilitating the coordinated regulation of plant growth, development, and responses to biotic and abiotic stresses. When operating this mechanism, numerous signalling cascades share common molecular components, which simultaneously influence multiple hormonal pathways. For example, DELLA proteins act as crucial nodes that connect GA signalling to other regulatory pathways (Box 4.1). They interact with auxin and ethylene to regulate root development, control seed dormancy and germination via ABA, and contribute to flood tolerance by integrating ethylene-mediated responses. Consequently, the accumulation and activity of DELLA proteins are dynamically regulated by a wide array of internal and external stimuli. Understanding how this limited set of shared regulators guides diverse physiological outcomes will provide insights into the integration of hormone signalling networks.

Box 4.1: The DELLA Proteins

DELLA proteins, which belong to the GRAS family of putative transcriptional regulators (named after GIBBERELLIN INSENSITIVE [GAI], REPRESSOR OF GA1–3 [RGA], and SCARECROW [SCR]), act as key growth repressors. These proteins are exclusively found in land plants, with their numbers varying among different species (Phokas and Coates 2021). Some species possess only a single DELLA protein, such as rice (SLENDER RICE 1, *Os*SLR1), barley (SLENDER 1, *Hv*SLN1), and tomato (PROCERA, *Sl*PRO), whereas others harbour multiple DELLA proteins. For instance, *Arabidopsis* contains five DELLA proteins: *At*GAI, *At*RGA, *At*RGL1, *At*RGL2, and *At*RGL3, which exhibit both distinct and overlapping functions.

The term "DELLA" is derived from the five conserved amino acids, aspartate, glutamate, leucine, leucine, and alanine (D-E-L-L-A), situated in the N-terminal domain. This motif exhibits a high degree of conservation across all plant species and is critically important for its regulatory function. DELLA proteins are negative regulators in the GA signal transduction pathway and belong to a class of growth-inhibitory proteins localized in the nucleus. Their primary function in vascular plants is to repress GA responses, which govern numerous key developmental processes, including seed germination, stem elongation, leaf expansion, and flowering. DELLA proteins act not only as integrators of environmental cues, such as light, temperature, and water, but also directly interact with key transcription factors, thereby playing a central role in various signalling routes. Because DELLA proteins lack a DNA-binding domain, they mainly execute their function by directly binding to transcription factors.

3 Hormonal Control of Plant Architecture

Plant architecture is predominantly determined by the combination of key morphological traits, including plant height, tiller number, tiller and leaf angles, grain number and size, primary root length, and crown and lateral root development. Phytohormones act as central regulators that integrate environmental signals with developmental programs, thereby playing crucial roles in modulating architectural features. Each structural trait is governed by a complex network of multiple hormones that interact either synergistically or antagonistically (Fig. 4.4). Moreover, microRNAs (miRNAs), transcription factors, and other regulatory proteins fine-tune plant architecture by modulating hormonal signalling cascades, highlighting the multi-layered control underlying this process (Guo et al. 2020).

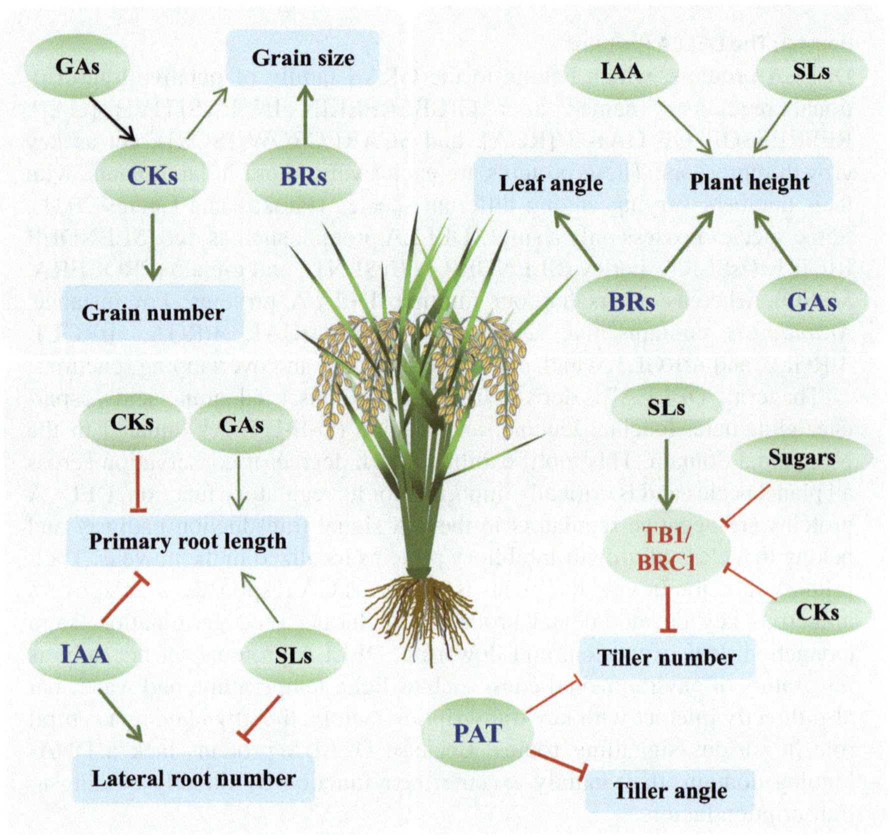

Fig. 4.4 Illustration of the contribution of phytohormones to shoot architecture, using rice as an example. The main hormones controlling each structural trait are represented. Plant height is primarily promoted by GAs, tiller number is suppressed by SLs, and tiller angle is negatively regulated by PAT through auxin redistribution. Additionally, BRs exert a strong promoting effect on leaf angle, while CKs and BRs have remarkable promoting influences on grain number and grain size, respectively. The regulation of branching is elaborately illustrated. The Teosinte Branched 1 (TB1) orthologs include OsTB1 in rice, SbTB1 in sorghum, and BRANCHED 1 (BRC1) in *Arabidopsis*, pea, and tomato. TB1/BRC1 acts as a central signalling integrator that suppresses axillary bud outgrowth. Its expression is induced by SLs, while CKs and sugars negatively regulate it. The green arrow represents positive regulation, the red line with perpendicular bars indicates negative regulation, and the black line shows the interaction between hormones. The thickness of the lines represents the extent of the effect that each hormone exerts on the target trait. ABA abscisic acid, BRs brassinosteroids, CKs cytokinins, GAs gibberellins, IAA indole-3-acetic acid, PAT polar auxin transport, SLs strigolactones. (Redrawn according to Wang et al. (2018) and Chen et al. (2022))

3.1 Plant Height

Plant height plays a crucial role in plant adaptation to diverse environments. In cereal crops, a shorter stature demonstrates a more advantageous response to nitrogen fertilization, as it alleviates excessive vegetative growth and *lodging*.

Consequently, shorter plants possess a greater potential to maintain higher grain yields, especially under a high rate of nitrogen fertilizer. Plant height is regulated by several hormones, including GAs, BRs, auxin, and SLs. Among these, GAs are notable for their crucial role in stem elongation and for their historical contribution to the *Green Revolution*, during which semi-dwarf, GA-insensitive cultivars of wheat and rice significantly enhanced yield and lodging resistance.

GAs promote internode elongation by stimulating both cell expansion and division, a phenomenon especially pronounced in grasses due to the intercalary meristems. The levels of bioactive GAs are tightly regulated through a feedback-controlled homeostatic mechanism. The link between GA biosynthesis and its physiological effects was first suggested by the abnormally high GA concentrations found in GA-insensitive dwarf mutants such as *Rht3* in wheat, *D8* in maize, and *gai* in *Arabidopsis*. Transcriptome studies conducted across multiple plant species have revealed that cellular GA levels specifically regulate the expression of key GA biosynthetic enzymes, namely, GA20-oxidase and GA3-oxidase, as well as the GA-deactivating enzyme GA2-oxidase. Genes encoding GA20-oxidase and GA3-oxidase are significantly downregulated upon exogenous GA application. Importantly, these transcriptional alterations take place within 1–3 h of GA treatment, significantly prior to any observable growth response. Conversely, GA 2-oxidase genes are promptly induced upon GA exposure. Collectively, these findings suggest that GA homeostasis in plants is maintained, at least in part, through the coordinated regulation of both anabolic and catabolic pathways.

The Green Revolution in the 1960s, which led to a significant increase in global cereal grain production, emphasizes the crucial role of GAs in regulating plant development (Fig. 4.5). A large international consortium, spearheaded by Norman Borlaug and Gurdev Khush, developed high-yielding semi-dwarf cultivars through mutations in the *Reduced height 1* (*Rht1*) gene in wheat and the *semidwarf 1* (*sd1*) gene in rice. Both the pivotal Green Revolution genes are integral components of the GA pathway. The *Rht1* locus in wheat encodes a DELLA protein. The dwarf alleles, primarily *Rht-B1b* and *Rht-D1b*, confer insensitivity to GA due to specific amino acid substitutions in the DELLA protein. This altered function of the DELLA protein suppresses GA-mediated growth responses, resulting in a reduction of plant stature from an initial height of 120–150 cm to 70–90 cm. The rice *sd1* gene, encoding a defective GA 20-oxidase, was originally identified in the Chinese cultivar Dee-Geo-Woo-Gen. The loss-of-function of GA 20-oxidase leads to reduced GA levels, thereby diminishing its positive effect on internode elongation. The shortened architecture mitigates lodging susceptibility, particularly under intensive management with abundant irrigation and N fertilizer. Additionally, a key characteristic of semi-dwarf wheat and rice plants is that the growth of vegetative organs (leaves and stems) is reduced more significantly than that of the reproductive organs (grains). In this way, semi-dwarf genes reconfigure resource allocation. Under the high-input conditions, the harvest index (HI) increases from 30%–35% to 45%–50%. As a result, a greater proportion of photosynthates is channelled towards grain development rather than stem elongation, leading to higher grain yield (Shen et al. 2025).

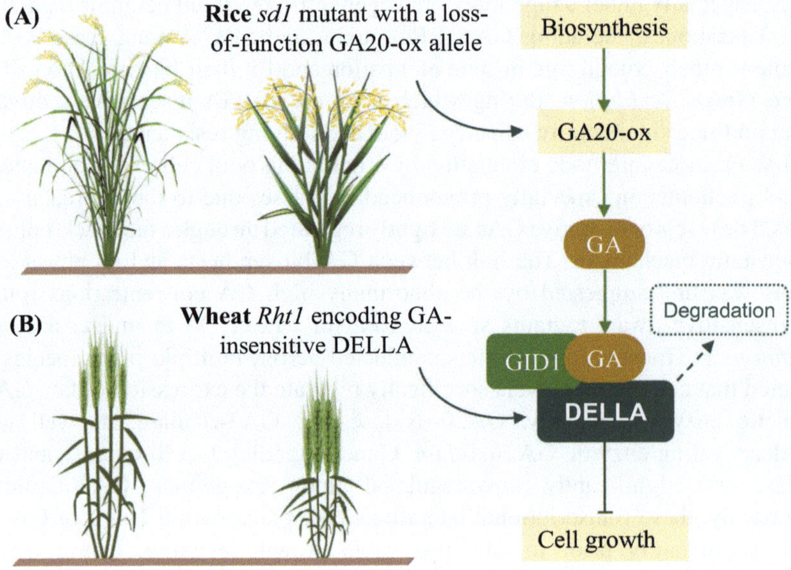

Fig. 4.5 The regulatory mechanism of the Green Revolution genes, *Rht1* and *sd1*. The GA signalling pathway involves both positive and negative regulatory components. The binding of active GA to its receptor protein (GIBBERELLIN INSENSITIVE DWARF 1, GID1) initiates the degradation of the transcriptional repressor DELLA. In wheat, *Rht1* represents a dominant gain-of-function mutation within the coding sequence of a DELLA repressor protein, resulting in partial insensitivity to GA. In contrast, the rice *sd1* gene harbours a loss-of-function mutation in a key gene encoding GA 20-oxidase, an enzyme critical for GA biosynthesis, thereby reducing GA levels and impairing stem elongation. (Redrawn from Liu et al. (2020))

Discovery of new alleles for genes in the GA signalling and biosynthesis pathway is an ongoing effort to assist breeding. Sub1A is an ethylene-responsive ERF transcription factor that enables rice to adopt the "quiescence" strategy for submergence tolerance (Fukao et al. 2006). Upon flooding, ethylene accumulates in plant tissues and strongly induces *Sub1A* expression. This activation leads to the accumulation of SLR1 and SLRL1, key repressors of GA signalling, thereby inhibiting GA-mediated shoot elongation. Concurrently, Sub1A upregulates GA 2-oxidase genes, which deactivate bioactive GAs and further suppress GA levels. By reducing both GA synthesis and responsiveness, *Sub1A* helps conserve carbohydrates and prevents energy-intensive growth during submergence. Moreover, *Sub1A* fine-tunes hormonal crosstalk by modulating BRs and ABA pathways, enhancing stress-responsive gene expression while restraining ethylene-driven elongation. Rice cultivars carrying the *Sub1A-1* allele exhibit shorter internodes under flood conditions, significantly improved survival rates, and maintain normal yields in non-flooded environments. Thus, the core function of *Sub1A* lies in suppressing GA synthesis and signalling to inhibit underwater shoot growth, enabling rice to withstand prolonged submergence without compromising agricultural productivity (Fukao and Bailey-Serres 2008).

In addition to the *sd1* gene, *Shortened Basal Internodes* (*SBI*) represents a novel lodging-resistance gene in rice. *SBI* encodes a GA 2-oxidase that specifically regulates the elongation of basal internodes by deactivating GA activity. It is predominantly expressed in the basal internodes of the plant (Liu et al. 2018). The recessive tall rice mutant *elongated uppermost internode* (*eui*) exhibits normal morphology until the final internode undergoes dramatic elongation at the heading stage. This stage-specific developmental effect of the *eui* mutation has been exploited in hybrid rice breeding to enhance the performance of male sterile cultivars during flowering. Map-based cloning revealed that the *EUI* gene encodes a previously uncharacterized cytochrome P450 monooxygenase, CYP714D1, which functions as a GA-deactivating enzyme. The identification of *EUI* as a GA catabolism gene provides further evidence that manipulating the GA metabolism pathway can be an effective strategy for improving crop performance (Zhu et al. 2006).

BRs function synergistically with GAs to regulate stem elongation, especially in seedlings. This interaction serves as the foundation for one of the most extensively used BR sensitivity bioassays: the bean second internode elongation assay. Mutants with diminished brassinosteroid levels in rice and tomato display stunted growth, which impairs both cell expansion and cell division. A well-known rice dwarf mutant, *dwarf2*, carries a mutation in a P450 family protein that is critically important for brassinosteroid biosynthesis. In tomato, the *DWARF* (*D*) gene encodes a P450 protein responsible for catalysing the oxidation of 6-deoxocastasterone to castasterone (CS), an active brassinosteroid and the direct precursor of brassinolide (BL), currently regarded as the most biologically potent BR. CS primarily functions in vegetative tissues, whereas BL predominantly regulates fruit development. Loss-of-function mutations in the *D* gene result in dwarfism in tomato.

Auxin is synthesized at the shoot apex and transported basipetally towards the roots through a process known as *polar auxin transport* (*PAT*). This transport occurs via specific transporters, including PIN auxin efflux carriers (PINs) and ATP-binding cassette subfamily B (ABCB) proteins, also referred to as P-glycoproteins (PGP) or multidrug resistance (MDR) proteins. PAT involves *canalization*,[4] which begins with the relocalization of PIN1 and subsequently promotes the differentiation of cells along the transport route into vascular tissue. Alterations in PAT underpin the dwarfing mechanisms that have been exploited in crop improvement. The reduced stature of the maize *brachytic2* (*br2*) mutant results from the loss of

[4] The canalization hypothesis was suggested in 1968 by Tsvi Sachs to account for the formation of vascular strands in response to wounding or auxin application. The hypothesis proposes that positive feedback between auxin movement through a cell and the cell's auxin conductivity leads to the gradual selection of narrow "canals" of polar auxin transport that will differentiate into vascular strands.

function in an ABCB/PGP/MDR protein, leading to decreased PAT (Multani et al. 2003). This reduction in PAT contributes to other agronomically desirable traits, such as more erect and darker stay-green leaves. Despite these beneficial characteristics, the *br2* mutation has not been commercially adopted due to its severe *pleiotropic* effects. A mutation in *dwarf3* (*dw3*), the sorghum ortholog of *br2* that also encodes an ABCB/PGP/MDR protein, has captured significant agronomic interest since the 1950s (Truong et al. 2015). The *dw3* mutant also reduces leaf insertion angles, thereby enhancing the vertical distribution of solar radiation within the canopy. Consequently, this optimizes canopy photosynthesis and increases nitrogen accumulation in leaves, collectively improving the overall performance of the sorghum crop.

3.2 Branching Pattern

Cereal crops have experienced divergent selective pressures on the branching pattern during the domestication process. Wheat and rice have been selected for a reduction in *apical dominance*, featuring multiple tillers that distribute grain production more uniformly and mature relatively synchronously. In contrast, maize, sorghum, and pearl millet have been selected through a distinct evolutionary pathway that strongly suppresses tillering. This phenomenon is characterized by enhanced apical dominance, which inhibits lateral branching and concentrates seed production on a single, large terminal inflorescence.

Shoot branching involves bud initiation, activation, expansion, and sustained branch growth. This process is regulated by a complex network involving phytohormones, sugars, environmental cues, and gene expression, with TB1/BRC1 acting as the key repressor of bud activation (Fig. 4.4). Specifically, phytohormones such as SLs, auxin, GAs, and ABA generally inhibit bud outgrowth, whereas BRs and CKs promote it.

SLs inhibit branching by inducing the expression of the TB1/BRC1. SLs' perception relies on a ternary complex comprising the SL receptor DWARF14 (D14), the F-box protein D3, and the transcriptional repressor D53. SLs trigger ubiquitination and degradation of D53, thereby activating the TB1/BRC1 pathway. Low N promotes phosphorylation and stabilization of D14, thus repressing rice tillering (Hu et al. 2024). In *Arabidopsis*, auxin indirectly inhibits bud outgrowth mainly by suppressing CKs biosynthesis while promoting SLs biosynthesis. Nevertheless, its effect on rice tillering is relatively limited. GAs signalling facilitates the proteasomal degradation of SLR1, which in turn triggers the degradation of MONOCULM 1 (MOC1), thereby repressing bud formation and outgrowth in rice. ABA, predominantly present in dormant buds, acts synergistically with SLs to inhibit bud outgrowth in rice.

CKs act antagonistically to auxin by suppressing apical dominance. CKs or their precursors are synthesized in the roots and transported acropetally via the xylem to arrested axillary buds, where they break dormancy. Emerging evidence suggests that CKs may influence PAT, increasing the expression of PIN transporters in the shoots.

Conversely, auxin modulates CK levels by repressing their biosynthesis, specifically through downregulation of the *ADENYLATE ISOPENTENYLTRANSFERASE* (*IPT*) gene family, thereby counteracting the promoting effect of nitrate on branching. Moreover, CKs function as primary regulators of sink establishment, a process critical for vigorous lateral shoot growth.

3.3 Tiller Angle

In cereal crops, tillers are specialized branches that originate from the axillary meristems of leaves and have the potential to develop into complete, fertile plants. The tillering of rice comprises two primary stages: an early spreading phase and a late compacting phase. During the initial spreading phase, plants adopt a loose and open architecture to maximize light capture and space utilization for tiller occurrence. In the subsequent compacting phase, leaves become more upright and densely arranged, thereby minimizing mutual shading and significantly enhancing canopy photosynthesis (Yu et al. 2007). The tiller angle, defined as the angle between the vertical line and the lateral tillers at maximum inclination, is important in optimizing light interception and promoting healthy interplant competition. Neither an overly spreading nor an excessively compact architecture is beneficial for the canopy function. Extremely compact crop plants are prone to disease, whereas plants with a prostrate growth habit occupy excessive space, thereby reducing grain yield per unit area. Therefore, maintaining an appropriate tiller angle is crucial for dense planting and, ultimately, for maximizing grain yield.

Gravitropism represents a critical environmental response in higher plants, as it directs root growth toward gravity (positive gravitropism) and shoot growth in the opposite direction (negative gravitropism). Tillers exhibit negative gravitropism, which is induced by the redistribution of auxin within the stem. Research on the tiller angle mutant *lazy1* and its suppressors demonstrates that auxin transport and redistribution are essential for determining tiller and leaf angles in rice. Moreover, SLs presumably widen the tiller angle by inhibiting auxin biosynthesis. At present, the gravitropic response is regarded as involving a series of sequential steps: gravity perception, signal transduction, and organ curvature resulting from the uneven distribution of auxin. Nevertheless, the regulatory mechanisms underlying gravitropism and tiller angle remain limited in staple crops (Wang et al. 2022).

3.4 Leaf Angle

Despite possessing the same LAI, diverse canopy architectures can emerge from the arrangement or stacking of foliage, giving rise to distinct light interception patterns. The penetration of radiation through the canopy is affected by multiple factors related to the 3D arrangement of leaves, including leaf size, leaf angle, spatial heterogeneity, leaf thickness, and leaf albedo. Leaf angle (LA) refers to the angle between the plant stem and the upper (adaxial) surface of the leaf blade. LA plays a critical role in determining plant architecture, varying with crop species, genotype, and environmental conditions. LA is one of the key factors that influence light interception efficiency and interspecific competition among adjacent plants, thereby affecting RUE and overall crop productivity.

Generally, cereal and grass crops have more erect leaves compared to dicotyledonous crops, with typical values less than 30° from the vertical. However, the early leaves of certain wheat cultivars are extremely prostrate, and highly fertilized grass crops can develop larger, horizontally oriented leaves at the top of the canopy. Under water stress conditions, leaves may wilt, resulting in a reduction in leaf angle. To address this variability, a quantitative description of the leaf angle distribution within a canopy can be obtained via cumulative frequency analysis of leaf angles. The leaf angles of most crop species follow one of the following idealized distributions: (1) erectophile, predominantly less than 30° (e.g. cereals); (2) plagiophile, predominantly between 30° and 60° (e.g. oilseed rape, soybean); and (3) planophile, predominantly greater than 60° (e.g. clover, sunflower).

High-density planting represents an effective strategy for increasing crop yield per unit land area. Erectophile leaves enable more efficient light interception within the canopy under dense planting, thereby enhancing photosynthetic efficiency, air circulation, and stress resistance. For instance, *osdwarf4-1*, a rice mutant deficient in BR biosynthesis, exhibits erect leaves and higher yield under dense planting (Sakamoto et al. 2006). Likewise, an isogenic single-cross maize hybrid carrying the *liguleless 2* (*lg2*) gene displays erect leaf morphology and can produce up to 40% more grain than its counterpart with horizontal leaves (Pendleton et al. 1968). Consequently, enhancing leaf erectness has become a target trait in crop improvement (Cao et al. 2022).

Numerous studies, especially on rice and maize, have reported a positive role of BRs in the regulation of LA formation. Tian et al. (2019) cloned *UPA1* (*Upright Plant Architecture 1*) and *UPA2*, two quantitative trait loci conferring upright plant architecture in maize. *UPA1* encodes brassinosteroid C-6 oxidase 1 (BRD1), the enzyme responsible for the final step of BRs biosynthesis. Elevated BR levels promote auricle expansion and increase leaf angle. *UPA2* is not a protein-coding gene but a distal regulatory element or enhancer located 9.5 kb upstream of the B3-domain transcription factor ZmRAVL1. A two-base insertion-deletion (InDel) present in the teosinte allele creates a strong binding site for DROOPING LEAF 1 (DRL1). DRL1 physically interacts with LIGULELESS 1 (LG1) and suppresses LG1-mediated activation of *ZmRAVL1*, thereby reducing *ZmRAVL1* expression and, consequently, BRD1 activity and leaf angle. In contrast, the maize-derived *UPA2* allele lacks these two bases, resulting in weakened DRL1 binding, increased expression of ZmRAVL1 and BRD1, and a wider leaf angle. The teosinte-derived allele was lost during domestication but can be reintroduced: near-isogenic lines carrying the wild-type *UPA2* allele or CRISPR-Cas9–edited *ZmRAVL1* exhibit a more erect leaf architecture and significantly enhanced grain yield under high-density planting conditions. Thus, the DRL1–LG1–ZmRAVL1–BRD1 regulatory cascade represents a precise molecular target for breeding maize cultivars tailored to dense cultivation systems.

3.5 Inflorescence Structure

The homeostasis of CK levels in the inflorescence meristem, governed by biosynthesis, activation, and degradation, plays a crucial role in regulating inflorescence architecture. Among them, deactivation is a key regulatory step that controls the abundance of biologically active CKs, as mediated by cytokinin oxidase/dehydrogenase (CKX), an enzyme that irreversibly cleaves the unsaturated isoprenoid side chain of CKs. Overexpression of CKX leads to reduced CKs content and delayed shoot development, attributable to significantly decreased cell proliferation rates in the SAM. Conversely, CKX overexpression promotes root growth and enlarges the RAM. Zhang et al. (2025) reported that the overexpression of GRAIN NUMBER ASSOCIATED (GNA) results in increased CKs levels, which in turn promote panicle branching and give rise to a larger panicle. This effect is mediated through the suppression of *OsCKX2* expression. In addition to CKs, GAs and the small peptide GRAIN NUMBER, GRAIN LENGTH AND AWN DEVELOPMENT 1 (GAD1) regulate grain number by promoting or repressing cytokinin activity in the panicle, respectively.

3.6 Root Architecture

The primary functions of the root system are to acquire nutrients and water from the soil and provide anchorage. Generally, plants produce only one or a limited number of roots during embryogenesis, and the majority of the root system develops subsequent to germination as the plant becomes established in the soil. The root angle plays a crucial role in the efficient absorption of nutrients and water from different soil layers. Steeper root angles allow access to mobile water and nitrogen from deeper soil layers, whereas shallow root angles assist in the capture of immobile phosphorus from the topsoil. A comprehensive understanding of the genetic regulation of root angle is, therefore, valuable for breeding crops with a high capacity to acquire mineral nutrients (Kirschner et al. 2024).

Root angle is determined by a balance between gravitropic and antigravitropic forces. Gravity perception predominantly occurs in the columella cells of the root cap. In these cells, specialized starch-containing plastids known as amyloplasts sediment in the direction of gravity, initiating the process of gravity sensing (Kirschner et al. 2024). Amyloplast sedimentation triggers a signal that is transmitted to the outer tissues via the meristematic zone. This signal results in an asymmetric distribution of auxin, causing asymmetric cell expansion in the elongation zone and ultimately leading to root bending as a gravitropic response. In cereal crops, numerous proteins have been identified as key players in the gravitropic response. However, only a few components of the antigravitropic response have been functionally characterized, including the regulation of asymmetric auxin distribution, ROS signalling, and cell wall stiffening in the elongation zone. Recent research suggests that these pathways may interact within a complex regulatory network, where mechanical cues modulate hormone gradients and oxidative bursts.

3.7 Growth Habit

The *growth habit* of a crop plant refers to its characteristic structural form and developmental pattern, such as whether it grows erect or prostrate and whether its growth is *determinate* or *indeterminate*. In determinate crops, such as wheat, rice, and maize, the transition from vegetative to reproductive growth is sudden and clearly defined. Once flowering commences, vegetative growth largely ceases. In contrast, indeterminate crops, including soybean, tomato, and pepper, maintain concurrent vegetative and reproductive development throughout a significant portion of their life cycle. After the initial floral transition, these plants undergo repeated and overlapping cycles of shoot elongation, leaf production, and flower/fruit formation, resulting in sprawling or bushy architectures that support continuous flowering and fruiting within a single growing season.

Repeated flowering begins when *florigen* concentrations reach a threshold sufficient to induce floral transition in the primary apical meristem. Meanwhile, newly formed vegetative buds are temporarily inhibited from flowering due to localized expression of *antiflorigen* signals. In the 1930s, it was proposed that the flowering hormone florigen is synthesized in leaves under inductive photoperiod conditions and then transported to the SAM, where it triggers flowering. Florigen is a globular protein encoded by the gene *FLOWERING LOCUS T* (*FT*), which belongs to the *CETS* gene family. This family is denominated after its orthologs *CENTRORADIALIS* from Antirrhinum, *TERMINAL FLOWER 1* (*TFL1*) from *Arabidopsis*, and *SELF-PRUNING* (*SP*) from tomato. Members of this family act as master regulators of developmental processes in both plants and animals. In contrast to other plant hormones, FT functions as the long-distance flowering signal that integrates multiple environmental cues and endogenous signals to indicate the transition to flowering (McGarry and Ayre 2021). FT is synthesized in the companion cells of the leaf vascular system and transported systemically to the apical and axillary meristems. In meristems, it interacts with transcription factor FLOWERING LOCUS D (FD) and *14-3-3 proteins* to form a hexamer, activating the expression of floral meristem identity genes, *LEAFY* (*LFY*) and *APETALA 1* (*AP1*) (refer to Sect. 4.5 in Chap. 7).

The flower-promoting activity of florigen is counteracted by antiflorigenic signals encoded by genes from the same gene family. In *Arabidopsis*, TFL1 functions as an antiflorigen that opposes the flower-inducing effects of FT. *TFL1* encodes a protein highly similar to FT in structure but antagonistic in function; moreover, a small number of amino acid substitutions can interconvert their functional identities. TFL1 maintains the SAM in a vegetative state. By binding to FD, TFL1 represses the expression of floral identity genes, thereby preventing flowering.

Mutations in the antiflorigen gene *SELF-PRUNING* in tomato (*sp*), along with its homologs in other crops such as soybean (*dt*) and cotton (*Gbsp*), result in precocious flowering of lateral shoots. This alteration leads to the transformation of continuously growing (indeterminate) shoot systems into compact (determinate) plants with synchronized flowering, both temporally and spatially. The tomato *sp* mutant exhibits a compact, determinate growth habit characterized by a sudden burst of flowering and fruit production, in contrast to the continuous, indeterminate, vine-like growth of classical cultivars and their wild ancestors. Similarly, in determinate and semi-determinate soybean genotypes, the apical meristem ultimately terminates vegetative growth by converting into a raceme (Fig. 4.6). In contrast, in indeterminate types, this transition is less abrupt and is modulated by competition for assimilates from developing seeds at existing nodes.

The flowering behaviour of crop species is governed by a dynamic balance between florigen and antiflorigen, forming a finely tuned, environment-sensitive regulatory system that coordinates both local developmental signals and long-distance systemic cues to modulate the timing of floral transition (Eshed and Lippman 2019). This delicate hormonal balance is crucial for flowering plants to reconcile two conflicting requirements: generating sufficient vegetative structures (source organs) to fulfil the high nutrient demands of developing flowers, fruits, and seeds (sink organs), while also ensuring the timely maturation of seeds.

Fig. 4.6 Determinate, semi-determinate, and indeterminate growth habits of the soybean plant. In determinate types, the main stem node accrual stops soon after flowering, leading to a terminal flower and a long pod-bearing raceme. In indeterminate types, nodes continue to form until seed filling, resulting in overlap of vegetative and reproductive developments. In semi-determinate types, node accrual continues past flowering but stops earlier than in indeterminate types. The difference between determinate and indeterminate types of soybeans is associated with contrasting alleles of a single gene locus (*dt1*). Indeterminate cultivars (*Dt1Dt1*) are well-suited to short-season, high-latitude environments due to their flexible growth period. In contrast, determinate cultivars (*dt1dt1*) are primarily cultivated in longer-season, low-latitude regions where growing conditions allow for more compact development

The selection of *FT* and *TFL1* alleles has resulted in more determinate growth patterns and earlier flowering in domesticated crops such as soybean, beet, barley, wheat, and sunflower (McGarry and Ayre 2021). The tomato *sp* mutant, which has a determinate growth habit and was discovered in a farmland in the 1920s, serves as an example of how genetic mutations can enhance crop productivity. A mutation in the *SP* gene transformed an indeterminate vine into a compact plant with uniform fruit set and synchronized ripening. This enabled higher planting densities and more convenient large-scale mechanical harvesting, facilitating extensive open-field tomato cultivation and revolutionizing agriculture in California's Central Valley. The *sp* allele has been extensively incorporated into commercial cultivars, and subsequent molecular research has shown that *SP* is homologous to *TFL1*. These genetic loci continue to be crucial focal points in contemporary crop improvement endeavours, as they illustrate that a single mutation can have a substantial impact on plant architecture and, in turn, the agricultural landscape.

4 Genetic Regulation of Plant Architecture

In addition to hormonal control, *genetic regulation* plays a pivotal role in shaping plant architecture. In this process, transcription factors (TFs) and hormones function in a hierarchical cascade. Hormone perception by receptor proteins triggers signal transduction, ultimately modulating the activity or expression of TFs. These TFs then bind to specific DNA sequences to directly regulate target genes involved in cell division, expansion, and differentiation, thereby mediating the morphogenetic processes. This hormone–TF nexus enables integration of environmental signals with developmental programs, allowing flexible adaptation of plant architecture. Engineering TFs offers an alternative strategy to manipulate hormone-regulated architectural traits without perturbing entire hormonal networks.

4.1 Transcriptional Factors in Crop Science

Transcription factors, the DNA-binding proteins, regulate gene expression by recognizing specific cis-acting elements within the promoter regions or intragenic sequences of target genes (Box 4.2). In plants, TFs act as master regulators governing key processes including metabolism, growth, and development. Moreover, they generate alternative transcripts through *intragenic promoters*, thereby enhancing the flexibility and robustness of cellular metabolism. The crucial role of TFs is evidenced by the large number of genes encoding them in plant genomes. For instance, the *Arabidopsis* genome contains approximately 2000 genes encoding TFs, making up roughly 6% of its total protein-coding genes. This abundance highlights the need for sessile plants to promptly regulate gene expression in response to a wide variety of environmental cues and to effectively coordinate complex physiological processes.

TFs are highly valuable targets for molecular crop breeding. Through the editing or overexpression of key TFs, researchers can simultaneously regulate entire downstream gene networks without directly modifying metabolic enzymes. Consequently, rapid improvements in crop yield, quality, and stress tolerance can be achieved while preserving core physiological processes. Wei et al. (2022) conducted a screening of TFs associated with photosynthesis in rice and identified OsDREB1C, a member of the DREB (Dehydration-Responsive Element-Binding) family. Notably, overexpression of OsDREB1C significantly enhances grain yield by 41.3%–68.3%, improves NUE, and accelerates flowering time. These findings underscore that the diverse array of TFs not only enhances plant adaptability to changing environments but also offers actionable targets for precision crop breeding.

In terms of plant architecture, the transcription factor IPA1 (Ideal Plant Architecture 1) enhances rice yield by suppressing unproductive tillers and enlarging panicle size. TB1 (Teosinte Branched 1) in maize restricts lateral branch development and promotes the initiation of female inflorescences. More recently, OsWRKY72, which is highly expressed in the leaf sheath and lamina joint, has been identified as a positive regulator of leaf angle through modulation of gravitropic responses by downregulating LAZY1 (Liu et al. 2024). Together, these findings demonstrate that plant architecture can be substantially optimized by modulating a single transcription factor.

Box 4.2: Diverse Transcription Factor Families

Transcription factors regulate gene expression by binding to specific cis-regulatory elements, thereby either promoting or inhibiting their interaction with components of the transcriptional machinery. This binding is typically sequence-specific, which ensures that each transcription factor targets a particular cis-element. For example, the auxin response factor (ARF) family of transcription factors specifically binds to the auxin response element (ARE).

Transcription factors generally possess a modular structure, comprising three principal domains: DNA-binding, transactivation, and nuclear localization signal domains. The DNA-binding domain enables the protein to recognize and bind specific cis-regulatory elements. Meanwhile, the transactivation domain interacts with components of the transcriptional machinery to facilitate regulation of transcriptional activity. Moreover, most transcription factors contain a nuclear localization signal, which directs the protein into the nucleus after its synthesis in the cytoplasm.

The diverse transcription factors in higher plants can generally be categorized into around 50 families, primarily based on sequence similarity. The nomenclature of these families often reflects conserved structural features. For example, basic helix–loop–helix (bHLH) proteins possess a DNA-binding domain composed of two α-helical regions connected by a loop. More commonly, however, family names reflect physiological functions, for instance, auxin response factors (ARFs). Additionally, some families are named after their first identified member. For example, AP2-like transcription factors, one of the largest transcription factor families in plants, were named after the APETALA2 (AP2) protein, a key regulator of floral development.

4.2 IPA1: A Central Regulator of Ideal Plant Architecture

Members of the SPL (SQUAMOSA PROMOTER-BINDING PROTEIN-LIKE) family play regulatory roles in a wide variety of developmental processes, including the transition between vegetative stages, the induction of flowering, branching, hormone signalling, and resistance to abiotic stresses. The *IPA1* gene encodes OsSPL14, which serves as a central regulator for the plant architecture and holds substantial potential for improving rice yield (Jiao et al. 2010; Miura et al. 2010). A systematic examination of *IPA1* expression levels and yield-related traits reveals that *IPA1* exerts opposing effects on tillering and panicle branching in a dosage-dependent manner, governed by miRNA-mediated post-transcriptional regulation and ubiquitin-mediated post-translational modification.

MicroRNAs (*miRNAs*) are a class of small non-coding RNAs naturally occurring in eukaryotes, which play crucial roles in gene regulation by either cleaving target mRNAs or inhibiting their translation. These regulatory molecules are extensively involved in diverse physiological processes, including growth, development, and stress responses. MicroRNAs such as *miR156* and *miR529* regulate *IPA1* at the post-transcriptional level by targeting its mRNA for degradation, thereby modulating its expression in a tissue-specific manner (Wang et al. 2018). *miR156* is predominantly expressed in seedlings, whereas *miR529* is mainly expressed in panicles, resulting in differential regulation of *IPA1* across tissues. In shoots, *IPA1* expression is suppressed by both *miR156* and DNA methylation. The IPA1 protein activates the expression of downstream target genes, including *OsTB1* and *D53*, which inhibit and promote bud outgrowth, respectively. In panicles, *IPA1* enhances the expansion of inflorescence meristems and increases the number of primordia by upregulating the expression of genes such as *DEP1*, which is suppressed by *miR529*.

Ubiquitination modifications, mediated by IPA1 INTERACTING PROTEIN 1 (IPI1), regulate *IPA1* at the post-translational level. IPI1 functions as a RING-finger E3 ligase. It promotes the degradation of IPA1 in panicles, while enhancing its stability in shoot apices. Furthermore, IPI1 can catalyse the ubiquitination of the IPA1-containing complex with distinct types of polyubiquitin chains, specifically adding K48-linked chains in panicles and K63-linked chains in shoot apices. This tissue-specific regulation of IPA1 protein stability enables precise modulation of its functional roles in determining plant architecture, particularly in key traits such as tiller number and panicle size (Wang et al. 2017).

In summary, the coordinated interplay between miRNA-mediated post-transcriptional regulation and ubiquitin-dependent post-translational modification precisely fine-tunes IPA1 activity in a tissue- and developmental stage-specific manner (Wang et al. 2018). This dual-layer regulatory architecture exemplifies the sophistication of plant gene regulation, where miRNAs govern transcript abundance, and ubiquitination controls protein turnover, thereby jointly shaping the spatiotemporal dynamics of IPA1 function. Critically, these context-dependent interactions not only deepen our understanding of IPA1-centered regulatory networks but also provide actionable genetic targets for breeding high-yielding rice cultivars.

4.3 TB1: A Key Suppressor of Shoot Branching

Archaeological evidence indicates that the domestication of maize commenced approximately 6000 years ago in central Mexico. Its wild progenitor has been identified as a large grass in Mexico and Nicaragua, known as teosinte. Teosinte differs significantly from cultivated maize in appearance, so it took scientists many years to recognize that the two plants are closely related. The most obvious difference between teosinte and maize is that teosinte is highly branched, whereas maize is not. Teosinte plants bear numerous long axillary branches ending in male inflorescences (tassels). In contrast, maize plants have fewer, shorter axillary branches ending in female inflorescences (ears).

Although maize and teosinte display significant morphological disparities, they exhibit minimal genetic variation. Both species possess 20 chromosomes with a highly conserved gene order. The remarkable morphological divergence between maize and its wild progenitor, teosinte, is largely attributable to allelic variation in five to six genomic regions. Each of these regions harbours major QTLs or regulatory genes that govern key domestication-related traits, such as plant architecture, inflorescence structure, and kernel size. Among these, altered expression of the *TEOSINTE BRANCHED 1 (TB1)* gene played a pivotal role in maize domestication. TB1, a member of the TCP transcription factor family, acts as a key regulator of shoot branching. The TCP family is named after its first three characterized members: TEOSINTE BRANCHED1 (TB1), CYCLOIDEA (CYC), and PROLIFERATING CELL FACTOR (PCF). These plant-specific TFs play essential roles in regulating leaf morphogenesis, floral patterning, shoot branching, hormone signalling, and stress responses.

The expression of TB1 is significantly upregulated in domesticated maize compared to its wild ancestor, teosinte, as a result of the insertion of a transposable element approximately 60 kb upstream of the gene (Hou et al. 2025). This insertion gives rise to a novel enhancer that promotes transcription. The elevated levels of TB1 protein bind to GGnCCC motifs (e.g. GGACCC or GGGCCC) in the promoters of key cell-cycle genes such as *PCNA2* and *MCM2*, suppressing their expression and thereby limiting cell proliferation in axillary meristems. Through comprehensive analyses of ChIP-seq, RNA-seq, and measurements of hormones and sugars in 1-mm axillary bud tissues, Dong et al. (2019) identified genetic pathways that are potentially regulated by TB1. They put forward a comprehensive model of the TB1-mediated tiller regulatory network. Initially, increased far-red light ratios trigger the expression of TB1. Subsequently, TB1 may decrease sugar accumulation by facilitating the expression of genes related to catabolism and sugar signalling. Moreover, TB1 serves as a central regulator of phytohormone homeostasis, positively modulating ABA and JA while inhibiting GA biosynthesis or signalling. The coordinated interaction between sugar availability and phytohormone dynamics is considered to maintain axillary bud dormancy and impede tiller outgrowth. Significantly, TB1 is anticipated to directly regulate key domestication genes such as *gt1*, *tru1*, and *tga1*, which played a crucial role in the morphological transformation from teosinte to modern maize, thus positioning TB1 as a master regulator within the domestication gene hierarchy.

The upregulation of TB1 expression, driven by a *transposon*-derived enhancer, was a crucial target during the early domestication of maize, ultimately shaping the plant architecture characteristic of modern maize. *TB1* is conserved across monocots and eudicots. Its orthologs include *OsTB1* (also known as *FINE CULM 1*, *FC1*) in rice, *SbTB1* in sorghum, and *BRANCHED 1* (*BRC1*) in *Arabidopsis*, pea, and tomato. These genes are expressed in axillary meristems and buds. Consistent with observations in maize, their overexpression inhibits branching, whereas loss-of-function mutations result in increased branching. Consequently, researchers initially hypothesized that *TB1*/*BRC1* is essential for inhibiting bud outgrowth and suppressing branching. However, more recent studies have shown that branching can still be suppressed in the absence of *BRC1*, and conversely, it can be activated even when *BRC1* is present.

5 Environmental Factors Influencing Plant Architecture

Plant architecture is flexible; it undergoes changes during growth from seedling to mature plant and in response to ambient conditions. Light duration and periodicity govern the transition from vegetative to reproductive growth, which is crucial in shaping plant architecture. Light intensity and spectral composition further modulate shoot architecture through the shade-avoidance response, a vital adaptation to intraspecific competition for light, particularly under dense planting. Even in modern maize cultivars, which are bred for strong apical dominance, lateral branching becomes prominent under low-density conditions. Conversely, high-density planting reduces photosynthetically active radiation and increases the far-red to red light ratio, thereby concurrently suppressing lateral bud outgrowth and promoting main stem elongation. Consequently, intense neighbour-induced shading severely restricts branching: many axillary buds fail to initiate, others arrest development before anthesis, and some undergo premature abortion.

In addition to light, temperature plays a crucial role in determining shoot architecture. During the seedling stage, low temperatures significantly affect the tiller angle of wheat, thereby enhancing its frost resistance (Marone et al. 2020). Grain number, plant height, and Heading date 7 (Ghd7) are key regulators for rice yield by modulating panicle branching and tillering. As a long-day-dependent negative regulator of flowering time, Ghd7 exerts a quantitatively variable influence on heading date and yield components. These effects are closely associated with *Ghd7* transcript levels and are further fine-tuned by environmental conditions and genetic backgrounds. Drought and high-temperature stress inhibit *Ghd7* expression, whereas low temperature promotes it, suggesting that *Ghd7* integrates dynamic environmental cues into architectural regulation and stress response pathways (Weng et al. 2014).

Nutrients play a pivotal role in determining plant architecture by influencing the allocation of resources between shoots and roots. In fertile soils, plants tend to allocate more resources to shoot growth rather than to root development, leading to an increased shoot-to-root ratio (SRR). The application of N and P fertilizers significantly promotes tillering in crops, a process strictly regulated by plant hormones. High N availability promotes the synthesis and transport of CKs from roots to shoots, stimulating axillary bud outgrowth and branching. Conversely, adequate P supply inhibits SL biosynthesis and thereby increases branching. In contrast, low N and low P conditions increase SLs content, thereby repressing lateral bud activation and reducing branching.

Nitrogen is also a crucial determinant of internode elongation. Research reveals that 75%–90% of the total N uptake in wheat and barley occurs during the stem elongation phase, highlighting a strong correlation between N assimilation and internode extension (Delogu et al. 1998). Significantly, semi-dwarf cultivars developed during the Green Revolution, such as rice with *sd1* and wheat with *Rht1*, demonstrate reduced sensitivity to N-induced height increases. The *sd1* and *Rht1* mutants frequently impair N uptake efficiency; however, this limitation can be addressed by overexpressing the transcription factor GROWTH-REGULATING FACTOR 4 (OsGRF4), which restores both N utilization and optimal growth (Li et al. 2018).

6 Crop Ideotype

Throughout the history of scientific development, novel *concepts* or research strategies occasionally emerge, generating significant enthusiasm and spurring substantial research endeavours. Every farmer aspires to cultivate the ideal plant: one that, when grown on the farmlands, would yield the maximum quantity and quality. This aspiration has propelled efforts to breed superior genotypes and inspired crop scientists to introduce the concept of an "*ideotype*" for identifying the optimal traits to integrate into such a perfect plant (Donald 1968).

6.1 The Concept of Ideotype

Plant breeding has long relied on traditional methods, such as "selection for yield" or "defect elimination". Nevertheless, breeding based on crop ideotypes offers a more strategic and forward-looking option. An ideotype refers to a theoretical model of an ideal plant, which is designed with a specific set of morphological, physiological, and genetic traits that collectively maximize the yield potential under defined environmental conditions. This approach improves plant breeding by providing a clear, science-based framework for determining which traits should be prioritized and how they should interact synergistically. In ideotype breeding, precise targets are set for each trait, including plant architecture, stress tolerance, and resource use efficiency, thus generating a comprehensive blueprint of the desired plant type. Unlike conventional methods that rely on the random appearance of favourable phenotypes through natural recombination, ideotype breeding enables a proactive and systematic pursuit of pre-determined characteristics, leveraging advanced genetic tools, genomic selection, and biotechnological innovations to accelerate progress (Ahn et al. 2023).

The term ideotype literally means a form that represents an idea. In its broadest sense, it refers to a biological model designed to exhibit predictable performance or behaviour under defined environmental conditions. More specifically, it denotes an idealized plant model engineered to produce higher yields or improved quality of grain, oil, or other valuable products when developed into a cultivar. Synonyms include model plant type, ideal plant type, and ideal model plant type. The original concept focused on the phenotypic traits of individual plants and the optimal combination of alleles or QTLs under specific growth environments. Over time, the definition has expanded to include integrated systems combining plant genotypes with appropriate management practices or cropping systems to achieve desired performance in specific ecological contexts, such as climatic, edaphic (soil-related), stress-tolerant, or disease- and pest-resistant ideotypes. Further development extended the concept to encompass not only individual plants but also the collective phenotypes of entire crop canopies. In this chapter, the term ideotype refers specifically to an optimized combination of morphological traits at the individual plant level.

The socio-economic conditions impose strong limitations on the acceptability of the ideotype. For example, mixing resistant and susceptible cultivars is highly effective in limiting the spread of aerial fungal diseases such as rusts or powdery mildews in wheat and barley. This limitation is partly due to physical barriers that prevent plant-to-plant infection. Such mixtures can therefore be considered as ideotypes for grain production with minimal or no need for fungicide use. However, they are often not adopted in practice because end-users are reluctant to process mixtures of grains whose precise composition they do not fully understand. Similarly, the practice of mixing conventional glutinous rice with modern hybrid cultivars in indica rice cultivation in southwestern China, aimed at preventing significant yield losses due to blast disease, would become unfeasible without the availability of low-cost manual labour to separately harvest waxy and non-waxy cultivars.

6.2 Weak Competitor

Evolutionary theory has augmented our comprehension of the traits that are likely to contribute to the success of crop plants compared to their wild counterparts (Gao et al. 2023). Natural selection favours the competitiveness of individual genotypes, even if it incurs a cost to group performance. In contrast, traits that minimize individual *competition* and maximize group performance are expected to boost crop productivity. Agronomists independently developed these concepts within the framework of crop ideotypes, which refer to the set of phenotypic traits that maximize productivity per unit of land area.

Donald (1968) initially noted the widespread trade-off between individual competitiveness and population productivity. Subsequently, he proposed that successful crop ideotypes likely represent weak competitors, thereby minimizing competition within the crop population. For instance, branching and plant height are traits that maximize individual competitiveness at the expense of group productivity. The ideotype should be "communal" to reduce interplant competition and reallocate resources previously invested in individual growth, thus enhancing resource use efficiency at the population level. A genotype that efficiently captures light via a costly leaf display in isolated plants may exhibit reduced light-use efficiency when grown in dense stands, due to mutual shading, which can decrease yield by 10%–20%. Consequently, individuals within the ideal populations do not display aggressive competitive behaviour, as illustrated in Fig. 4.7. These concepts laid the foundation for the development of short and compact plant architectures in rice and wheat, which are achievements of great significance for the yield increases during the Green Revolution.

Weak competition serves as a crucial characteristic of the ideotype, which posits that high-yielding, superior crop plants should be weak competitors. Nevertheless, it has led to certain semantic confusions in the existing literature. As an alternative, Zhang et al. (1999) proposed the use of "growth redundancy" to clearly denote that surplus growth in resource-harvesting structures can be sacrificed to enhance yields. The authors also associated Donald's ideotype with the well-known *tragedy of the*

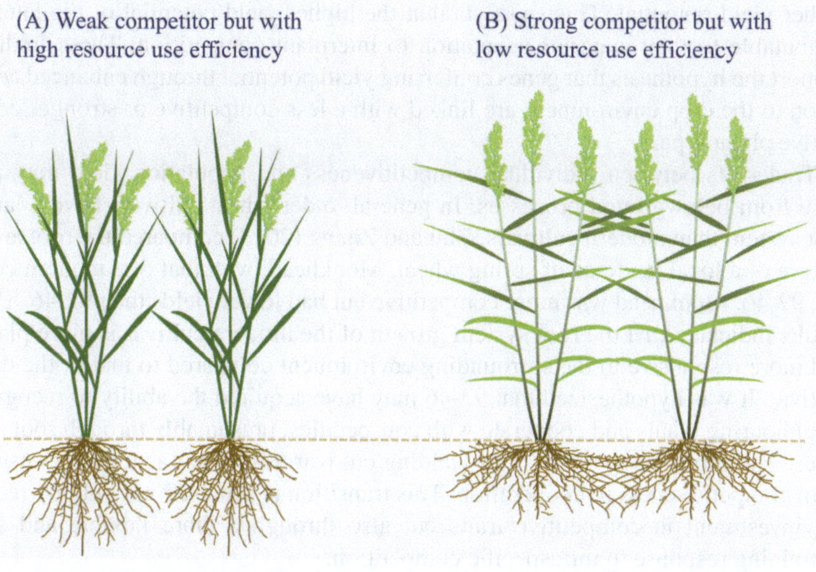

Fig. 4.7 The weak and strong competitors. (**A**) The less competitive cultivar exhibits restraint in both above-ground and below-ground growth, efficiently utilizing resources within its spatial boundaries. (**B**) In contrast, the more competitive cultivar adopts an aggressive strategy, extending its leaves and roots to capture additional light, mineral nutrients, and water from neighbouring areas. To maximize the utilization of natural resources, particularly light, high-yielding modern crops are generally characterized as weak competitors or strong cooperators. These crops typically exhibit erect and compact architectural traits, which also align with the requirements for mechanical planting and harvesting

commons[5]: If all members of a crop population invest less in aggressive competition, they can collectively benefit; yet individual restraint would be exploited. The most prominent example of this is the significant reduction in plant height during the Green Revolution. Shorter but higher-yielding plants are undoubtedly weak competitors in a light-limited canopy, offering much-needed empirical evidence for Donald's view.

Reynolds et al. (1994) conducted field experiments using modern semidwarf spring wheat cultivars representing a spectrum of yield potential to compare their adaptability to varying levels of interplant competition for resources. Competition levels were manipulated by adjusting planting density in eight-row plots at the flag-leaf ligule emergence stage, the phenological stage when interplant competition is assumed to be most intense. To mitigate competition for light, adjacent rows were bent away from the central two rows, thereby enhancing light penetration to the lower leaves of the central rows. Concurrently, underground (root) competition was reduced by uprooting plants adjacent to the central two rows, achieving simultaneous alleviation of both light and root competition. Although average yield increases were 25% under the light-only treatment and 40% under the root-and-light treatment, a significant genotypic effect was observed: genotypes with lower yield potential exhibited greater yield responses to competition reduction than those with higher yield potential. This suggests that the higher yield potential of the latter is attributable to their superior adaptation to interplant competition. These findings support the hypothesis that genes conferring yield potential through enhanced adaptation to the crop environment are linked with a less competitive or stronger cooperative phenotype.

Trade-offs between individual competitiveness and population yield may also arise from belowground processes. In general, older wheat cultivars have a larger root system than modern cultivars. Zhu and Zhang (2013) compared the root architecture of a local landrace of spring wheat, Monkhead, with that of a modern cultivar, 92-46. Monkhead was more competitive but had lower yields than 92-46. Their results indicated that the root system growth of the modern cultivar is more plastic and more responsive to the surrounding environment compared to that of the older cultivar. It was hypothesized that 92-46 may have acquired the ability to recognize neighbouring plants and cooperate with conspecifics, presumably through root exudates. Therefore, the modern high-yielding cultivar 92-46 may exhibit a transition from competitiveness to cooperation. This transition is achieved not only by reducing investment in competitive traits but also through a more flexible and self-restraining response to intraspecific competition.

[5] In economics, the tragedy of the commons is the depletion of a shared resource by individuals, acting independently and rationally according to each one's self-interest, despite their understanding that depleting the common resource is contrary to the group's long-term best interests.

6.3 Crop Ideotypes

Ideotype is a popular concept among plant breeders, who use it to identify the optimal combinations of traits within a specific genotype to achieve predetermined yield levels based on socio-economic conditions. In 1968, Donald introduced the concept of designing an optimal wheat plant architecture or ideotype, which includes a short and sturdy stem, few small erect leaves, a large erect ear with awns, and a single *culm* (Fig. 4.8). The ideotype approach has been utilized in breeding

Fig. 4.8 The wheat ideotype proposed by Donald in 1968. The plant exhibits a dwarf stature, characterized by a single, sturdy culm (stem) that supports a large, erect ear bearing an awn, along with a relatively small number of upright leaves. The root system is highlighted, indicating a high proportion of seminal roots, which are crucial for early-stage nutrient uptake

programs at the International Rice Research Institute (IRRI) and in China to enhance rice yield potential. Initial generation of new plant type (NPT) lines, developed from tropical japonica at IRRI, exhibited poor yield due to limited biomass production and inadequate grain filling. Several second-generation NPT lines outyielded the first-generation lines as well as the indica check. The proposed NPT exhibits the following key characteristics: low tillering capacity (3–4 tillers when directly seeded); fewer unproductive tillers; 200–250 grains per panicle; a plant height ranging from 90–100 cm; thick and sturdy stems; dark green, upright, and thick leaves; a vigorous root system; a growth duration of 100–130 days; and an elevated harvest index (Peng et al. 2008).

China's super hybrid rice breeding project has developed a large number of F1 hybrids by integrating the ideotype approach with intersubspecific heterosis. These hybrids have achieved grain yields of 12 t ha^{-1} in on-farm trials, 8%–15% higher than those of the control cultivars. The success of super hybrid rice can, to some extent, be attributed to the incorporation of beneficial traits from the NPT. For instance, both breeding strategies emphasized larger panicle size, limited tillering capacity, and robust lodging resistance. More importantly, super-hybrid rice breeding prioritized the architecture of the top three leaves and optimal panicle positioning within the canopy to support heavy panicles. The success of super hybrid rice breeding in China and advances in NPT breeding at IRRI demonstrate that the ideotype approach is effective in breaking through the rice yield ceiling.

7 Revolutionizing Ideotype

Plant architecture has been shaped drastically to meet human needs and preferences through the processes of crop domestication and breeding. Most progress in this field has been achieved using conventional breeding techniques, including trait-based selection following controlled crosses. When a desired trait is controlled by a single gene, it can be effectively introgressed into crop cultivars via traditional breeding methods. However, most agronomic traits, including the plant architecture, are complex traits governed by multiple genetic loci and diverse environmental conditions. This complexity poses significant challenges for the genetic modification of target architectural traits.

Early efforts in ideotype breeding for rice and wheat were largely unsuccessful due to constraints imposed by *pleiotropy* and *linkage drag* arising from crosses with genetically inferior germplasm (Ahn et al. 2023). Linkage drag is a genetic constraint that results from the close physical linkage between beneficial and deleterious alleles. It hinders crop improvement by co-transferring undesirable traits into elite germplasm, leading to reduced yield or quality and extended breeding cycles. Modern tools, including high-density molecular markers, recombination-enhancing approaches, genome editing, and strategic hybridization, offer robust strategies to identify, reduce, or fully overcome linkage drag. Recently, the "*de novo domestication*" approach has been proposed to overcome the limitation of linkage drag by enabling precise genome editing of wild species to domesticate them from scratch, as a substitute for the conventional approach of transferring alien genes from wild species into domesticated crops (Zsögön and Peres 2018).

Modern rice cultivars are uniformly diploid, and polyploidization has long been pursued due to the potential benefits of enhanced genome buffering, increased vigour, and improved stress resilience. However, an effective and reproducible approach has yet to be established. Yu et al. (2021) adopted a practical strategy, namely, the de novo domestication of wild *allotetraploid* rice. By selecting the wild allotetraploid *O. alta* (CCDD) genotype polyploid rice 1 (PPR1), researchers established an efficient tissue culture and genome editing system and generated a high-quality genome assembly. Leveraging domestication-related genes from cultivated diploid rice, they identified the corresponding homologs in the *O. alta* genome, thereby providing critical information for targeted editing. Subsequently, CRISPR/Cas9, base editing, and multiplex editing technologies were employed to enhance PPR1 by targeting a set of genes associated with domestication traits and agronomic significance in diploid rice. This approach generated edited lines with enhanced characteristics: diminished seed shattering, shorter awns, moderate plant height, enlarged grain size, sturdy stems, and an adjustable heading date. This work delineates a well-defined pathway for developing novel crop cultivars with improved plant architecture by integrating the benefits of polyploidy, functional genomics from cultivated crops, and rapid breeding enabled by genome-editing technologies.

8 Concluding Remarks and Perspectives

Ideotype breeding is a strategic approach that involves pre-defining optimal trait combinations and introducing them into a target crop species to enhance yield potential. The successful execution of ideotype breeding relies on two crucial elements: (1) the identification of genes that govern key structural traits related to yield and (2) the ability of target genetic modification to produce predictable phenotypic improvement. Recent research has significantly deepened our understanding of the genetic and hormonal regulation of plant architecture, predominantly through investigations on a restricted set of model eudicot species such as *Arabidopsis*, tomato, and pea, as well as monocot species like rice and maize. Even though many genes have been cloned and characterized, their application in ideotype breeding is still limited, mainly because of pleiotropic effects or linkage drag.

Recent advancements in biotechnology–information technology (BIT) integration have significantly accelerated the precise design and manipulation of architectural traits, thereby facilitating ideotype breeding (Ahn et al. 2023). Gene-editing platforms, such as CRISPR-Cas systems, enable precise modification of genes, thereby facilitating accelerated pyramiding of multiple gene loci. High-throughput phenotyping platforms allow real-time acquisition of detailed 3D plant structural data across thousands of field plots. Simultaneously, AI models are capable of automatically segmenting plant organs, reconstructing high-resolution crop plants, and predicting growth dynamics, thereby converting raw image data into quantitative architectural descriptors. These descriptors are then incorporated into crop models and *digital twin* systems to project the architectural phenotypes across diverse climatic, edaphic, and management scenarios. This closed-loop system significantly shortens the breeding cycle: candidate genetic edits can be designed, implemented, phenotyped, and performance-evaluated via simulation within weeks rather than years. Overall, crop breeding is undergoing a transformative shift, from traditional selection methods towards a highly integrated BIT pipeline. This convergence marks the dawn of a new era in rational, data-driven ideotype design, offering a scalable approach to meeting rising global food demand in the face of intensifying climate pressures.

Summary

1. Plant architecture refers to the spatial arrangement of plant organs, playing a crucial role in the resource use efficiency, the yield potential, and the compatibility of crops with mechanized planting technologies.
2. The architecture of cereal crops is primarily governed by genetic factors that regulate plant height, tiller number, angles of tillers and leaves, grain size and number, length of primary roots, and the abundance of crown and lateral roots.
3. Shoot architecture exhibits significant divergence across various crop species, yet it is governed by shared developmental processes and principles rooted in meristems and phytomers.
4. The architecture of plants is regulated by complex networks involving hormones, miRNAs, TFs, and genes. At the core is the delicate balance of phytohormone distribution over time and space.
5. The antagonistic interactions between florigen (FT) and antiflorigen (TFL1 or SP) play a crucial role in regulating determinate and indeterminate growth patterns in plants.
6. The transcription factor IPA1 reduces unproductive tillers and increases the grain number per panicle in rice, while TB1 limits branch outgrowth and initiates the formation of female inflorescences in maize.
7. The ideotype is a plant model that, when developed as a cultivar, is expected to exhibit improved crop productivity in terms of the quantity or quality of grain, oil, or other valuable products.
8. The ideotype should be designed to promote communal growth by minimizing interplant competition for natural resources, thereby improving resource use efficiency at the population level.

Important Terms

BRC1; DELLA; de novo domestication; Florigen/Antiflorigen; Genetic regulation; Green Revolution; Growth habit; Hormonal regulation; Ideotype; IPA1; Meristem; NPT; Plant architecture; Phytomer; *Rht1;* SAM; *sd1;* TB1; TFs.

Review Questions

Multiple Choice

1. ( ) was the first growth hormone to be studied in plants, which was discovered in early studies of coleoptile bending during phototropism.
   (a) GAs
   (b) CKs
   (c) Auxin
   (d) SLs

2. ( ) is the primary hormone that plays a key role in inhibiting shoot branching, promoting cambial activity and secondary growth, and supporting interactions within the rhizosphere.
   (a) GAs
   (b) CKs
   (c) ABA
   (d) SLs

3. Which of the following statements about the concept of ideotype is INCORRECT? ( )
   (a) The term "ideotype" literally means "a form denoting an idea".
   (b) It has been expanded to encompass the combination of plant genotypes and management actions or cropping systems in order to achieve specific performances.
   (c) It has been extended to include the phenotypes of crop canopies rather than just single plants.
   (d) The socio-economic conditions impose little limitations on the acceptability of the ideotype.

True or False

1. Phytohormones are produced in one cell and modulate cellular processes in another cell by interacting with specific proteins ( )
2. Root angle regulation involves competing gravitropic and antigravitropic offset mechanisms ( )

Blank Filling

1. Shoot __________ meristem gives rise to all other shoot meristems and organs, thus forming a new phytomer.
2. __________ function as key integrator of environmental and developmental signals, thus playing essential roles in the regulation of plant architecture.
3. Although comprehensive measurements of crop leaf angle are limited, most species fall into one of three idealized distributions: (1) __________ (<30°, e.g. grasses and cereals), (2) plagiophile (30–60°, e.g. sugar beet, rape, soybean), and (3) __________ (>60°, e.g. clover, bean, sunflower).

Short Answer

1. Illustrate how domestication and artificial selection improve crop yield by changing the plant form.
2. Key environmental factors that impact plant architecture.
3. The role of GA signalling in crop improvement during the Green Revolution.
4. Examples showing the transcriptional control of shoot structure.
5. The core concept of the ideotype by Donald (1968).

Sketching

Draw a generic eudicot and a monocot, clearly labelling the key structural features and indicating their major regulatory phytohormones.

Essay Items

1. Donald initially noticed the widespread trade-off between individual competitiveness and overall crop productivity. He then suggested that successful crop types were likely weaker competitors, which could help reduce interplant competition within the crop stand. Reynolds et al. (1994) carried out field experiments using modern semi-dwarf spring wheat cultivars with varying yield potentials. Their goal was to evaluate how well these cultivars adapted to different levels of competition among plants for resources. Please read this paper (Field Crops Research, Volume 37, Issue 3, 1994, Pages 149–160) and provide a summary in your own words.

2. Key structural features of plants, such as plant height and branching, are shaped by the coordination of multiple phytohormones. This raises an important question: does this multi-hormonal control ultimately benefit crop productivity and fitness? Let's explore this idea by using plant height as an example to understand how hormonal crosstalk balances plant growth and resilience.

References

Ahn E, Botkin J, Curtin SJ et al (2023) Ideotype breeding and genome engineering for legume crop improvement. Curr Opin Biotechnol 82:102961. https://doi.org/10.1016/j.copbio.2023.102961

Cao YY, Zhong ZJ, Wang HY et al (2022) Leaf angle: a target of genetic improvement in cereal crops tailored for high-density planting. Plant Biotechnol J 20(3):426–436. https://doi.org/10.1111/pbi.13780

Chen RZ, Deng YW, Ding YL et al (2022) Rice functional genomics: Decades' efforts and roads ahead. Sci China Life Sci 65(1):33–92. https://doi.org/10.1007/s11427-021-2024-0

Delogu G, Cattivelli L, Pecchioni N et al (1998) Uptake and agronomic efficiency of nitrogen in winter barley and winter wheat. Eur J Agron 9(1):11–20. https://doi.org/10.1016/S1161-0301(98)00019-7

Donald CM (1968) The breeding of crop ideotypes. Euphytica 17:385–403. https://doi.org/10.1007/BF00056241

Dong Z, Xiao Y, Govindarajulu R et al (2019) The regulatory landscape of a core maize domestication module controlling bud dormancy and growth repression. Nat Commun 10(1):3810. https://doi.org/10.1038/s41467-019-11774-w

Dresselhaus T, Balboni M, Berg L et al (2025) How meristems shape plant architecture in cereals–Cereal Stem Cell Systems (CSCS) Consortium. Plant Cell 37(7):koaf150. https://doi.org/10.1093/plcell/koaf150

Eshed Y, Lippman ZB (2019) Revolutions in agriculture chart a course for targeted breeding of old and new crops. Science 366:6466. https://doi.org/10.1126/science.aax0025

Fukao T, Bailey-Serres J (2008) Submergence tolerance conferred by *Sub1A* is mediated by SLR1 and SLRL1 restriction of gibberellin responses in rice. Proc Natl Acad Sci USA 105(43):16814–16819. https://doi.org/10.1073/pnas.0807821105

Fukao T, Xu K, Ronald PC et al (2006) A variable cluster of ethylene response factor–like genes regulates metabolic and developmental acclimation responses to submergence in rice. Plant Cell 18(8):2021–2034. https://doi.org/10.1105/tpc.106.043000

Gallei M, Luschnig C, Friml J (2020) Auxin signalling in growth: Schrödinger's cat out of the bag. Curr Opin Plant Biol 53:43–49. https://doi.org/10.1016/j.pbi.2019.10.003

Gao LX, Kantar MB, Moxley D et al (2023) Crop adaptation to climate change: an evolutionary perspective. Mol Plant 16(10):1518–1546. https://doi.org/10.1016/j.molp.2023.07.011

Guo W, Chen LM, Herrera-Estrella L et al (2020) Altering plant architecture to improve performance and resistance. Trends Plant Sci 25(11):1154–1170. https://doi.org/10.1016/j.tplants.2020.05.009

Hou J, Liu M, Yang K et al (2025) Genetic variation for adaptive evolution in response to changed environments in plants. J Integr Plant Biol 67(9):2265–2293. https://doi.org/10.1111/jipb.13961

Hu QL, Liu HH, He YJ et al (2024) Regulatory mechanisms of strigolactone perception in rice. Cell 187(26):7551–7567. https://doi.org/10.1016/j.cell.2024.10.009

Jiao YQ, Wang YH, Xue DW et al (2010) Regulation of *OsSPL14* by OsmiR156 defines ideal plant architecture in rice. Nat Genet 42(6):541–544. https://doi.org/10.1038/ng.591

Kirschner GK, Hochholdinger F, Salvi S et al (2024) Genetic regulation of the root angle in cereals. Trends Plant Sci 29(7):814–822. https://doi.org/10.1016/j.tplants.2024.01.008

Li S, Tian YH, Wu K et al (2018) Modulating plant growth-metabolism coordination for sustainable agriculture. Nature 560(7720):595–600. https://doi.org/10.1038/s41586-018-0415-5

Liu C, Zheng S, Gui JS et al (2018) *Shortened Basal Internodes* encodes a gibberellin 2-oxidase and contributes to lodging resistance in rice. Mol Plant 11(2):288–299. https://doi.org/10.1016/j.molp.2017.12.004

Liu SL, Zhang M, Feng F et al (2020) Toward a "Green Revolution" for soybean. Mol Plant 13(5):688–697. https://doi.org/10.1016/j.molp.2020.03.002

Liu L, Zhao LR, Liu YW et al (2024) Transcription factor OsWRKY72 controls rice leaf angle by regulating *LAZY1*-mediated shoot gravitropism. Plant Physiol 195(2):1586–1600. https://doi.org/10.1093/plphys/kiae159

Marone D, Rodriguez M, Saia S et al (2020) Genome-wide association mapping of prostrate/erect growth habit in winter durum wheat. Int J Mol Sci 21(2):394. https://doi.org/10.3390/ijms21020394

McGarry RC, Ayre BG (2021) Cotton architecture: examining the roles of *SINGLE FLOWER TRUSS* and *SELF-PRUNING* in regulating growth habits of a woody perennial crop. Curr Opin Plant Biol 59:101968. https://doi.org/10.1016/j.pbi.2020.10.001

Miura K, Ikeda M, Matsubara A et al (2010) *OsSPL14* promotes panicle branching and higher grain productivity in rice. Nat Genet 42(6):545–549. https://doi.org/10.1038/ng.592

Multani DS, Briggs SP, Chamberlin MA et al (2003) Loss of an MDR transporter in compact stalks of maize *br2* and sorghum *dw3* mutants. Science 302(5642):81–84. https://doi.org/10.1126/science.1086072

Murray JAH, Jones A, Godin C et al (2012) Systems analysis of shoot apical meristem growth and development: integrating hormonal and mechanical signalling. Plant Cell 24(10):3907–3919. https://doi.org/10.1105/tpc.112.102194

Pendleton JW, Smith GE, Winter SR et al (1968) Field investigations of the relationships of leaf angle in corn (*Zea mays* L.) to grain yield and apparent photosynthesis. Agron J 60(4):422–424. https://doi.org/10.2134/agronj1968.00021962006000040027x

Peng SB, Khush GS, Virk P et al (2008) Progress in ideotype breeding to increase rice yield potential. Field Crop Res 108(1):32–38. https://doi.org/10.1016/j.fcr.2008.04.001

Phokas A, Coates JC (2021) Evolution of DELLA function and signaling in land plants. Evol Dev 23(3):137–154. https://doi.org/10.1111/ede.12365

Reynolds MP, Acevedo E, Sayre KD et al (1994) Yield potential in modern wheat varieties: its association with a less competitive ideotype. Field Crop Res 37(3):149–160. https://doi.org/10.1016/0378-4290(94)90094-9

Sakamoto T, Morinaka Y, Ohnishi T et al (2006) Erect leaves caused by brassinosteroid deficiency increase biomass production and grain yield in rice. Nat Biotechnol 24(1):105–109. https://doi.org/10.1038/nbt1173

Shen L, Qi X, Guo Z (2025) Wheat. Curr Biol 35(21):R1037–R1043. https://doi.org/10.1016/j.cub.2025.09.007

Tian JG, Wang CL, Xia JL et al (2019) Teosinte ligule allele narrows plant architecture and enhances high-density maize yields. Science 365(6454):658–664. https://doi.org/10.1126/science.aax5482

Truong SK, McCormick RF, Rooney WL et al (2015) Harnessing genetic variation in leaf angle to increase productivity of Sorghum bicolor. Genetics 201(3):1229–1238. https://doi.org/10.1534/genetics.115.178608

Wang J, Yu H, Xiong GS et al (2017) Tissue-specific ubiquitination by IPA1 INTERACTING PROTEIN1 modulates IPA1 protein levels to regulate plant architecture in rice. Plant Cell 29(4):697–707. https://doi.org/10.1105/tpc.16.00879

Wang B, Smith SM, Li JY (2018) Genetic regulation of shoot architecture. Annu Rev Plant Biol 69:437–468. https://doi.org/10.1146/annurev-arplant-042817-040422

Wang WG, Gao HB, Liang Y et al (2022) Molecular basis underlying rice tiller angle: current progress and future perspectives. Mol Plant 15(1):125–137. https://doi.org/10.1016/j.molp.2021.12.002

Wei SB, Li X, Lu ZF et al (2022) A transcriptional regulator that boosts grain yields and shortens the growth duration of rice. Science 377(6604):386. https://doi.org/10.1126/science.abi8455

Weng XY, Wang L, Wang J et al (2014) *Grain Number, Plant Height, and Heading Date7* is a central regulator of growth, development, and stress response. Plant Physiol 164(2):735–747. https://doi.org/10.1104/pp.113.231308

Yu BS, Lin ZW, Li HX et al (2007) *TAC1*, a major quantitative trait locus controlling tiller angle in rice. Plant J 52(5):891–898. https://doi.org/10.1111/j.1365-313X.2007.03284.x

Yu H, Lin T, Meng XB et al (2021) A route to *de novo* domestication of wild allotetraploid rice. Cell 184(5):1156–1170. https://doi.org/10.1016/j.cell.2021.01.013

Zhang DY, Sun GJ, Jiang XH (1999) Donald's ideotype and growth redundancy: a game theoretical analysis. Field Crop Res 61(2):179–187. https://doi.org/10.1016/S0378-4290(98)00156-7

Zhang JH, Lin QB, Wang X et al (2025) The DENSE AND ERECT PANICLE1-GRAIN NUMBER ASSOCIATED module enhances rice yield by repressing *CYTOKININ OXIDASE 2* expression. Plant Cell 37(1):koae309. https://doi.org/10.1093/plcell/koae309

Zhu L, Zhang DY (2013) Donald's ideotype and growth redundancy: a pot experimental test using an old and a modern spring wheat cultivar. PLoS One 8(7):e70006. https://doi.org/10.1371/journal.pone.0070006

Zhu YY, Nomura T, Xu YH et al (2006) *Elongated Uppermost Internode* encodes a cytochrome P450 monooxygenase that epoxidizes gibberellins in a novel deactivation reaction in rice. Plant Cell 18(2):442–456. https://doi.org/10.1105/tpc.105.038455

Zsögön A, Peres LEP (2018) Molecular control of plant shoot architecture. Plant Cell 34(2):703–705. https://doi.org/10.1093/plcell/koab296

Source and Sink Relation

5

Critical Factors Behind Assimilate Allocation Among Vegetative and Reproductive Organs

Contents

Z. Liu, *Crop Physiology*, https://doi.org/10.1007/978-981-92-2960-4_5

Abstract

Numerous strategies have been developed to genetically enhance either source or sink strength independently to improve crop yield. However, many of these efforts have yielded limited success due to the intricate and dynamic interplay between source and sink organs. This chapter provides a comprehensive summary of recent advances in understanding source–sink interactions. It begins by defining key concepts such as source, sink, and flow, followed by an in-depth examination of the physiological and molecular mechanisms that regulate source–sink interactions. Emphasis is placed on the transporters and signalling molecules that facilitate nutrient allocation between sources and sinks. Finally, this chapter discusses the crucial role of balanced source–sink relationships in enhancing crop productivity and identifies key unresolved questions to guide future research.

Keywords

Source–sink relation · Transporter · Sugar signalling · Crop yield

Did You Know

- For most C3 plants, photosynthetic rate may decrease by up to 50% when the O_2 concentration is doubled from its ambient level of 21%, while this rate can increase by as much as twofold when the O_2 concentration is reduced to below 2%.
- Sucrose level in the mesophyll cytosol generally ranges from 10 to 50 mM. In contrast, it is significantly higher in the phloem, reaching 400–1400 mM. For example, maize has a phloem sucrose level of 900 mM, rice 574 mM, barley 1030 mM, and wheat 251 mM.
- Amino acids are the second most abundant metabolites in phloem sap, accounting for 5%–15% of total solutes and reaching up to 360 mM in maize. The dominant amino acids in phloem sap are glutamate, glutamine, aspartate, and asparagine. In certain species, such as *Arabidopsis*, citrus, and maize, proline, alanine, or glycine can also be prevalent.
- The allocation of photosynthetically derived carbohydrates likely represents the most critical difference distinguishing cultivated crops from their wild counterparts. For example, cultivated potato plants allocate ~80% of their dry weight to tubers, ten times more than their wild relatives.
- In some plant species, such as cassava, drought-induced stress triggers leaf abscission; when favourable growing conditions return, stored carbon is redirected from the roots to fuel the development of new foliage.
- Gaseous air pollutants, such as nitrogen dioxide (NO_2), can enter the leaves via stomata. They may dissolve in the aqueous phase of the apoplastic space and then be transformed into nitrate (NO_3^-), thereby enabling transport across the plasmalemma into the cells.

Learning Objectives

After reading this chapter, students will be able to

- Understand the concepts of source, sink, and flow
- Explain the physiological mechanisms underlying source–sink interactions
- Identify key indicators for assessing source–sink relationships
- Evaluate the ongoing debate regarding whether crop productivity is limited by source or sink capacity
- Understand the challenges of genetic modification in source–sink relationships

Introductory Remarks

Crop yield formation is mainly the process of carbon (C) assimilation driven by solar energy intercepted by the canopy. Sucrose, the product of photosynthetic assimilation, serves a distinctive function as the molecule for distributing C throughout the plant. It can be utilized locally for the growth and development of leaves, stored as starch in chloroplasts, or exported to other parts of the plant, such as internodes and seeds. The allocation of assimilated C to various organs is termed C allocation. Meanwhile, the more extensive processes that involve the partitioning of nitrogen (N), phosphorus (P), potassium (K), and other minerals in conjunction with C are collectively referred to as assimilate or nutrient allocation (Halford 2010).

Regarding nutrient allocation, the terms "source production" and "sink utilization" are commonly used. A *source* organ is defined as the net producer or exporter of nutrients such as photosynthates, amino acids, and minerals, while a *sink* organ is regarded as the net importer or utilizer. A long-standing debate has persisted among crop scientists regarding whether crop productivity is limited by source or sink constraints; that is, whether yield is restricted by photosynthetic assimilation or the utilization of photosynthates. A growing body of evidence indicates that both source and sink factors jointly regulate plant development and growth. Therefore, achieving efficient coordination between source and sink organs is of significant importance for both crop breeding and management.

Although the impacts of external conditions, including high temperature and diseases, on crop productivity are well-understood, insufficient attention has been given to internal factors, such as the nutrient allocation at the whole-plant level (Rosado-Souza et al. 2023). Recently, advancements in genomics have improved our understanding of assimilate partitioning through the modification of target genes related to source activity or sink strength. Nevertheless, our understanding of source–sink interactions remains incomplete, and our capacity to regulate these processes directionally is still limited. This chapter conducts an in-depth exploration of the mechanism of assimilate allocation between sources and sinks, aiming to elucidate the intricate physiological basis underlying crop yield and quality.

Concept Map

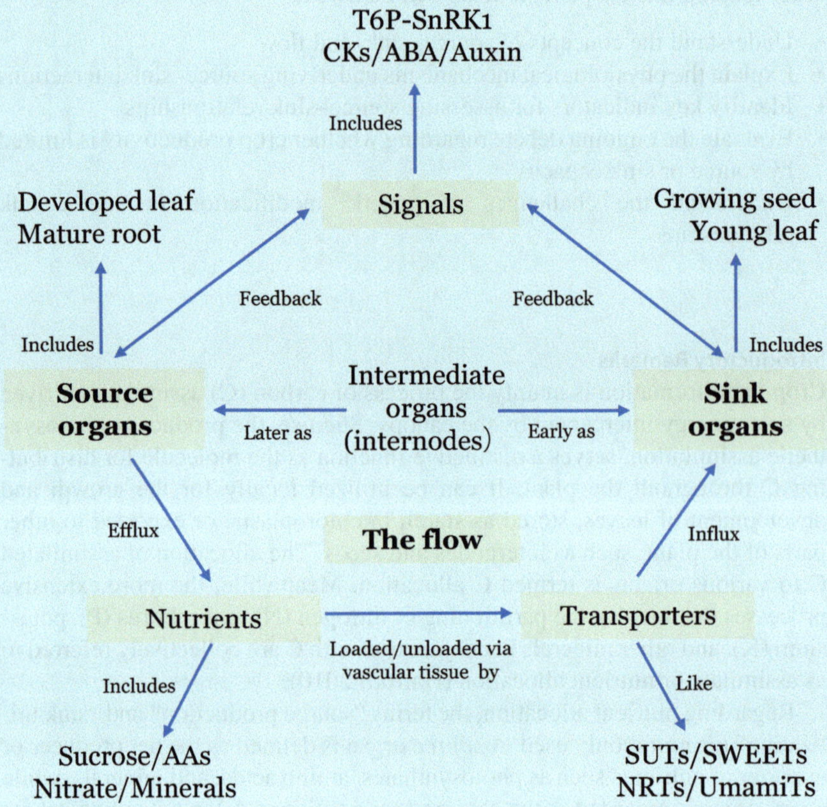

Annotation: Nutrients, whether in inorganic forms such as nitrate or in organic forms such as sucrose and amino acids, are essential for plant development and growth, as they provide the fundamental building blocks and energy required for the formation of plant tissues. The allocation of nutrients between source and sink organs is crucial for determining crop productivity. Source organs, such as mature leaves and roots, produce and export nutrients, whereas sink organs, such as developing seeds, import and utilize them. Additionally, there are intermediate organs that initially act as sinks but later transition to functioning as sources. The movement of nutrients from source to sink is facilitated by transporters located in the plasma membrane of the phloem, such as SUT and SWEETs for sugars, and UmamiTs and NRTs for nitrogen compounds. Signalling molecules, including T6P, CKs, and auxin, play a regulatory role in modulating the interaction between source and sink organs.

1 Basic Concepts of Source–Sink Theory

The fundamental concept of source–sink relations in plant physiology was originally proposed by Mason and Maskell (1928), focusing on the mechanisms of *nutrient* allocation between vegetative and reproductive organs. The essential nutrients for plant life include carbon (C), nitrogen (N), oxygen, water, and a variety of soil-derived minerals, such as P, K, and various micronutrients (White et al. 2016).

1.1 Classification of Source and Sink Organs

Generally, a *source* is a nutrient producer and exporter, a *sink* is a nutrient importer and consumer, while the *flow* is the translocation of nutrients from source to sink via vascular tissues (Fig. 5.1). However, there is an *intermediate organ* that undergoes a source–sink transition during its life cycle. For instance, the stems in cereal crops can shift from initially acting as a sink before flowering to ultimately serving as a source for grain filling.

1.1.1 Source Organs

Source organs primarily consist of mature leaves and roots, along with photosynthetically active non-leaf tissues, such as stems, inflorescences (e.g., ears or panicles), green floral parts, seed pods, and fruits (Yu et al. 2015). Mature leaves serve the function of photosynthetic carbon assimilation, producing sucrose that is actively loaded into the phloem and then transported to sink tissues. Meanwhile, a part of it is temporarily stored in the leaf, in the form of starch, for energy utilization during the night when photosynthesis ceases. Starch functions as a daytime C sink and transforms into a nocturnal C source. Typically, 25%–30% of the photosynthates are temporarily stored as starch during daylight hours; 7% are transformed into soluble sucrose; and the remainder fuels essential metabolic activities associated with growth and respiration in leaves (MacNeill et al. 2017).

Mature roots absorb water and minerals from the soil, which are subsequently translocated to the shoot for utilization. In addition, mineral nutrients can also be taken up by aerial organs, such as the leaves, fruits, and stems (Box 5.1). In addition, storage organs, such as seeds and tubers, store starch, lipids, proteins, and minerals, which are broken down during germination. The resulting small molecules, such as sucrose, glucose, fructose, amino acids, fatty acids, phosphate, and other minerals, are then utilized as the primary source to support seedling growth.

Although leaves are regarded as the major site of photosynthesis, other green non-foliar organs and tissues, including floral organs, bracts, and stems, are also capable of producing a significant amount of photosynthates. The physiology of non-foliar photosynthetic organs and their contribution to overall plant carbon fixation are attracting increasing scientific interest. Wheat ears and rice panicles have been shown to be highly photosynthetic, contributing 10%–60% to grain yield (Lawson and Milliken 2023). Photosynthesis occurring in the pods plays a critical role in determining both seed yield and quality in the Brassicaceae family. In

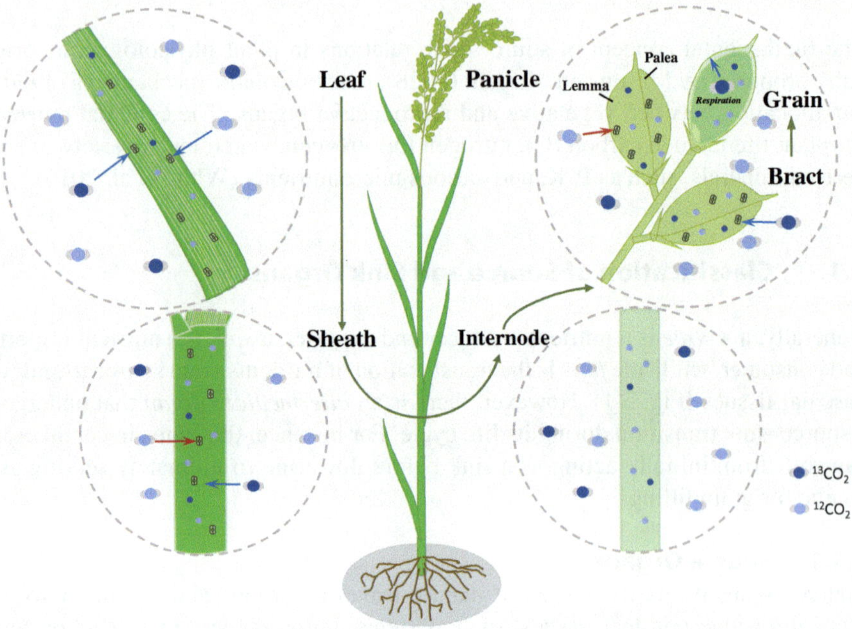

Fig. 5.1 Source, sink, and intermediate organs of the rice plant at the grain-filling stage. The leaf and bracts serve as the source organs, while the grains function as the sink organs; meanwhile, the sheath and internode act as intermediate organs that store transient starch and contribute to grain filling. The stable C isotopes of $^{13}CO_2$ and $^{12}CO_2$ are presented to show the discrimination against the heavier ^{13}C at various organs. In leaves, the lighter ^{12}C is always preferred, while the heavier ^{13}C is discriminated against during the process of photosynthesis. The internodes exhibit a higher enrichment of ^{13}C compared to the leaves, suggesting that there exists a substantial fractionation effect during the loading and unloading of sucrose, as well as the conversion between sucrose and starch in the phloem. The bracts contain chlorophyll and possess stomata on their surface. These features enable them to not only fix atmospheric CO_2 but also refix the CO_2 generated by grain respiration, thereby providing assimilates for the grains. However, due to the lower permeability of CO_2 diffusion in the bracts and the lower ratio of intracellular to atmospheric CO_2 concentration (C_i/C_a), the $\delta^{13}C$ of assimilates is always higher (with a smaller negative value) than that of the leaves. For further information on stable C discrimination, see Box 8.1 in Chap. 8

Camelina sativa, in planta ^{13}C labelling experiments revealed that photosynthates from the proximal pod contributes 33%–45% of the carbohydrates stored in seeds. Remarkably, the photosynthetic activity of the pod alone is sufficient to support as much as 79% of seed DM accumulation, even when foliar photosynthesis is eliminated (Koley et al. 2022).

As summarized by Araus et al. (2021), photosynthetic rates in non-foliar organs of certain C3 crop species ranged from 0.6 to 30.3 μmol m^{-2} s^{-1}, varying with tissue type and growing conditions. Due to their proximity to grains, ear bracts possess a higher capacity than leaves and awns for refixing CO_2 released from grain respiration. This can mitigate respiratory losses and increase ear WUE, as it operates independently of gas exchange with the atmospheric environment. Wheat and barley ears can recycle 55%–75% of the grain-respired CO_2, thereby contributing substantially to the grain-filling process (Sanchez-Bragado et al. 2020). Although the photosynthetic activities of ears are generally lower than those of functional leaves, these organs possess a comparably large surface area. Therefore, when the entire area is considered, the photosynthetic strength of the ears can be comparatively higher.

The contribution of ear photosynthesis to grain yield becomes more prominent when leaf photosynthetic activity is diminished by abiotic stress. For instance, Zhang et al. (2011) reported that nonfoliar organs of wheat accounted for 27%–62% of the total green area per culm, and this proportion increased significantly as drought severity intensified. Under water scarcity, the photosynthetic capacity of flag leaves declines, along with reduced activity of the photosynthetic enzymes. In contrast, the photosynthetic activity of ears and awns remains unaltered, thus supporting grain filling and maintaining yield under drought stress (Tambussi et al. 2021).

Box 5.1: Foliar Fertilization

For over a century, agriculture has harnessed the natural ability of leaves to absorb substances delivered in liquid form. The direct application of nutrients to foliage is a standard agronomic practice employed by farmers for the cultivation of a diverse range of crops. This approach is especially valuable when the availability of nutrients in the soil is inadequate, when prompt correction of observable nutrient deficiencies is necessary, or when the nutrient requirements increase significantly at critical developmental stages.

Leaves can absorb nutrients through the cuticle, stomata, and micro-pores present on their surface. The cuticle, which is a waxy layer, serves as a barrier to minimize water loss while enabling the diffusion of nutrients. Although nutrients can enter through the stomata, the hydrophobic waxy layer lining these pores hinders the absorption of water-soluble nutrients, thereby restricting their diffusion into the leaf tissues. Micro-pores, which are smaller than a nanometer and more numerous on the shaded side of the leaf, offer an alternative pathway for nutrient absorption, especially under conditions of high moisture. Nutrient absorption is enhanced when the nutrient solution spreads uniformly across the leaf surface, a process that is facilitated by surfactants through the reduction of surface tension.

Foliar fertilizers are widely used globally, yet their effectiveness varies significantly because of the complex physicochemical, physiological, and environmental factors involved (Fernández et al. 2021). To date, our understanding of the composition and structure of leaf surfaces, i.e., the cuticle, and the mechanisms underlying foliar absorption remains incomplete. This knowledge gap is primarily attributable to the complexity arising from the cuticle's nature as a composite membrane, comprising both hydrophilic constituents (polysaccharides) and hydrophobic components (waxes and cutin), with functionally diverse molecular groups arranged heterogeneously at the nanoscale.

1.1.2 Sink Organs

Plant sink organs can be broadly classified into two functional categories: growth sinks and storage sinks (Yu et al. 2015). Growth sinks, which encompass meristems, young leaves, and emerging roots, are metabolically active sites that vigorously absorb carbohydrates and nutrients to support cell division, expansion, and differentiation. In contrast, storage sinks, such as developing seeds, fruits, and tubers, are specialized in importing sugars and amino acids and transforming them into stable storage reserves, including starch, sucrose, proteins, and lipids.

The duration of storage varies significantly: some reserves (e.g., transient starch in chloroplasts) are remobilized within minutes or hours, while others (e.g., starch in endosperms) endure for months or years. This temporal distinction serves as the foundation for the structural and functional divergence between transient and long-term storage compounds. For instance, starch granules in cereal and legume seeds act as extended assimilate reservoirs that support germination. These granules are characterized by a highly structured, multilayered arrangement and notable biochemical stability. Conversely, transient starch in leaf mesophyll chloroplasts is smaller in size, has lower crystallinity, and undergoes dynamic restructuring over the diurnal cycle, enabling efficient mobilization to promote sucrose production during dark periods.

1.1.3 Intermediate Organs

Intermediate organs function either as a source or as a sink, depending on the physiological status and developmental stage of the crop plant. The parenchyma cells of internodes and leaf sheaths typically serve as temporary storage sites for C and N prior to the formation of definitive sink organs. Once sink organs have formed, these tissues switch roles and act as sources, supplying nutrients for seed development. Temporary storage sinks are generally located adjacent to axial transport pathways in stems and roots. Photosynthates are stored temporarily when supply from source organs exceeds demand by sink tissues; conversely, they are remobilized from temporary storage when source-derived nutrient production and export are insufficient to meet sink demand.

Carbon storage in the stems of many temperate grasses mainly consists of fructans, which are water-soluble fructose polymers with the potential to resist low temperatures. Typically, fructans and sucrose are co-stored in the stems of wheat, barley, and oats. Cereals typically do not accumulate a substantial quantity of starch in their stems. However, rice is an exception as it stores starch in its stems and lacks the ability to synthesize fructans. Stem storage of carbohydrates serves as a crucial buffer for supplying photosynthates to cereal ears during the grain-filling stage, especially under drought conditions (Engels et al. 2012). In temperate cereals such as wheat, fructans accumulate in the stems before anthesis and during the early post-anthesis period, subsequently contributing to grain filling. In wheat, the contribution of carbohydrate reservoirs to grain weight and yield depends on both genotype and growing conditions. Under drought stress during the grain-filling stage, fructan remobilization can account for up to 50% of the grain yield. In maize, significant quantities of N are temporarily sequestered in the stem, and approximately half of the grain nitrogen content is derived from this pool.

1.1.4 Strength and Activity

Source tissues predominantly serve the function of absorbing and assimilating nutrients from the external environment. The term *source strength* typically refers to the net rate at which a particular nutrient, such as carbon dioxide (CO_2), nitrate (NO_3^-), or ammonium (NH_4^+), is absorbed by the source organ, measured in moles per second (mol s^{-1}) (White et al. 2016):

$$\text{Source strength} = \text{Source size} \times \text{Source activity}$$

where source size refers to the physical size of the source organ (g), and source activity denotes the specific uptake rate of nutrient from the environment (mol g^{-1} s^{-1}).

In sink tissues, the primary biological processes involve the consumption of nutrients for constructing new tissues or maintaining the function of the existing tissues. Similarly, *sink strength* is defined as the net rate at which a specific organ takes up a particular nutrient, such as sucrose or an amino acid (mol s^{-1}):

$$\text{Sink strength} = \text{Sink size} \times \text{Sink activity}$$

where sink size is the physical size of sink organ (g), and sink activity represents the specific uptake rate of the nutrient from the source (mol g^{-1} s^{-1}).

1.2 Functional Transition of Leaves from Sink to Source

Throughout its lifespan, a leaf undergoes a functional shift, from being a sink for minerals and photosynthates in the early stage to becoming a net source at maturity (Engels et al. 2012). During the early developmental phase, cells at the leaf tip cease proliferation and begin differentiation, leading to the formation of mature structures such as the epidermis and stomata. This maturation process proceeds from the leaf tip toward the base, establishing a basipetal (distal-to-proximal) differentiation wave. Cells ahead of this wave continue to proliferate and expand actively, whereas those behind it gradually complete their differentiation. This spatiotemporal heterogeneity partitions the leaf into three distinct functional zones: the distal mature zone (at the leaf tip), where cells are fully differentiated; the middle expansion zone, characterized by predominant cell enlargement; and the proximal division zone (at the leaf base), where sustained mitotic activity enables continuous longitudinal leaf growth. Thus, leaf maturation exhibits spatial heterogeneity across the leaf blade, commencing at the apex and progressing in a basipetal wave of differentiation. Consequently, the sink-to-source transition follows a basipetal pattern along the leaf's longitudinal axis.

During an intermediate phase, bidirectional phloem transport occurs within a single leaf: some vascular bundles import photosynthates, while others export them. In sugar beet, the shift from net import to net export occurs when the leaf has attained approximately 40%–50% of its final area and photosynthetic capacity. Likewise, in other dicotyledonous species, this shift generally occurs when leaves are 30%–60% expanded. The sink-to-source transition in leaves involves biochemical, physiological, and anatomical changes. Specifically, the shift towards sucrose synthesis is associated with alterations in enzyme activities: a decline in acid invertase and sucrose synthase, both involved in sucrose catabolism, and an increase in sucrose phosphate synthase activity, which catalyses sucrose biosynthesis. Additionally, this transition entails modifications in plasmodesmatal connectivity between mesophyll and epidermal cells, leading to reduced symplasmic continuity. Nevertheless, the regulatory mechanisms governing nutrient import and export during leaf maturation remain poorly understood.

1.3 The Flow Connecting the Source and the Sink

The flow is defined as the translocation of nutrients from the source to the sink. In crop plants, source and sink organs are spatially separated, necessitating efficient long-distance transport of photosynthates and other nutrients through the phloem. Phloem sap consists of 88%–90% water and 10%–12% dry matter, which comprises water-soluble components, including sugars, amino acids, organic acids, inorganic ions, hormones, RNAs, proteins, and a diverse range of secondary metabolites. Sugars account for up to 80% of the osmotic content of the sap. Collectively, these solutes maintain the sieve tube sap at a pH of 7.0–7.5 and generate high osmotic pressure.

1.3.1 The STE-CC Complex

The phloem is composed of two distinct cell types: *sieve tube elements* (*STEs*) and their associated *companion cells* (*CCs*), which originate from the asymmetric division of a common progenitor cell. In flowering plants, STEs serve as the primary functional units of the phloem tissue, responsible for long-distance transport of nutrients throughout the plant. Upon maturation, STEs align end-to-end to form elongated conducting tubes that transport leaf-derived photosynthates. Each STE is physiologically associated with one or more CCs. CCs are characterized by a dense, granular cytoplasm containing a prominent nucleus, numerous mitochondria, abundant rough endoplasmic reticulum, a high density of ribosomes, and multiple small vacuoles. These features reflect their highly metabolically active role in nutrient transport. The elevated organelle content and extensive plasmodesmatal connections empower CCs to energize, regulate, and sustain long-distance nutrient transport within the phloem. CCs generally perform at least three fundamental functions: (1) providing ATP to meet the energy requirements of STEs, (2) synthesizing crucial macromolecules required by the enucleate STEs, and (3) facilitating assimilate loading in source tissues and unloading in sink tissues. Intercellular molecular exchange between STEs and CCs is mediated by specialized, branched plasmodesmata. This mechanism ensures coordinated physiological activity and enables the two cell types to function as an integrated structural and functional unit, termed the sieve tube element–companion cell (STE–CC) complex.

1.3.2 The Apoplastic and Symplastic Pathways

Two distinct mechanisms for phloem loading have been identified: apoplastic loading and symplastic loading. In the apoplastic pathway, sucrose is transported from mesophyll cells adjacent to the phloem into the apoplast and is subsequently actively loaded, against the concentration gradient, into the STE-CC complex by energy-dependent transporters located in the plasma membrane. The detailed interpretation of the sugar transporters involved is provided in the following Sect. 2.2.2. In symplastic phloem loading, sucrose is transported from mesophyll cells to CCs via plasmodesmata. Subsequently, it undergoes energy-dependent reactions to form higher-molecular-weight (HMW) sugars, typically oligosaccharides such as raffinose, stachyose, and verbascose. The precise mechanism of symplastic loading remains incompletely understood. Nevertheless, one proposed model posits that the HMW sugars synthesized in CCs are too large to diffuse back through plasmodesmata into mesophyll cells, resulting in polymer trapping and thereby confining their movement exclusively to the phloem.

After being loaded into the phloem, sucrose is transported to sink tissues, where it can enter destination cells through symplastic and apoplastic pathways. In the symplastic pathway, sucrose moves passively through plasmodesmata, driven by its concentration gradient. It first accumulates in the STE–CC complex and then diffuses directly into the cytoplasm of sink cells, without crossing any plasma membrane. This bypasses membrane transporters, allowing intact sucrose to enter sink cells unaltered. Once inside, sucrose is rapidly cleaved by invertases and sucrose synthase into glucose and fructose: hexose units that fuel biosynthetic processes, including starch and cell wall synthesis.

The apoplastic pathway for phloem unloading is more prevalent across plant species than the symplastic pathway. It plays a particularly critical role during embryogenesis, as developing embryos lack direct symplastic connections with maternal tissues. In this context, sucrose can either be imported directly into embryonic cells via transporters or first hydrolyzed to glucose and fructose by extracellular invertases in the apoplast, as detailed in Sect. 2.2.2. Importantly, carbohydrates serve not only as metabolic intermediates but also as signalling molecules that regulate the expression of numerous developmental and stress-responsive genes. Consequently, understanding the functional implications of distinct sucrose transport modes, especially their spatial regulation, enzymatic control, and downstream signalling, is increasingly recognized as essential for unravelling nutrient-mediated developmental control in plants.

1.3.3 The Pressure-Flow Hypothesis

According to the pressure-flow hypothesis, sucrose in the phloem sap is transported via mass flow rather than osmosis (Fig. 5.2). This movement of water does not violate the laws of thermodynamics, as it is a pressure-driven mass flow process, which is distinct from osmosis governed by a water potential gradient. The passive,

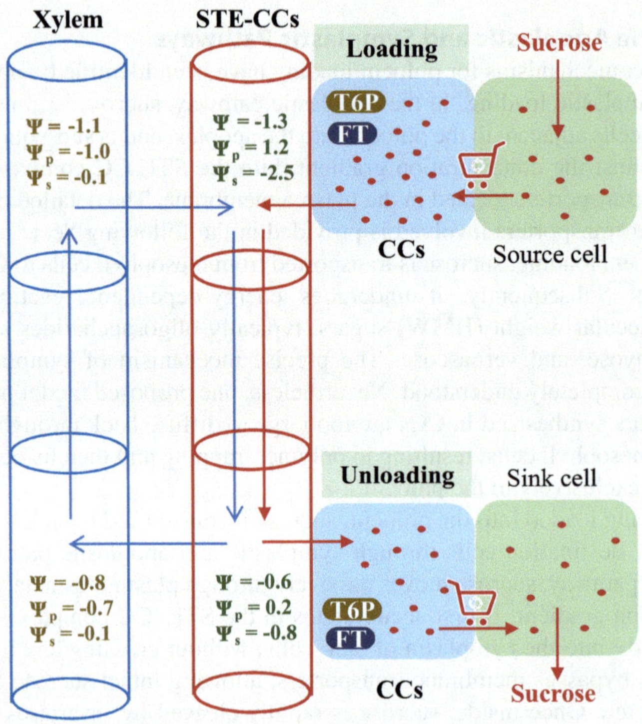

Fig. 5.2 The pressure-flow model for phloem loading and unloading of sucrose. The water potential of a solution (Ψ) can be decomposed into individual components, typically expressed as $\Psi = \Psi_s + \Psi_p$. In this context, Ψ_s and Ψ_p represent the contributions of solutes and pressure, respectively, to the free energy of water. In source tissues, an elevation in the sugar concentration within sieve elements leads to a low (negative) solute potential (Ψ_s), consequently reducing the overall water potential (Ψ). This gradient prompts the influx of water into the sieve elements, causing an increase in turgor pressure (Ψ_p). At the sink, phloem unloading lowers sugar concentration within sieve elements, resulting in a less negative (i.e., higher) solute potential (Ψ_s). Consequently, when the water potential (Ψ) in the phloem exceeds that in the xylem, water moves out of the phloem along the water potential gradient, resulting in a reduction of turgor pressure within the sieve elements at the sink. It is important to note that the companion cell in the leaf serves as the primary site for the biosynthesis of the sugar-signalling molecule trehalose-6-phosphate (T6P) and the florigen protein FT. Consequently, it functions as an integrator of metabolic and developmental signals in regulating source–sink relationships. CCs companion cells, STE sieve tube element. (Redrawn from Taiz et al. (2015) and Crang et al. (2020))

long-distance translocation of solutes within sieve tubes is ultimately dependent on active phloem loading in source organs and phloem unloading in sink organs.

In source tissues, sugars are actively transported from leaf mesophyll cells into the phloem via specific sugar transporters. This energy-dependent process establishes a substantial concentration gradient of sucrose between the cytosol of mesophyll cells and the phloem sap. Sucrose concentrations in the mesophyll cytosol typically range from 10 to 50 mM, whereas phloem sap concentrations reach approximately 1 M. As a result, the *osmotic potential* within sieve elements is markedly lower than that in the adjacent xylem. This differential in osmotic potential drives the influx of water from the xylem into the phloem, generating considerable hydrostatic pressure within the phloem network.

At the terminal end of the translocation pathway, phloem unloading decreases the sugar concentration in the sieve elements, thereby increasing the solute potential (i.e., making it less negative) within the sieve elements of sink tissues. As the phloem water potential rises relative to that of the xylem, water diffuses out of the phloem along the water potential gradient and re-enters the surrounding xylem. The resulting decrease in turgor pressure in the sieve elements of the sink establishes a pressure gradient that drives bulk flow of phloem sap from source to sink.

During the translocation from one sieve tube to another, no membranes are crossed, and sugars move at the same velocity as the water. Owing to this mechanism, mass flow can occur from a source organ with a lower water potential to a sink with a higher water potential, depending on the metabolic requirement of the source and sink tissues. Furthermore, sucrose is transported via open phloem pathways rather than through sealed channels within sieve tubes (Thomas 2017). The sugar concentration and turgor pressure are controlled by release and retrieval mechanisms on the plasma membrane of STE–CC complexes, driven by the proton-motive force. These mechanisms are differentially regulated along the phloem, controlling sugar and water movement in specific zones. In the collection phloem, loading dominates; in the release phloem, unloading is primary.

2 Physiology of Source–Sink Interaction

Source–sink interaction refers to the exchange of nutrients between source and sink organs, regulated by a highly complex signalling network in response to the cellular carbon and nitrogen status, agronomic practices, and fluctuating environmental conditions. It is extremely complex, and currently, only a limited part of the underlying mechanisms is understood. This section conducts a systematic examination of the interaction from three critical perspectives: (1) the nutrients transported in the vascular tissues, (2) the transporters that promote nutrient transport, and (3) the signals that coordinate metabolic activities between source and sink organs.

2.1 Nutrients Transported in the Vascular Tissues

2.1.1 Sugars

The phloem predominantly transports the non-reducing disaccharide sucrose, the main form of carbon allocated from source to sink tissues. In specific plant species that utilize the polymer-trap loading mechanism, such as sweet watermelon and cucumber, the primary sugars transported are raffinose-family oligosaccharides (RFOs), including raffinose, stachyose, and verbascose. These RFOs play a role in osmoprotection and reduce back-diffusion due to their relatively high molecular weight. Moreover, certain plants from the Rosaceae and Cucurbitaceae families primarily transport sugar alcohols, such as sorbitol and mannitol, through the phloem. Reducing monosaccharides, such as glucose, fructose, or mannose, are generally excluded from phloem transport because their reactive aldehyde or ketone functional groups may initiate unwanted chemical reactions; instead, these sugars are promptly phosphorylated or converted into sucrose prior to loading into the phloem.

2.1.2 Nitrogenous Compounds

Plants predominantly absorb inorganic nitrogen in the form of nitrate and ammonium. Nevertheless, the uptake of amino acids and other organic nitrogen compounds via roots is also ecologically significant in certain species. Moreover, legumes can fix atmospheric N_2 through symbiotic bacteria in root nodules. In soybeans, ammonia, the direct product of symbiotic nitrogen fixation, is initially incorporated into glutamine, aspartate, and glycine. These amino acids provide both the N and C skeletons required for purine ring biosynthesis. The resulting purines are subsequently oxidized to xanthine, then to uric acid, and finally to allantoin and allantoic acid (the primary ureides), which are then transported to the shoots. During the reproductive stage, ureides constitute 60%–75% of the soluble nitrogen in the xylem sap of soybean.

Amino acids serve as the main N carriers moving from source to sink (Lalonde et al. 2004). Glutamate, glutamine, aspartate, and asparagine are the four most abundant amino acids in phloem sap, collectively accounting for about 70% of all free amino acids and reaching concentrations of 100–200 mM. Conversely, their concentrations in xylem sap are approximately ten times lower, with basic and acidic amino acids being the dominant components. The long-distance transport of organic nitrogen in plants is at least as intricate as that of sugars. This is due to the high diversity of N-containing compounds transported in the phloem sap, including the 20 proteinogenic amino acids, various amino acid analogues such as GABA, oligopeptides, and numerous uncharacterized N-containing compounds. Moreover, unlike sucrose, amino acids are transported in both the phloem and xylem, requiring complex communication between these two vascular tissues.

2.2 Transporters for Sugars and Nitrogenous Compounds

Plants allocate different tasks among organs specialized for specific functions. While roots are primarily specialized in taking up mineral nutrients and assimilating nitrate into amino acids, leaves are mainly responsible for transforming CO_2 into sugars. To facilitate the exchange of nutrients among these organs, two transport systems have evolved: the xylem transports nitrate, amino acids, and ureides from the roots to the shoots, whereas the phloem transports sugars from the leaves to sink organs such as seeds and roots. The transport of these nutrients through the vascular tissues necessitates crossing multiple membranes, which is mainly mediated by transport proteins.

2.2.1 Transporters in Plants

Transporters are integral membrane proteins that facilitate the selective transport of ions, metabolites, hormones, and other molecules across cellular membranes, thereby integrating intracellular and extracellular signalling pathways. These proteins play essential roles in plant growth and development by mediating: (1) nutrient uptake—transporters for nitrate, ammonium, and phosphate are responsible for the uptake of essential nutrients from soils; (2) cellular homeostasis—ion channels and H^+-coupled symporters regulate intracellular pH, osmotic balance, and membrane potential; (3) hormone distribution—specific transporters mediate the redistribution of auxin, ABA, and other phytohormones to coordinate physiological responses; and (4) stress adaptation—numerous transporters are upregulated under conditions of drought, salinity, heavy metal toxicity, or pathogen attack, contributing to detoxification, osmotic adjustment, and signal transduction.

Plant transporters are predominantly classified according to their energy source and transport mechanism: (1) channels and pores, which serve as passive pathways for water (e.g., aquaporins), ions, and small molecules; (2) primary active transporters, the ATP-dependent pumps, among which the ATP-binding cassette (ABC) superfamily is the largest, participating in processes such as detoxification, hormone translocation, and lipid trafficking; (3) secondary active *cotransporters*, which harness electrochemical gradients to drive substrate transport, including the NRTs, SUTs, and MSTs family members; and (4) transmembrane electron carriers, which facilitate substrate movement by coupling it to electron transport across membranes. These carriers link redox reactions occurring on opposite sides of a membrane, typically coupling substrate oxidation to proton translocation or the reduction of extracellular electron acceptors. For example, NADH dehydrogenase, the primary electron carrier in the respiratory chain, transfers electrons and simultaneously pumps protons across the membrane, thereby coupling NADH oxidation to proton gradient formation and energy storage.

The cotransporter is an integral membrane protein that mediates the simultaneous transport of two distinct solutes across a biological membrane. The movement of one solute (the "driving ion") is powered by its electrochemical gradient, and this energetically favourable process is coupled to the translocation of a second solute against its own electrochemical gradient (the "driven substrate"). Because the energy required for this transport is derived from the electrochemical gradient of the driving ion rather than from direct ATP hydrolysis, cotransport is classified as a form of secondary active transport. Cotransporters are functionally divided into two subclasses based on the direction of solute movement: *symporters*, in which both solutes move in the same direction, and *antiporters*, in which the solutes move in opposite directions. For example, transporters such as SUTs, STPs, NRTs, and AAPs function as symporters, whereas TSTs operate as antiporters, as described below.

2.2.2 Sugar Transporters

There are three major families of sugar transporters: MSTs (monosaccharide transporters), SUTs/SUCs (sucrose transporters or sucrose carriers), and SWEETs (Sugar Will Eventually be Exported Transporters). These transporters play distinct roles in sugar transport, especially in terms of their structure and functional participation in sugar loading and unloading processes. Both SUTs and MSTs belong to the major facilitator superfamily (MFS) and possess 12 transmembrane domains (TMDs). In contrast, SWEETs have seven TMDs. In terms of transport mechanisms, MSTs and SUTs act as proton-coupled sucrose cotransporters, while SWEETs are classified as *uniporters* that mediate passive sugar transport along concentration gradients without requiring additional energy input, such as a proton motive force or pH gradients.

Most monosaccharide transporters (MSTs) are predominantly expressed in sink tissues, where they facilitate the cellular uptake and distribution of simple sugars, including glucose and fructose. Phylogenetically, MSTs are classified into seven distinct clades: Early Response to Dehydration-like 6 (EDRL6), Sugar Transport Protein or Hexose Transporter (STP/HT), Tonoplast Sugar Transporter (TST), Plastidic Glucose Translocator (pGlcT), Inositol Transporter (INT), Polyol/Monosaccharide Transporter (PMT), and Vacuolar Glucose Transporter (VGT). Among these, STPs and HTs are widely investigated due to their central role in sugar acquisition. They are integral plasma membrane proteins that catalyse proton-coupled symport of hexoses into the cytosol. In contrast, TSTs reside on the tonoplast and function as proton-coupled antiporters, exporting protons out of the vacuole while importing sucrose or hexoses into its acidic lumen.

SUTs are integral plasmalemma proteins that function as H^+/sucrose symporters, utilizing the energy stored in transmembrane proton electrochemical gradients to drive sucrose uptake across biological membranes. In photosynthetically active source tissues, these transporters actively import sucrose from the apoplastic space into the STE–CC complexes, a key step in phloem loading essential for the systemic distribution of photosynthates. In contrast, within non-photosynthetic sink tissues, SUTs operate in reverse mode, mediating sucrose efflux from phloem sap into surrounding parenchyma cells, where it serves as a carbon and energy source for growth and metabolism. Thus, SUTs serve dual, context-dependent roles: enabling phloem loading in source tissues and facilitating phloem unloading in sinks, thereby underpinning whole-plant C allocation. In *Arabidopsis*, SUT family members are phylogenetically classified into three major clades: Clade I comprises AtSUC1, AtSUC2, AtSUC5, and AtSUC6–AtSUC9; Clade II contains only AtSUC3; and Clade III consists solely of AtSUC4. While most SUTs/SUCs exhibit strict specificity for sucrose, notable functional exceptions exist: AtSUC5 also transports biotin, and AtSUC9 preferentially mediates glucoside uptake but lacks significant sucrose transport activity.

SWEETs serve as reversible sugar carriers, enabling the movement of sugars both out of and into cells, thereby governing C partitioning across various biological membranes. Unlike proton-coupled transporters, SWEETs operate independently of energy input and do not rely on ion gradients; instead, they function as passive uniporters that shuttle sucrose, glucose, and fructose across the plasmalemma as well as intracellular membranes such as the tonoplast. This bidirectional flexibility underpins their critical involvement in other physiological processes, including pathogen-induced sugar release. SWEETs mediate sucrose loading into and unloading from the STE–CC complex, where subsequent active transport is driven by the sucrose transporter SUT1 and energized by proton-pumping H^+-ATPases. Phylogenetically, members of the SWEET family are classified into four major clades. In *Arabidopsis*, Clades I and II mainly facilitate the transport of hexoses, while Clade III (comprising AtSWEET9 to AtSWEET15) is specialized in sucrose translocation. The members of Clade IV, namely AtSWEET16 and AtSWEET17, are localized in the tonoplast and exhibit broad substrate specificity, preferentially transporting fructose but also being capable of transporting glucose and sucrose.

Collectively, these three transporter families coordinate the distribution of carbohydrates from source to sink, regulating sugar import, export, and intracellular allocation to support plant growth, reproduction, and adaptation to environmental stress. SUTs act as active sucrose importers driven by the proton motive force. They primarily mediate sucrose loading into the phloem for long-distance translocation and its delivery to developing seeds or fruits. MSTs serve as high-affinity H^+-symporters for monosaccharides, providing hexoses to rapidly growing sink tissues such as roots, pollen, and seeds. They are often upregulated under stress conditions that demand rapid sugar uptake. In contrast, SWEETs function fundamentally differently as facilitative uniporters, allowing passive diffusion of sugars along their concentration gradient. These transporters play vital roles in sugar efflux processes such as phloem unloading and nectar secretion, as well as in inter-organ and plant–microbe carbohydrate exchange. Since they are independent of the proton gradient, SWEETs are bidirectional and are often utilized by pathogens to access host-derived C resources.

An illustrative example demonstrating the coordinated function of three types of transporters involved in sugar loading into the developing endosperm is presented in Fig. 5.3. In cereal grains, the translocation of sugars from maternal tissues, such as the nucellus projection (in rice and wheat) and the placental chalaza (in maize), to the starchy endosperm occurs via two complementary pathways (Liu et al. 2022): direct sucrose transport and indirect hexose transport. Together, these two pathways ensure that cereal seeds can efficiently allocate sucrose-derived C to the starchy endosperm, either as intact sucrose or as its hexose constituents, depending on developmental stage and physiological conditions.

1. Direct Sucrose Transport. Sucrose is released from the STE-CC complex into the apoplast via plasmalemma-localized SWEETs and subsequently taken up by SUTs/SUCs located on the endosperm plasmalemma. This process delivers intact sucrose into the cytosol for starch biosynthesis and is particularly critical during the rapid grain-filling phase, when high C demand necessitates efficient assimilate delivery.
2. Indirect Hexose Transport. Following SWEET-mediated sucrose efflux, extracellular sucrose can be decomposed into glucose and fructose in the apoplastic

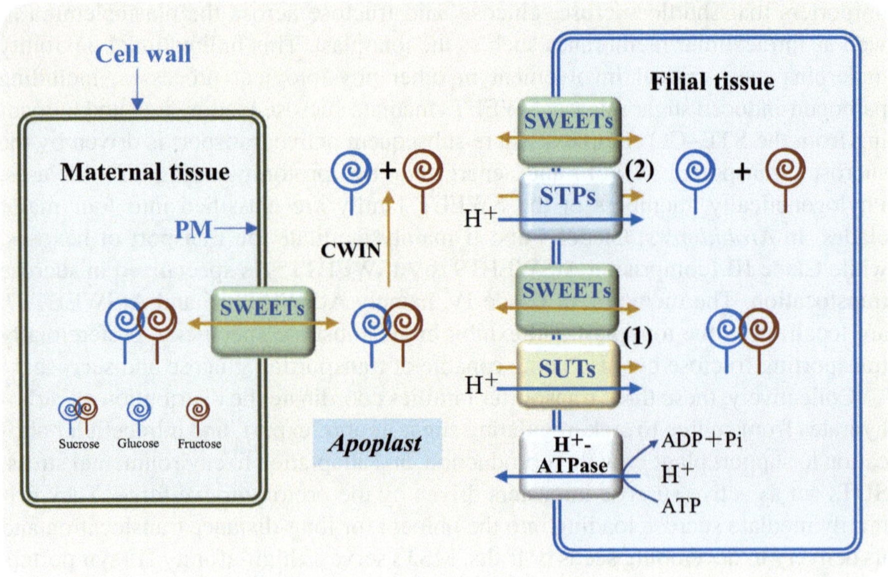

Fig. 5.3 Schematic models for routes of sugar loading into cereal endosperms. Sucrose unloaded from the maternal vasculature is either (1) directly loaded into the endosperm through SWEETs and SUTs or (2) initially hydrolyzed into glucose and fructose by cell wall invertase (CWIN). Subsequently, the resulting hexoses are transported into the endosperm via SWEETs or monosaccharide transporters (MSTs), such as members of the STP family of hexose/H^+ symporters. PM plasma membrane, STP sugar transport protein. (Redrawn from Liu et al. (2022) and Eom et al. (2015))

space by cell-wall invertase (CWIN). These hexoses are then transported into endosperm cells via proton-coupled MSTs, such as members of the STP family. This route fine-tunes the apoplastic sucrose to hexose ratio, alleviates osmotic stress, and supplies flexible C sources for energy production, biosynthetic reactions, and sugar-signalling cascades.

2.2.3 Nitrogen Transporters

The transport of N from the root to the shoot takes place in the xylem, whereas the partitioning of N from source leaves to sinks takes place in the phloem. In the sequence of N transfer within the plant, plasmalemma-localized transporters play a crucial role in regulating N uptake by the roots, root-to-shoot translocation, leaf-to-sink transport, and seed loading (Fig. 5.4). It should be noted that most of these transporters function as symporters, except for UmamiTs, which operate as uniporters, and AVT1, which functions as an antiporter. This section describes a subset of the N transporters; for more detailed information on the entire set of transporters, refer to the review by Tegeder and Masclaux-Daubresse (2018).

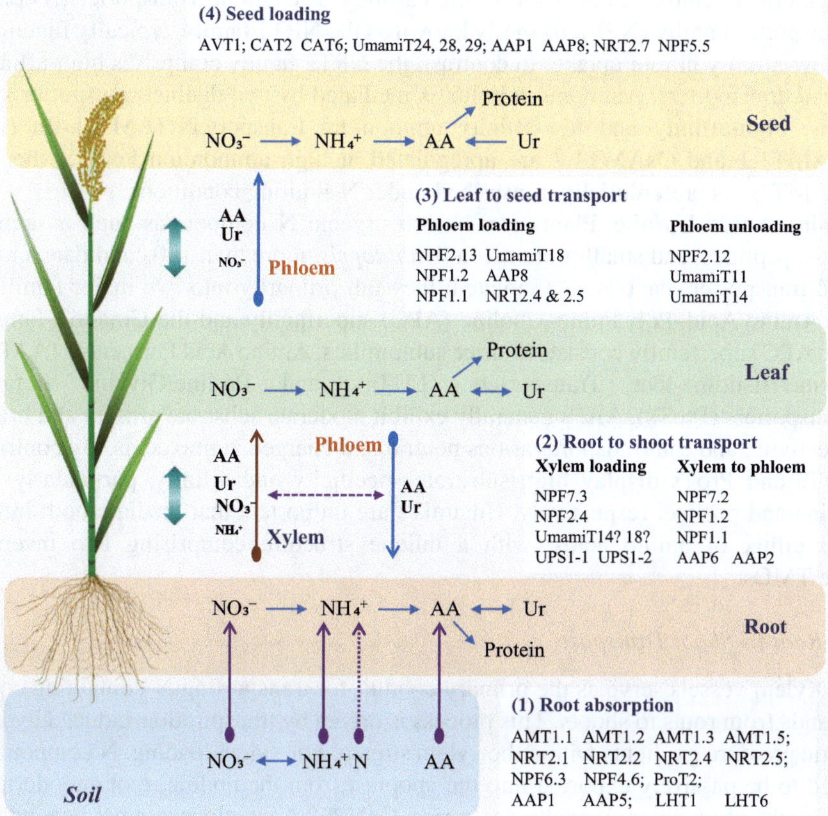

Fig. 5.4 Schematic overview of nitrogen (N) transport processes at the whole-plant level. The root–shoot translocated nitrogenous compounds include inorganic N such as nitrate (NO_3^-), ammonium (NH_4^+), and di-nitrogen (N_2), as well as organic N such as amino acids (AAs) and ureides (Ur). The listed N transporters are involved in (1) root absorption, (2) root to shoot transport, (3) leaf to seed transport, and (4) seed loading. Green two-way arrows signify feedback controls exerted by source–sink interaction. The smaller font size of xylem NH_4^+ and phloem NO_3^- indicates their lower levels in comparison to other transported N compounds. AA amino acids, AAP amino acid permease, AMT ammonium transporter, AVT amino acid vacuolar transporter, CAT cationic amino acid transporter, LHT lysine/histidine-like transporter, NRT/NPF nitrate transporter/peptide transporter family, ProT proline and glycine betaine transporter, UmamiT usually multiple acids move in and out transporter, UPS ureide permease, Ur ureides. (Redrawn from Tegeder and Masclaux-Daubresse (2018))

1. *Nitrogen Absorption by the Root*

Roots possess multiple transporters for inorganic and organic nitrogen, each exhibiting distinct substrate binding affinities and selectivity. This functional diversity enables plants to dynamically adjust N acquisition in response to fluctuating soil N availability across different growth stages.

Nitrate and Ammonium Uptake Nitrate absorption from the soil is facilitated by transporters exhibiting either low or high affinity. The Nitrate Transporter 1/Peptide Transporter Family (NPF), formerly known as the NRT1 family, typically functions in low-affinity nitrate uptake. In contrast, the NRT2 family comprises high-affinity nitrate transporters. Ammonium influx is mediated by two distinct transporter systems: high-affinity and low-affinity ammonium transporters (AMTs). In rice, OsAMT1.1 and OsAMT1.2 are upregulated at high ammonium levels, whereas OsAMT1.3 is preferentially transcribed under N-limiting conditions.
Amino Acids Uptake Plants can absorb organic N compounds such as amino acids, peptides, and small proteins. In *Arabidopsis*, more than 100 candidate amino acid transporters have been annotated; they fall primarily into two major families: the Amino Acid–Polyamine–Choline (APC) superfamily and the UmamiT family. The APC superfamily consists of three subfamilies: Amino Acid Permeases (AAPs), Lysine/Histidine-like Transporters (LHTs), and Proline/Glycine Betaine Transporters (ProTs). AAPs generally exhibit moderate substrate affinity and broad selectivity, and can transport various neutral and charged amino acids. By contrast, LHTs and ProTs display high substrate specificity and affinity, particularly for lysine and proline, respectively. UmamiTs are uniporters that mediate both influx and efflux of amino acids, with a unique structure comprising two inverted five TMDs.

2. *Root-to-Shoot Transport*

Xylem vessels serve as the primary conduit for transporting N-containing compounds from roots to shoots. This process is driven by transpiration-induced hydrostatic pressure gradients within the xylem stream. For xylem loading, N compounds need to be passively exported into the apoplast from the nodule, root endodermis, pericycle, or vascular parenchyma. In rice, OsNPF2.4 functions as a pH-dependent, low-affinity nitrate transporter located in the plasmalemma of epidermal cells, xylem parenchyma, and phloem CCs. It is crucial for nitrate uptake and root-shoot transport (Xia et al. 2015).

Beyond long-distance transport, the xylem actively participates in N recapture along its pathway, contributing to the establishment of N reserves in parenchyma tissue in roots, stems, and leaves. Moreover, it facilitates xylem-to-phloem transfer to directly supply fast-growing sinks. This transfer requires retrieval of N from the xylem into parenchyma cells via symporters such as NPF7.2 and AAP6, followed by movement towards the phloem and apoplastic loading.

3. *Leaf-to-Seed Transport*

Sink organs typically do not import N compounds via the xylem because of their low transpiration rates. The STE–CC complex in the phloem constitutes the long-distance conduits for the import of nitrogen into sink organs. The nitrate levels are comparatively low in the phloem, with the ratios of nitrate to amino acids generally ranging from 1:100 to 1:10. Consequently, the transport of nitrogen from leaves to sink organs predominantly occurs in the organic forms of amino acids or ureides.

Phloem Loading in Leaf Phloem loading commences with active apoplastic transport mediated by transporters in CCs. In minor veins, the key transporters include NPF2.13 and NRT2.5; in major veins, the process depends on NRT2.4, NPF1.2, and NPF1.1. Amino acids are first exported from mesophyll cells into the apoplastic space by exporters such as UmamiT18. Subsequently, they are loaded into the STE–CC via symporters, including AAPs, CATs, and LHTs.

Phloem Unloading in Seed Amino acid unloading is mediated by UmamiT efflux systems at the terminus of the maternal vasculature. Both UmamiT11 and UmamiT14 are located across the plasmalemma of cells near the protoxylem and phloem STEs, where they are responsible for the import of amino acids into seeds. Additionally, NPF2.12/NRT1.6 is situated in the vascular tissue of the silique, importing nitrate for embryo development and temporary N storage.

4. *Seed Loading and Storage*

Amino acids represent the predominant form of N transferred from maternal tissues to developing embryos. The export of amino acids from the seed coat into the apoplastic compartment is facilitated by two transporters, namely UmamiT28 and UmamiT29. Once amino acids enter the apoplast, they are subsequently absorbed by cotyledonary cells, primarily through AAPs and CAT6, and then delivered to the growing embryo.

A portion of the nitrate is sequestered within vacuoles for temporary storage and subsequently utilized for protein synthesis. AVT1, an amino acid antiporter localized to cotyledonary vacuoles, facilitates the uptake of neutral amino acids, namely tyrosine, glutamine, asparagine, isoleucine, and leucine. CAT2 is recognized as an effective transporter for basic amino acids. Additionally, UmamiT24, a tonoplast-localized transporter, is hypothesized to regulate the vacuolar loading of amino acids for temporary storage during seed development.

2.3 Signals Coordinating the Source–Sink Communication

Information is the reflection of the dynamic states and changes of entities in the objective world, capturing their mutual relationships and reciprocal influences. It unveils the intrinsic nature of how these entities evolve and influence one another. In source and sink organs, there exists the "flow" information reflecting the state of nutrient movement and changes in terms of the composition and concentration of sugars and amino acids. A *signal* in a biological system can be conceptualized as the transmission and interpretation of information that carries biological significance, whether at the genetic, physiological, or biochemical level. Biological signals embrace a variety of molecules, such as hormones, sugars, and small peptides, among others. *Signalling*, the cascade of signal transmission that integrates the "flow" information with metabolic and developmental events, regulates nutrient flow pattern and metabolic direction, thus being crucial for optimizing source–sink relations. Understanding these mechanisms is important for unravelling the molecular basis of source–sink communication while also developing genetic manipulation strategies for crop improvement. However, key components that underlie these signalling pathways are only now being discovered.

2.3.1 Sugar Signalling

Sugars, including sucrose, glucose, fructose, and trehalose, along with their derivatives, not only function as metabolic components of cells but also display hormone-like regulatory functions. Sugars govern the fundamental processes, such as embryogenesis, germination, vegetative growth, reproduction, senescence, and responses to diseases and environmental stimuli. Sugars upregulate genes related to source–sink transport, anabolism in the sink organs, and seed ontogenesis, while downregulating those related to leaf photosynthesis and reserve remobilization under abiotic stress. Conversely, sugar starvation exerts the opposite effect.

Sucrose, the main form of carbon transported to the sink organs, is regulated by a specific signalling pathway mediated by trehalose 6-phosphate (T6P). T6P, the sucrose sensor, regulates cellular anabolism or catabolism depending on local sucrose availability (Fig. 5.5). Its effect on plant metabolism and development is thought to occur via inhibition of SnRK1 (SUCROSE NON-FERMENTING 1 (SNF1)-RELATED KINASE). SnRK1 is a multiprotein complex orthologous to AMP-activated protein kinase (AMPK) in animals and SNF1 in yeast. It is activated by energy deprivation (AMP) and inhibited uniquely by the sucrose signal (T6P) in plants. SnRK1 plays a crucial role in preventing cell and organismal death caused by energy exhaustion by conserving carbon and energy: it suppresses anabolism and ATP/carbon consumption while activating catabolic processes such as storage reserve degradation.

Cellular T6P levels are usually low; however, they respond sensitively to sucrose availability. When a large amount of sucrose is transported into sink organs, T6P levels rise sharply, leading to inhibition of SnRK1 activity and consequently increased sink activity. T6P inhibits SnRK1 activity both temporally and spatially in developing wheat grains. During the grain-filling stage, SnRK1 is activated by reduced T6P concentrations in developing embryos and pericarps, whereas it is inactivated by high sugar and T6P concentrations, which promote endosperm starch biosynthesis. Reducing T6P through genetic modification or marker-assisted selection, or elevating T6P via chemical intervention, leads to substantial yield improvements in maize, rice, and wheat (Paul et al. 2017).

FLOWERING LOCUS T (FT), recognized as florigen, serves as a conserved regulator of floral induction. It is synthesized within the leaves and subsequently transported to the SAM, where it initiates the formation of the inflorescence meristem (see Sect. 4.1.4 in Chap. 7). The role of FT appears to extend beyond the control of flowering. Notably, an FT homologue, SP6A, in potato interacts with the sucrose efflux transporter SWEET11 to prevent the leakage of sucrose into the apoplast, thereby facilitating symplastic phloem loading of sucrose. This finding may signify a link between sugar metabolism and photoperiod pathways that govern

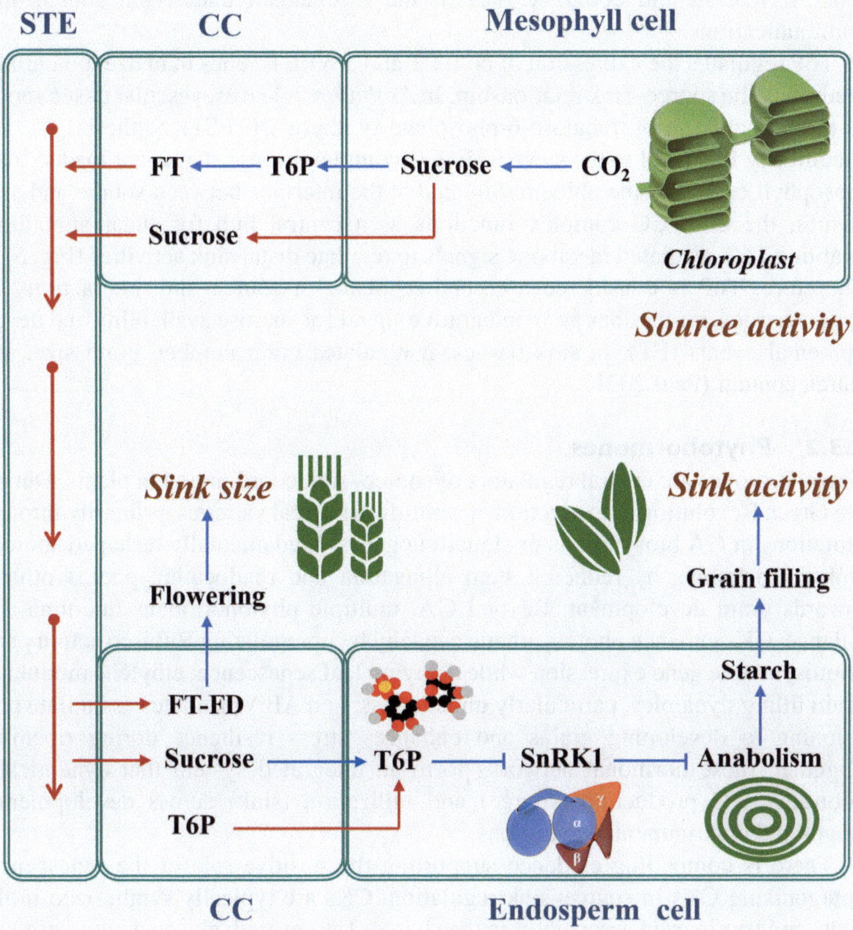

Fig. 5.5 The T6P–SnRK1 signalling pathway as a central hub linking source activity and sink strength. In mammals, blood glucose levels are tightly regulated by two key hormones: insulin and glucagon. In flowering plants, a functionally analogous regulatory system operates through the sugar-signalling molecule trehalose 6-phosphate (T6P). Like insulin, T6P serves as a negative feedback regulator of sucrose metabolism. T6P senses the sucrose level and acts as a central signalling molecule that inhibits the energy-sensing kinase SnRK1. SnRK1 is a heterotrimeric kinase complex composed of a catalytic α subunit and two regulatory subunits, β and γ, and functions as a central regulator of metabolic and energy homeostasis (Crozet et al. 2014). When T6P levels are high, SnRK1 activity is suppressed. As a result, anabolic genes, such as ADP-glucose pyrophosphorylase, are activated to direct excess carbon into starch accumulation in sink tissues. The companion cells (CCs) serve as the primary site for the biosynthesis of T6P and florigen (FT), thereby acting as an integrative hub of metabolic and developmental signalling. FD FLOWERING LOCUS D, SnRK1 (SUCROSE NON-FERMENTING 1 (SNF1)-RELATED KINASE), STE sieve tube element. (Redrawn from Paul et al. (2017))

tuber formation and could be part of the mechanism underlying source–sink communications.

T6P regulates the expression of both FT and SWEET genes in maize, potentially mediating the source–sink relationship. In *Arabidopsis* leaves, vascular tissue serves as the primary site of trehalose-6-phosphate synthase 1 (TPS1) synthesis. TPS1 is specifically localized to the sieve STE–CC complex, where sugars are loaded from mesophyll cells into the phloem. Situated at the interface between source and sink tissues, the STE–CC complex functions as a central hub for sugar signalling, enabling T6P-mediated metabolic signals to regulate distal sink activities (Fig. 5.2). Therefore, T6P is considered a crucial regulator of source–sink interactions. In source tissues, it functions as an integrative signal for sucrose availability and developmental events (FT). In sink tissues, it regulates grain number, grain size, and starch content (Paul 2021).

2.3.2 Phytohormones

Phytohormones are central regulators of source–sink coordination in plants. During the Green Revolution, the selection of semi-dwarf cereal varieties, primarily through mutations in GA biosynthesis or signalling genes, fundamentally reshaped source–sink relationships by reducing stem elongation and reallocating photosynthates towards grain development. Beyond GA, multiple phytohormones fine-tune this balance: CKs enhance photosynthetic capacity by upregulating Rubisco activity and photosynthetic gene expression while delaying leaf senescence; ethylene modulates grain filling dynamics, particularly under stress; and ABA promotes assimilate partitioning to developing grains and enhances stress resilience during ripening. Together, these hormonal networks form an integrated system that dynamically coordinates C production (source) and utilization (sink) across developmental stages and environmental conditions.

There is compelling evidence supporting the positive role of the senescence-antagonizing CKs in source–sink regulation. CKs are typically synthesized in the roots and transported acropetally to the shoots. Tissues with elevated cytokinin levels function as strong metabolic sinks. The onset of post-flowering leaf senescence can be attributed to the reallocation of CKs from the leaves to the developing seeds, thereby enhancing the seeds' efficiency as nutrient sinks relative to senescing leaves. Premature leaf senescence significantly diminishes photosynthetic activity and severely reduces crop yield. Cytokinin oxidase/dehydrogenase (CKX), encoded by multigene families, is the primary enzyme responsible for degrading CKs via irreversible cleavage of their N^6 side chains to yield adenine or adenosine. Modulating CKX can alter cytokinin homeostasis. *OsCKX11* acts antagonistically to CKs in regulating senescence, thereby delaying leaf senescence and coordinately modulating source–sink dynamics in rice (Zhang et al. 2021).

Although the significance of long-distance auxin and sugar transport for plant growth and development has long been acknowledged, the mechanistic crosstalk between auxin signalling and sucrose distribution remains poorly understood. Zhao et al. (2022) identified a hierarchical regulatory module in rice comprising OsARF18, OsARF2, and OsSUT1, through which auxin orchestrates carbon allocation across source–sink interfaces. Specifically, OsARF2 activates *OsSUT1* transcription by directly binding to sugar-responsive elements (SuREs) in its promoter region. Conversely, OsARF18 exerts dual repressive control: it inhibits *OsARF2* expression by binding to an auxin-responsive element (AuxRE) in the *OsARF2* promoter, and simultaneously suppresses *OsSUT1* expression by targeting SuREs within the *OsSUT1* promoter.

3 Indicators of Source–Sink Relation

Various physiological indicators, such as the harvest index and shoot-to-root ratio, serve as proxies for source–sink dynamics. However, their correlation with final crop yield is not consistently proportional. Indeed, altering any single metric may produce variable, context-dependent outcomes, underscoring the complexity and plasticity of source–sink relationships.

3.1 Shoot to Root Ratio

The coordination between the shoot and the root is of utmost importance for plants to thrive. The shoot captures solar energy via photosynthesis, whereas the root absorbs water and minerals from the soil. The vascular system, composed of xylem and phloem, provides structural support to the plant body and facilitates the transport of nutrients and signalling molecules between the shoot and the root (Fig. 5.6).

Hormones have been comprehensively investigated as key regulators of shoot–root communication. Root-derived ABA and CKs are transported in an acropetal direction to the shoot through the xylem. Conversely, auxins and JAs move in a basipetal direction from the shoot to the root via the phloem, together with photosynthates. The case of GAs is more intricate, as they are synthesized in both the shoot and the root, particularly in rapidly growing tissues, such as young shoot apices, developing seeds, root tips, and expanding leaf vascular systems. After synthesis, GAs are distributed throughout the plant via bidirectional transport.

The bidirectional hormonal signalling circuit enables efficient and instantaneous communication of physiological status between the shoot and the root. For example, when the roots detect sufficient external water, they can transmit an upward GA signal to the above-ground parts, indicating that it is an appropriate time for rapid vertical growth. When the shoots exhibit high photosynthetic intensity and strong productivity, they can send a downward GA signal to the roots, suggesting that it is time for the roots to penetrate deeper into the soil to acquire more minerals. In addition, small peptides, proteins, and RNA molecules have also been found to be mobile in the vascular tissues, coordinating nutrient transport and stress signalling between the shoot and root (Li et al. 2021).

The shoot to root ratio (SRR) has long been acknowledged as a crucial physiological characteristic that influences crop productivity. A multitude of investigations, encompassing empirical field trials and mechanistic modelling methods, have

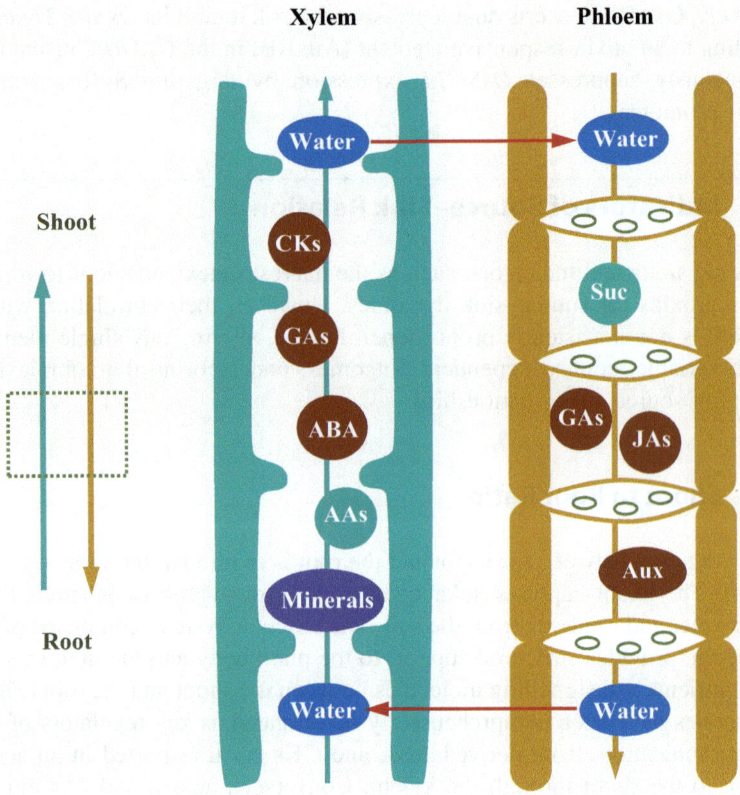

Fig. 5.6 Exchange of nutrients and hormones between shoots and roots. Hormones function as long-distance signals between the shoot and the root. (1) Root-derived hormones are primarily transported acropetally through the xylem. Abscisic acid (ABA) is synthesized in the roots and translocated upward via the transpiration stream to the shoot, where it regulates stomatal closure and other physiological responses in aerial tissues. Cytokinins (CKs), produced in the root, are also exported upward through the xylem. Gibberellins (GAs) are predominantly synthesized in rapidly growing regions of the plant, such as shoot apical meristems and root tips. GAs can be transported basipetally or acropetally via the phloem or xylem, depending on developmental signals. (2) Shoot-derived hormones primarily move downward through the phloem. Auxins, such as indole-3-acetic acid (IAA), are synthesized in young leaves and the shoot apex and subsequently loaded into the phloem for transport to the root, where they regulate root branching and elongation. Jasmonates (JAs) are synthesized mainly in the chloroplasts and peroxisomes in leaves. They are transported basipetally via the phloem to the root, modulating root growth and stress responses. Aux auxins, AAs amino acids, Suc sucrose

been carried out to determine its optimal value for maximizing yield (Chang and Zhu 2017). For example, in rice, artificially limiting tillering to preserve only the primary stem, a manipulation that effectively reduces SRR, has been demonstrated to enhance stomatal conductance, photosynthesis, total DM production, and panicle size under well-watered conditions (Nada and Abogadallah 2016). Conversely, studies on wheat in water-limited environments indicate that reducing the root system, thereby increasing SRR, leads to significant improvements in both water use efficiency and final grain yield (Ma et al. 2010). Furthermore, breeding progress in wheat has led to modern high-yielding varieties showing a consistent increase in SRR, especially in rainfed Mediterranean agroecosystems (Siddique et al. 1990). Collectively, these context-dependent responses highlight the complexity of how SRR regulates yield determination across different species and environmental scenarios.

3.2 Harvest Index

The *harvest index* (HI) is defined as the ratio of the biomass of the harvested product to the total aboveground crop biomass, typically expressed as a percentage. Biomass typically refers to the aboveground parts, except when the harvested organ is situated underground, like potato tubers and sweet potato roots. HI is not an inherent and immutable characteristic of a species; rather, it varies according to genotypes, management practices, and environmental conditions, which reflect the allocation of photosynthates among reproductive, storage, and vegetative organs. A clear understanding of the allocation patterns across plant organs facilitates the development of more precise breeding and agronomic strategies tailored to each crop species (Sadras et al. 2024).

For the three major cereal crops, the HI of contemporary high-yielding cultivars generally ranges from 0.4 to 0.6. This range reflects decades of intensive breeding efforts, which have led to a substantial increase in the HI compared to the historical levels of around 0.3. In temperate regions, wheat and barley typically attain HI values within the range of 0.4–0.5, and certain elite cultivars can approach 0.6. Under intensive irrigated conditions, rice shows an HI of approximately 0.4–0.55. Modern maize hybrids regularly reach HI values between 0.4 and 0.55. Many legume and oilseed crops present lower HI values, generally falling within the range of 0.2–0.4. Among them, soybean allocates a relatively larger proportion of biomass to seed production, leading to a comparatively high HI, often close to 0.4. In contrast, oilseed rape has a significantly lower HI, usually around 0.3–0.35. Crops with underground harvested organs, such as sugar beet (with an HI of 0.5–0.7) and potato (with an HI of up to 0.8), also exhibit relatively high HIs. Fodder crops can achieve exceptionally high HI values, up to 0.9, as nearly all aboveground biomass is harvested and utilized.

When implementing crop breeding, it is crucial to consider the inherent HI ceiling determined by the harvestable organ. In cereal crops, yield improvements are often restricted by the contribution of biomass transferred from stems to grains. Evidence indicates that the increases in HI are approaching a plateau and are improbable to surpass 0.6. Nevertheless, root, tuber, and leaf crops may still have greater potential for enhancing HI through the optimization of sink strength. Additionally, HI is influenced by multiple factors, including rainfall patterns, nitrogen fertilization, and temperature. It has been noted that both HI and yield decline under conditions of heightened drought or salt stress. However, there may be circumstances where yield and total biomass production increase while HI decreases. As a key indicator of the synchronization between sources and economic sinks, its fundamental characteristics are still not comprehensively understood. For example, the exact way in which resources conserved during stem development are reallocated to support grain filling remains poorly understood.

The genetic and physiological foundation of crop yield is intricate; nevertheless, the significant increments in cereal yields during the second half of the twentieth century were attained via relatively uncomplicated strategies, mainly by augmenting the HI. In rice and wheat, the yield potential was elevated by breeding GA-insensitive cultivars with a higher HI, along with enhanced nitrogen input. These methods demonstrated efficacy in cereals but were not directly applicable to other crop varieties, such as legumes, owing to a variety of biological and agronomic limitations (Ahn et al. 2023).

First, cereals demonstrate a determinate growth habit that is characterized by robust apical dominance, accompanied by a clear transition from the vegetative to the reproductive phase. Conversely, legumes present an indeterminate growth pattern, featuring a substantial overlap between the vegetative and reproductive phases, which restricts the efficacy of dwarfing genes in enhancing the HI. Secondly, cereal grains are borne on terminal reproductive structures that emerge at the SAM. Mutations influencing GA signalling generally modify the overall plant height or internode elongation, yet they have minimal impacts on the fundamental architecture of the terminal inflorescence. This makes the inflorescence relatively insensitive to GA mutations that predominantly affect vegetative growth, allowing breeders to adjust plant height without significantly reducing the grain number. In contrast, legume grains develop within pods that are sequentially formed at multiple nodes along the stem. The nodal distribution of pods implies that alterations in GA signalling may concurrently affect both vegetative and reproductive development, potentially affecting pod number, placement, and overall yield. Finally, cereal seeds are primarily composed of carbohydrates, whereas legume seeds contain more than just carbohydrates. The accumulation of proteins and lipids involves distinct biosynthetic and remobilization pathways. Therefore, breeding or management strategies must consider both carbon and nitrogen supply to optimize the balanced seed composition and achieve a higher HI.

3.3 Dynamics of Dry Matter in Internodes

In crop plants, an *internode* refers to the stem segment located between two consecutive nodes, which are the sites of leaf, branch, or flower emergence. Internode development in cereal crops can be divided into four distinct phases.

1. Tissue Differentiation: The newly formed internode undergoes a tightly regulated sequence of cell division and differentiation, giving rise to the primary tissue systems of the stem, namely, epidermis, vascular tissues (xylem and phloem), and ground tissues (parenchyma and sclerenchyma). Vascular bundle differentiation plays a pivotal role in determining internode structural strength and conductive capacity, thereby influencing the stem's mechanical support and efficiency in nutrient transport in later developmental phases.
2. Elongation and Initial Thickening: Rapid longitudinal growth is driven by cell division in the intercalary meristem at the base of the internode, followed by extensive cell elongation; in the upper region, a transition from cell division to elongation occurs simultaneously with the initiation of cell wall thickening.
3. Dry Matter (DM) Accumulation: Sclerenchyma cells deposit large amounts of cellulose, hemicellulose, and lignin in their thickened walls, while parenchyma cells accumulate soluble carbohydrates, including starch, fructans, and sucrose.
4. Dry Matter Export: Stored carbohydrates are hydrolyzed into sucrose, which is then loaded into the phloem and transported to sink organs such as developing leaves and reproductive structures. The four stages are not strictly distinct and often exhibit overlap during the actual growth process. For instance, a limited degree of DM accumulation may commence as early as the elongation stage.

DM translocation during the material export phase constitutes a central component of the source–sink interaction, directly impacting final yield and quality. Xu et al. (2021) sampled internodes separately and increased sampling intensity from 21 days before anthesis (DBA) to 60 days after anthesis (DAA), with weekly intervals before anthesis and 10-day intervals thereafter. Through two-year field and pot experiments involving both japonica and indica rice cultivars under two nitrogen fertilization regimes, they identified a consistent pattern of internode DM dynamics during grain filling across contrasting growing conditions, as depicted in Fig. 5.7. Accordingly, they proposed that internode DM can be considered as an integrative index for the source and sink relationship.

Phase I A steady upward trend in internode DM until 10 DAA. Because of the swift extension of the stem, photosynthates continuously accumulate in the internodes from anthesis to 10 DAA. Therefore, the internode functions as a competing sink organ during the early grain-filling stage, a period critical for endosperm development (Lawlor and Paul 2014).

Phase II A significant reduction in internode DM between 10 and 30 DAA. This can be explained by the high demand from the grain-filling process. By the end of this phase, up to 95% of DM in superior grains (at maturity) and at least 80% in inferior grains has been accumulated, indicating that the grain-filling process is

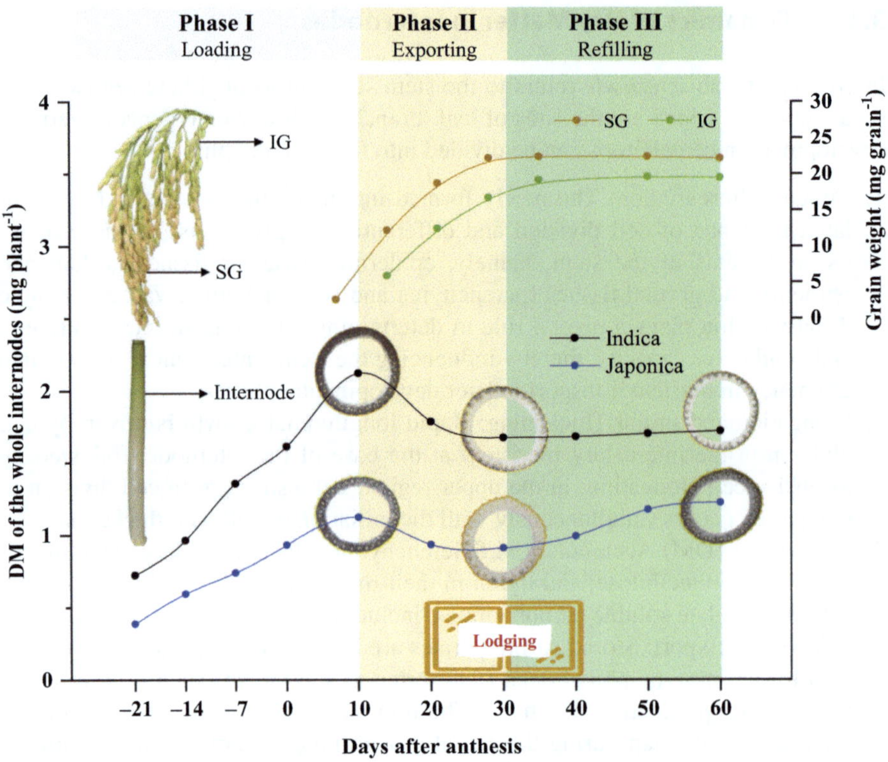

Fig. 5.7 Diagram of the dynamics of internode dry matter (DM) from 21 days before anthesis to maturity. It should be noted that indica rice differs from japonica rice in the recovery of DM in internodes, partly due to the reduced photosynthetic capacity of leaves that undergo early senescence. The embedded diagram illustrates the histochemical localization of starch in transverse sections of the internode, where bluish-purple staining indicates starch deposition revealed by I_2-KI treatment. The icon of a golden window spanning 20 to 40 DAA indicates the lodging susceptibility window of japonica rice. Inferior grains (IG), located on the distal secondary branches of the lower rachis in the panicle, typically exhibit delayed flowering, reduced assimilate allocation, slower grain filling, and smaller grain size. In contrast, superior grains (SG), situated on the apical primary branches of the upper and middle rachis, flower earlier, receive greater nutrient supply, and generally attain higher grain weight and improved quality. (Redrawn from Xu et al. (2021))

nearly 90% complete. Similarly, leaves and sheaths also undergo substantial DM loss during this phase. The hollow stem, combined with the heavy panicle, significantly increases the risk of lodging. Thus, the last 10 days of this phase and the first 10 days of the following phase constitute a critical window for lodging susceptibility, particularly in japonica rice.

Phase III A notable recovery in internode DM. Superior grains are completely filled, while inferior grains exhibit a slower rate of accumulating the remaining 20% of DM. Consequently, the recovery of DM during the late grain-filling stage can be ascribed to the surplus photosynthates from leaves that surpass the assimilate requirements of growing grains. Furthermore, the degree of internode DM recovery is more significant in japonica rice compared to indica rice, which might be partially explained by the earlier occurrence of leaf senescence in indica cultivars.

As depicted in Fig. 5.7, when the 60-day rice grain-filling phase is divided into two 30-day periods, it is evident that by the end of the first 30 days, superior grains have already accumulated approximately 95% of their final DM, whereas inferior grains have reached at least 80% of theirs. This indicates that only ~10% of the grain yield is formed during the second month, making this period less significant for rice yield formation. Nevertheless, the final month is still vital for the timely establishment of subsequent-season crops like wheat, implying a possible opportunity for agronomic intervention during the late growth phase in highly intensive and productive double-cropping systems. So far, limited attention has been paid to shortening the rice growing period, especially the late grain-filling phase. In contrast, there has been a growing trend to extend the duration of rice growth to maximize yield potential, as evidenced by the release of three benchmark cultivars across different decades in the Yangtze River Delta region of China (Xu et al. 2021). The lower-yielding cultivar WYJ-3, released in the 1990s, matures approximately 1 week earlier than the high-yielding cultivar NJ-8, released in the 2020s. This implies that the selection for increased yield may have unintentionally led to a loss of early maturity during crop improvement. Consequently, breeding programs aimed at developing high-yielding cultivars should explicitly incorporate earliness as a key selection criterion and adjust their strategic priorities accordingly. For agronomists, plant growth regulators serve as a valuable tool for intervention at the late growth phase. Ethylene-releasing agents, such as ethephon, have been shown to effectively promote earlier maturity in cotton cultivation.

4 Relevance to Crop Breeding and Management

The source–sink relationship represents a fundamental physiological mechanism influencing crop yield formation, thus offering crucial scientific insights to guide both breeding programs and agronomic management practices.

4.1 The Critical Periods of Yield Formation

For *grain crops*, the yield is determined by multiplying the quantity of grains or seeds per unit area by their mean weight. Accumulating evidence indicates that the variation in grain yield is more dependent on the grain number than the grain weight. The significance of the grain number has triggered numerous studies on various crops, aiming to identify the *critical periods* in the crop cycle when the grain number is susceptible to source attenuation.

The critical periods for yield formation are defined as specific developmental phases during which environmental stresses, such as frost, heat, and drought, exert the most significant influence on the determination of grain number. These periods have been identified predominantly by adjusting the source–sink ratio through shading treatments applied during different developmental stages. Divergent responses have been observed across crop species, primarily attributed to the variations in their phenology and growth habits (Savin and Slafer 2019). In determinate crops like rice, wheat, and maize, grain number is generally determined prior to the flowering time, whereas grain weight is established subsequently, with only a slight overlap between these two developmental stages. Conversely, in semi-determinate and indeterminate species, such as legumes, in which the basal nodes have already reached the pod-setting stage while the upper nodes are only beginning anthesis, seed filling continues beyond anthesis. Consequently, there is a substantial overlap between the periods of seed number growth and seed weight increase. Therefore, in the case of cereals, the critical period begins during the late vegetative stage, which includes stem elongation, extends through the early anthesis phase, and ends shortly after flowering. For legumes, due to their indeterminate growth pattern, the critical period shifts to the pod-setting stage and persists until the early seed-filling stage.

During the grain filling, cereals respond slowly to changes in the source–sink ratio. Therefore, management strategies aimed at enhancing grain yield should ensure that crops pass through their most sensitive developmental stages prior to anthesis, without coinciding with periods of extreme stress. Moreover, high-yielding practices should focus on optimizing plant growth and development during these critical phases by means of appropriate irrigation and balanced mineral nutrition. It is important to note that the actual yield is determined not only by the potential sink capacity established during the critical period but also by the post-anthesis assimilate supply from the leaves. Post-anthesis abiotic and biotic stresses can accelerate leaf senescence and reduce assimilate production, leading to yield losses due to inadequate grain filling.

4.2 Source or Sink Limitation?

The development and enlargement of sink organs may be constrained either due to the inadequate availability of photosynthates from source tissues (source limitation) or because of the limited capacity of the sinks to absorb and metabolize these compounds (sink limitation). For over a century, scholars have engaged in a vigorous debate regarding whether crop yield is primarily constrained by source or sink limitation. Decades of research on elevated CO_2 have demonstrated that enhancing source activity by improving the photosynthetic rate frequently fails to lead to proportional increases in yield. Similarly, modifying sink strength does not invariably result in higher yields. Hence, it is essential to assess on a case-by-case basis whether the limiting factor for crop yield is associated with source or sink organs.

Current evidence indicates that wheat grain weight is seldom restricted by the supply of assimilates during the grain-filling stage, even under sub-optimal growing conditions, such as the Mediterranean climates. As a result, wheat exhibits significant sensitivity to source reduction solely under highly specific and ecologically rare scenarios, which do not represent most global wheat-growing regions. Parallel investigations on other temperate cereals, including barley and triticale, under experimentally controlled source–sink ratios consistently reveal similar patterns: grain yield shows a weak response to source limitation but a strong response to sink manipulation. This consistency implies predominant sink limitation on the yield formation process. Although co-limitation by both source and sink has been reported, contemporary high-yielding genotypes of temperate cereals are sink-limited once flowering is completed.

The significance of source strength in determining potato yield is demonstrated by genetically manipulating the tubers' ability to metabolize sucrose and accumulate starch (Smith et al. 2010). In one experiment, expression of yeast invertase in the tuber apoplast enhanced sucrose catabolism, thereby augmenting sink strength. In these invertase-upregulated potatoes, the number of tubers was reduced compared with that of the WT, and the tubers were larger. In another experiment, sink starch biosynthesis was reduced by downregulating ADP-glucose pyrophosphorylase (AGPase) activity, producing a greater number of smaller tubers compared with the WT. While changes in sink strength significantly influenced the competitive dynamics among individual tubers, they had a minimal impact on the overall tuber biomass accumulation. Under conditions of low sink strength attained through reduced AGPase activity, plants produced numerous small tubers. Conversely, under high sink strength triggered by increased invertase activity, intense competition for available sucrose resulted in the development of fewer yet larger tubers. Nevertheless, the overall tuber weight was only slightly affected. It decreased by merely 30% compared to the WT under low sink strength and remained unchanged under high sink strength. Therefore, in this scenario, source organs played a more dominant role than sink organs in determining the allocation of sucrose to potato tubers.

5 Genetic Modification

Since the advancements of the Green Revolution in the 1960s, agriculture has not yet witnessed another substantial increase in crop yield. As traditional breeding approaches reach a plateau, there is an urgent necessity to explore frontier technologies derived from fundamental plant biology to enhance *genetic gains*. Researchers aiming to improve crop yield have concentrated on three critical aspects: increasing photosynthetic assimilation in the leaf (source activity), enhancing the loading and unloading of sucrose in the phloem (transport or flow), and improving carbohydrate storage in sink organs (sink activity). In the past, most studies concentrated on the first aspect. Nevertheless, in recent decades, attention has been increasingly directed towards the other two aspects (Wen et al. 2022).

5.1 Source Activity

5.1.1 Improving Leaf Photosynthetic Activity

Photosynthesis stands as one of the most precisely regulated biological processes, characterized by multiple interwoven and finely balanced regulatory layers. This enables plants to function and thrive under highly variable environmental conditions. Due to this multilevel complexity, optimizing photosynthesis has remained a longstanding challenge. Much of current research focuses on identifying natural genetic variations that enhance photosynthetic efficiency. However, the inherently limited genetic variation in photosynthetic enzymes and related components hinders efforts to optimize photosynthesis, particularly in major crops.

Fernie et al. (2020) presented a comprehensive overview of various strategies that directly or indirectly impact photosynthesis. These strategies encompass: (1) alterations to individual proteins within the CBC or cytochrome complex; (2) enhancements to C-concentrating mechanisms; (3) approaches for introducing synthetic bypasses for photorespiration (see the Sect. 5.1.2); (4) strategies designed to optimize photosynthesis in dense crop stands, where self-shading results in a decrease in photosynthetic activity and rapid variations in light intensity; and (5) methods for enhancing the responsiveness of stomata, which often fail to adapt to fluctuating light.

Under intense solar irradiance, leaves protect themselves from photodamage by converting the excess absorbed light energy into thermal energy. Nevertheless, the transition from this protective heat-dissipation mode to active photosynthetic energy utilization is relatively slow, specifically slower than the movement of clouds overhead. As a result, even after a leaf is exposed to shaded conditions, residual nonphotochemical quenching persists, temporarily reducing photosynthetic performance. De Souza et al. (2022) developed a biotechnological approach to enhance the recovery process mediated by nonphotochemical quenching in soybean. Field experiments have revealed that transgenic plants demonstrated significantly enhanced photosynthetic performance under dynamic light environments, leading to an increase in seed yield of up to 33%. Significantly, this yield increase took place without any discernible alterations in seed protein or oil concentrations. These findings emphasize the potential of optimizing photosynthetic efficiency as a feasible and sustainable approach to improve agricultural productivity.

5.1.2 Reducing the Photorespiration Loss

Plants have evolved a diverse range of biochemical strategies to transform atmospheric CO_2 into stable organic molecules, with sucrose being the primary end-product. Most plant species, including staple crops such as rice, wheat, tomato, and potato, depend on the C3 photosynthetic process for this carbon assimilation. In C3 carbon fixation, CO_2 is initially incorporated into 3-phosphoglycerate (3-PG), a three-carbon compound, through the catalytic action of Rubisco. However, Rubisco also possesses oxygenase activity, resulting in the competition between O_2 and CO_2 for the active site. This leads to photorespiration, a process that generates the toxic intermediate 2-phosphoglycolate (2-PG), which must be metabolically recycled into 3-PG through an energy-consuming pathway. Overall, photorespiration significantly reduces the efficiency of carbon fixation in C3 plants, resulting in photosynthetic losses exceeding 30% and yield reductions surpassing 50% (Walker et al. 2016). As a result, photorespiration is generally regarded as an energetically wasteful metabolic pathway and is thus considered a prime candidate for genetic and biochemical interventions aimed at enhancing crop yield.

Synthetic biology, a rapidly advancing interdisciplinary field, shows great potential for engineering novel biochemical pathways in plants with precise and predictable outcomes, thus revolutionizing traditional breeding strategies (Batista-Silva et al. 2020). Employing synthetic biology approaches, several photorespiratory bypasses have been constructed to efficiently metabolize 2-PG and concentrate CO_2 in chloroplasts, thus markedly suppressing photorespiration. South et al. (2019) introduced synthetic glycolate metabolic pathways into tobacco chloroplasts to suppress their glycolate efflux, leading to a 20% increase in photosynthetic quantum yield. Field trials demonstrated that biomass productivity increased by more than 40% in the homozygous transgenic lines. Likewise, transgenic rice lines incorporating an innovative photorespiratory bypass pathway demonstrated substantial improvements in photosynthetic performance, along with 23%–31% increases in total biomass and 18%–25% higher nitrogen content in leaves relative to the WT (Shen et al. 2019).

Beyond its function of diverting 2-PG from the C fixation pathway, photorespiration is also essential for regulating N metabolism. Suppressing secondary N assimilation remains a critical challenge in genetically engineering photorespiratory pathways. Chen et al. (2025) developed a C and N metabolism-coupled photorespiratory bypass, namely the GCBG bypass, in rice chloroplasts. The GCBG bypass modified the levels of key metabolites and their derivatives related to natural photorespiration, thus enhancing net photosynthesis. Metabolomic analyses indicated that oxaloacetate derived from glycolate had a significant impact on amino acid metabolism and carbohydrate biosynthesis. The GCBG plants showed an average yield increase of 19.0% compared to the WT under field conditions. Furthermore, the enhanced N uptake in GCBG plants compensated for 44.1% of the yield loss observed under N-deficient conditions. This study offers a promising strategy for breeding elite crops with simultaneously improved photosynthetic performance and NUE.

5.1.3 Exploiting the Photosynthesis of Non-Leaf Organs

Over the past several decades, a wide variety of approaches have been investigated to enhance crop productivity by improving leaf-level photosynthetic performance. However, as emphasized by Araus et al. (2021), most of these interventions have yielded only marginal or limited agronomic benefits. Even when measurable improvements in photosynthetic capacity occur, they are unlikely to increase grain or biomass yield by more than 10%–15%. Furthermore, historical analyses indicate that the observed improvements in photosynthetic performance primarily reflect changes at the whole-plant scale. These changes are driven not only by leaf-level alterations but also by coordinated enhancements in both photosynthetic (e.g., leaves) and non-photosynthetic (e.g., stems, petioles) organs.

Although the precise role of non-leaf tissues in carbon fixation and grain yield remains incompletely characterized, their potential to enhance crop productivity has received relatively little attention. A growing body of evidence indicates that reproductive organs, such as flowers, fruits, and seeds, possess functional photosynthetic capacity. Photosynthesis occurring in the ears of wheat and barley has been shown to substantially support grain development, even under suboptimal environmental or management conditions. As a result, ear photosynthesis is increasingly regarded as a valuable physiological trait for cereal improvement. Targeting photosynthetic activity in non-foliar organs thus emerges as a novel and promising breeding strategy. This concept is further supported by Simkin et al. (2020), who reported elevated photosynthetic rates not only in leaves but also in the ears of transgenic wheat plants engineered to overexpress the CBC enzyme SBPase. Significant variability has been observed in the ears and panicles of C3 cereal crops, offering avenues for leveraging this inherent diversity in modern breeding programs. Moreover, genetic differences in ear-level resilience to water deficit have been empirically confirmed, suggesting a valuable yet underutilized trait for developing climate-resilient wheat ideotypes (Li et al. 2017).

5.2 Sink Activity

Although genetic modification has clearly enhanced photosynthetic efficiency, further gains are likely to require concurrent modifications to sink tissues to fully unlock the potential of improved source capacity. Only when sink strength matches source capacity can these efficiency gains be effectively translated into increased yield (Batista-Silva et al. 2020; Rossi et al. 2015).

Cell wall invertase (CWIN) determines the capacity of C unloading from the maternal vascular tissues to the grains. Overexpression of CWIN leads to an increase in grain size, number, and starch content, and consequently boosts the grain yield in rice and maize (Li et al. 2013a). Lou et al. (2024) developed high-efficiency prime-editing tools to precisely insert a 10-bp heat-shock element (HSE) into the promoters of CWINs in rice. The insertion of HSE enables CWINs to exhibit heat-responsive upregulation in both controlled and field environments, thereby enhancing C allocation to the grain. This results in a 25% increase in grain yield and recovers up to 41% of heat-induced grain losses. In addition to invertases, enhancing sucrose synthase SUS activity represents a valuable strategy for increasing starch accumulation in sink organs. Despite conflicting results reported in *Arabidopsis*, cotton exhibits a strong correlation between seed and fibre growth with SUS activities, while the *ectopic expression* of *StSUS4* in maize led to enhanced starch accumulation in seeds and a more favourable ratio of amylose to amylopectin (Li et al. 2013b).

Downregulation of glucan, water dikinase (GWD), the primary enzyme responsible for starch phosphorylation specifically in the endosperm, leads to lower starch phosphate levels while simultaneously enhancing grain size and overall yield in wheat (Ral et al. 2012). Elevated expression of histone acetyltransferase (HAT1), which enlarges spikelet hulls by promoting cell division, enhances plant biomass and grain weight in rice (Song et al. 2015). Ectopic expression of a sugar-inducible transcription factor, SUSIBA2, promotes carbon allocation to aboveground biomass rather than to roots, thereby increasing starch content in stems and seeds and consequently improving overall yield. Additionally, it reduces root exudates that positively contribute to methane emissions in rice paddies (Su et al. 2015).

Research on maize has significantly advanced our understanding of how plant hormones regulate sink activity. ABA functions as a central regulator of sink strength: it enhances phloem unloading into developing kernels and directly upregulates key genes involved in starch biosynthesis, thereby improving both grain yield and quality. The transcription factor ZmbZIP29, a basic leucine zipper protein, shows high expression during the early stages of endosperm development in maize and functions cooperatively with ZmABI19 to modulate the expression of O2, a key regulator of endosperm starch accumulation. According to Yang et al. (2022), ABA treatment significantly enhances the transcript levels of *ZmbZIP29*, *ZmABI19*, and *O2*. Moreover, under ABA signalling, ZmbZIP29 and ZmABI19 act synergistically to activate transcription from the *O2* promoter. Phosphoproteomic profiling by LC-MS/MS revealed site-specific phosphorylation events at Thr57 in ZmABI19, Thr75 in ZmbZIP29, and Thr387 in O2. Subsequently, SnRK2.2 was confirmed as the kinase directly responsible for this phosphorylation. These results demonstrate that ABA-triggered phosphorylation serves as a pivotal regulatory switch for the expression of critical endosperm-filling genes in maize.

5.3 Nutrients Transport

Manipulating nutrient transporters provides an alternative approach to regulate the source–sink relationship. The apoplastic pathway in the leaf vasculature loads sucrose into the phloem STE-CC complex, a process mainly mediated by SUTs/SUCs. Results from antisense inhibition of *SUT1* in tomato, knockout or upregulation of *SUT1* in rice, and overexpression of *SUT1* in pea collectively suggest that SUTs/SUCs play a crucial role in phloem loading and thereby influence crop yield (Wang et al. 2015; Wen et al. 2022).

In *Arabidopsis*, the *amino acid permease AAP8*, specifically localized in leaf phloem tissues, mediates the uptake of amino acids into the phloem stream, facilitating their systemic allocation throughout the plant. Loss-of-function *aap8* mutants exhibit impaired phloem loading of amino acids, leading to significant reductions in silique development and seed set (Santiago and Tegeder 2016). In rice, overexpressed *OsAMT1.1*, a central ammonium transporter in the *OsAMT1* family, markedly enhances root ammonium influx capacity. This genetic enhancement translates into improved vegetative growth and increased grain yield, under both favourable and limiting NH_4^+ availability (Ranathunge et al. 2014).

The phloem in the vascular bundles of the panicle neck plays a vital role in delivering photosynthates from source leaves to developing grains, and its efficiency is closely linked to grain yield. By conducting a *genome-wide association study (GWAS)* using 386 rice accessions, Zhai et al. (2023) successfully identified and characterized *large vascular bundle phloem area 4* (*LVPA4*), a gene critical for regulating panicle vascular tissue development. Phenotypic analyses were performed on the near-isogenic line NIL-*LVPA4*^{LT} in the high-yielding indica cultivar Teqing, as well as on overexpression lines with the Lemont allele of *LVPA4*. Both lines exhibited a marked increase in LVPA, accompanied by notable increases in flag leaf size. NIL-*LVPA4*^{LT} showed a 7.6%–9.6% yield increase, primarily attributed to a highly developed vascular system that efficiently transports photosynthates to the developing grains.

In the chalazal zone of the ovule, where the maternal phloem vessels terminate, a novel structure termed the phloem end structure (PES) has been identified as a regulatory checkpoint for the import of nutrients from the maternal plant into the seeds. As reported by Liu et al. (2025), the accumulation of callose physically seals the PES in unfertilized, fully developed ovules, thus rendering this nutrient gateway inactive. Nevertheless, callose is enzymatically degraded upon fertilization, leading to the structural remodelling of the PES and the activation of nutrient transfer to the nascent seed. A mutant in an ovule-expressed β-1,3-glucanase gene (*AtBG_ppap*) exhibited incomplete callose degradation after fertilization and produced smaller seeds, seemingly due to the partially closed state of the PES. In contrast, an *AtBG_ppap* overexpression line produced larger seeds because of continuous callose degradation, fully opening the PES gate to enable nutrient loading into the seed. Intriguingly, the rice ovule possesses a structure highly like the PES. As a proof of concept, Liu et al. (2025) generated a line overexpressing the *Arabidopsis*-derived *AtBG_ppap* gene under the control of the 35S promoter. This overexpression line exhibited a 9% higher yield compared to the WT, indicating potential conservation of this structure-dependent mechanism of nutrient flow at least among angiosperms.

6 Agronomic Intervention

Agronomic practices such as irrigation, fertilization, and the application of plant growth regulators significantly influence source–sink dynamics by modulating the developmental and metabolic processes in plant organs. For instance, nitrogen fertilization during the panicle differentiation stage promotes spikelet formation, leading to larger panicles or spikes. Similarly, implementing alternate wetting and drying after flowering enhances grain filling efficiency, improves the HI, and increases overall yield, largely due to the maintenance of robust root systems. When integrated with other improved management practices, these targeted interventions can substantially boost crop productivity.

As detailed in the preceding section, T6P is an endogenous sugar signal in plants that regulates source–sink relationships and thereby enhances crop yield. However, it cannot be directly introduced into crops or fully genetically controlled. The leaf cuticle is a thin polymeric layer composed of cutin, a polyester derived from long-chain fatty acids, intricately integrated with pectin and waxes. Due to the presence of carboxyl groups on the fatty acid chains of cutin, the outer surface of most leaves carries a net negative charge. This negative surface charge effectively repels other negatively charged entities, such as airborne microbes and charged solutes, functioning as an electrostatic barrier. T6P is a highly polar molecule bearing a negative charge. When applied to the leaf surface, it undergoes electrostatic repulsion like the forces that restrict pathogen adhesion, thus impeding its access to the cutin matrix. Moreover, the cutin layer is both hydrophobic and relatively thick, which further restricts the diffusion of charged and hydrophilic molecules through the polymer network. As a result, negatively charged T6P molecules encounter significant resistance when attempting to traverse the cutin layer, making direct uptake by epidermal cells inefficient without the assistance of carriers, surfactants, or physical modification of the cuticle.

Recently, Griffiths et al. (2025) developed a plant-permeable, sunlight-activated T6P signalling precursor, DMNB-T6P, for ready uptake into leaf tissues. The scalable synthesis of DMNB-T6P facilitated field trials conducted over a four-year period in Argentina, where current wheat yields are close to the global average and precipitation patterns are typical of global agriculture. Spaying DMNB-T6P ten days after anthesis at a dose of 1–2 mM led to yield increases in all four years (ranging from 9.3% to 12.7%), regardless of whether the years were wet or dry. Importantly, no additional fertilizer was applied, and grain protein concentration remained stable despite higher grain yields. In addition, Griffiths et al. (2025) estimated the synthesis cost of DMNB-T6P to be $300 per ton, which translates to a few cents per hectare when applied at optimal rates. If this treatment leads to an average yield increase of 10.4%, the global wheat yield (averaging 3.6 tons per hectare) would rise by 0.37 tons per hectare, generating an additional $116 per hectare (based on February 2023 wheat prices). This corresponds to a global value increase of $25.6 billion (assuming 221 million hectares of wheat were grown in 2021). This significant advancement in agricultural practices offers a promising avenue for sustainably increasing global cereal yields amid ongoing climate change.

7 Concluding Remarks and Perspectives

Due to the complex interplay between source and sink and their susceptibility to various whole-plant factors, manipulating a single trait or cellular process related to either source, flow, or sink often leads to non-linear responses. This complexity might account for the limited success achieved in current biotechnological efforts aimed at improving crop yields through the manipulation of cellular-level source or sink functions.

Over the past several decades, functional genomic approaches have substantially enhanced our understanding of the genetic and molecular mechanisms underlying source–sink interactions. However, several crucial questions regarding how source and sink organs metabolically regulate each other remain unanswered (Fernie et al. 2020). (1) What strategies can most effectively enhance photosynthetic capacity without eliciting inhibitory feedback from sink tissues? Additionally, what factors govern the kinetics of sucrose loading into the phloem, specifically the availability of mesophyll sucrose or the activity of specific transporters? (2) Do long-distance signals from sink organs regulate phloem loading, and how is sucrose transport coordinated with that of amino acids and minerals? (3) Which molecular mechanisms and signalling pathways coordinate the functional balance between sources and sinks, and how do these regulatory components respond to abiotic and biotic stresses?

In addition to these metabolic mechanisms that regulate source–sink interactions, an additional layer of complexity, specifically the developmental aspects, also needs to be considered (Rosado-Souza et al. 2023). T6P has emerged as a central integrator of sugar availability and developmental regulation in plants. In the model species *Arabidopsis* and likely across all angiosperms, T6P is predominantly synthesized in the vasculature, a strategically positioned site that enables the coordination of systemic sugar status with whole-plant developmental responses. Accumulating evidence indicates that T6P is capable of intercellular movement and may be transported between tissues, a hypothesis supported by the expression patterns of enzymes responsible for T6P synthesis and the localization of T6P targets, such as SnRK1. Acting both locally and systemically, T6P may further coordinate source metabolic activity with sink development by interacting with key regulatory components such as FT (Fichtner 2025). However, key mechanistic questions remain: What are the molecular mechanisms underlying intercellular T6P movement? Moreover, how does T6P function as a signalling hub that coordinates metabolic homeostasis with developmental transitions in plants?

Summary

1. The source organ, such as a leaf, produces and exports nutrients, while the sink organ, like a grain, imports and consumes them. Additionally, there is an intermediate organ, such as the internode of cereals, which functions as either a source or sink depending on its physiological status and developmental stage.
2. The source–sink interaction refers to the relationship between assimilate production in source organs and assimilate utilization in sink organs, which is regulated by a highly complex signalling network, where hormones, sugars, and environmental cues converge to govern developmental processes and adapt to stress.
3. Interactions between source and sink organs can be dissected through three aspects: nutrients transported through vascular tissues, transporters that facilitate nutrient mobilization into and out of the vascular tissues, and signalling pathways that regulate nutrient transport and developmental events.
4. T6P functions as a cellular sucrose sensor that dynamically regulates plant metabolism in response to local sucrose levels. Current evidence supports a model in which T6P directly inhibits SnRK1 kinase activity, thereby suppressing the expression of genes involved in carbohydrate catabolism and stress-responsive pathways.

5. While it has been acknowledged that a co-limitation exists between the source and sink of assimilates, contemporary temperate cereals are predominantly limited by their sinks.
6. The close interaction between source and sink, combined with multiple factors operating at the whole-plant level, leads to a non-linear response when modifying any single trait or process related to source, flow, or sink functions. This high level of nonlinearity likely accounts for the limited success attained by current genetic engineering strategies in enhancing crop yields.

Important Terms

AAP; AMT; Flow; Information; Intermediate organs; MST; NRT1; Signal; Signalling; Sink; SnRK1; Source; Strength; SUT; SWEET; T6P; Transporter; UmamiT.

Review Questions

Multiple Choice

1. Regarding the intermediate organ, which of the statements below is NOT right (   )

   (a) Carbon storage in the stems of many temperate grasses primarily consists of glucans.
   (b) Generally, fructans and sucrose are stored together in the stem, as seen in wheat, barley, and oats.
   (c) Cereals generally do not store significant amounts of starch in their stems, except for rice.
   (d) Stem storage of carbohydrates serves as an important buffering system for supplying photosynthates to cereal ears during grain-filling, especially under drought.

2. A crop species with a harvest index of 0.3 is likely to be (   )

   (a) Oilseed rape
   (b) Wheat
   (c) Rice
   (d) Maize

3. Regarding the functional transition from sink to source during leaf growth, which of the statements below is NOT right (   )

   (a) The long-distance transport of sucrose is exclusively mediated by the phloem, coupling the sink-to-source transition with a corresponding shift from phloem unloading (import) to phloem loading (export).

(b) There is a phase during which bidirectional phloem transport occurs within a single leaf, with some vascular bundles importing photosynthates while others are exporting them.
(c) The transition progressing acropetally along the leaf.
(d) The transition entails alterations in plasmodesmata frequency and structural organization within both mesophyll and epidermal cells.

4. According to the pressure-flow hypothesis, which of the statements below is NOT right ( )

(a) Sucrose in the phloem sap is transported via mass flow rather than through osmosis.
(b) The pressure gradient driving mass flow is established by phloem unloading in leaves and phloem loading in sink organs.
(c) During the transport process within sieve tubes, sucrose is translocated through the phloem via partially permeable pathways rather than fully sealed conduits.
(d) The movement of sucrose from source to sink tissues occurs through symplasmic and/or apoplasmic pathways, driven by the metabolic needs and storage requirements of these tissues.

5. The sink–source transition in growing leaves is characterized by biochemical, physiological, and anatomical changes. This process is accompanied by the ( )

(a) Increase in invertase activity
(b) Increase in sucrose synthase activity
(c) Decrease in sucrose phosphate synthase activity
(d) Reduced symplasmic continuity between mesophyll cells and epidermal cells

6. ( ) is responsible for the transport of glutamine and aspartate from the leaf to the seed of soybean plants.

(a) AMT1.1
(b) NRT1.1
(c) AAP8
(d) UPS

7. With regard to the classification of transporters in plants, the sugar transporter SWEETs belong to the ( ).

(a) Symporter
(b) Antiporter
(c) Cotransporter
(d) Uniporter

8. With regard to the classification of transporters in plants, the amino acid transporter UmamiTs belong to the (   ).
   (a) Symporter
   (b) Antiporter
   (c) Cotransporter
   (d) Uniporter

True or False

1. The most common sugar transported in the phloem is sucrose, although depending on the plant species, raffinose or stachyose may also be transported (   )
2. Sucrose transport from source to sink organs occurs in both the phloem and xylem, in contrast to amino acids, which requires sophisticated exchange mechanisms between these two vascular tissues (   )
3. Studies across different crops demonstrated that grain or seed number per unit area (GN m^{-2}) is the dominant yield component (   )
4. Companion cells are distinguished by their large nuclei and prominent nucleoli, a high density of mitochondria, abundant ribosomes, and a densely packed cytoplasm, all of which reflect their highly active metabolic state (   )

Blank Filling

1. Because of the complex interplay between source and sink and their susceptibility to various whole-plant factors, manipulating a single trait related to either source, flow, or sink often leads to ______________ responses.
2. Three families of transporters, ______________, ______________, and ______________, have been identified as key players in the intercellular transport of sugars.

Short Answer

1. Compare the leaf with non-foliar organs, such as the ear or panicle, in terms of source activity.
2. Explain the role of intermediate organs in crop yield formation.
3. Describe the forms of nutrients when transported in the vascular systems.
4. Diagram the role of T6P-SnRK1 in regulating the plant response to sugar supply.
5. Justify the indication of HI in the source–sink relation.
6. Use one crop as an example to show how the source or sink limits the grain yield.

Essay Items

1. Read the article titled "Critical period for yield determination across grain crops" published in *Trends in Plant Science* (2024, 29: 329–342) and summarize its main points using your own words. Subsequently, assess the significance of the "critical period" in relation to the breeding or management of crop species you are familiar with.
2. Over the past three decades, a diverse array of transgenic plants has substantially deepened our understanding of source–sink interactions. Nevertheless, only limited progress has been made in the genetic engineering endeavours targeting the enhancement of crop yields through the manipulation of cellular-level source or sink functions. Please analyse the reasons for this issue by taking a specific crop as an example.

References

Ahn E, Botkin J, Curtin SJ et al (2023) Ideotype breeding and genome engineering for legume crop improvement. Curr Opin Biotechnol 82:102961. https://doi.org/10.1016/j.copbio.2023.102961

Araus JL, Sanchez-Bragado R, Vicente R (2021) Improving crop yield and resilience through optimization of photosynthesis: panacea or pipe dream? J Exp Bot 72(11):3936–3955. https://doi.org/10.1093/jxb/erab097

Batista-Silva W, da Fonseca-Pereira P, Martins AO et al (2020) Engineering improved photosynthesis in the era of synthetic biology. Plant Commun 1(2):100032. https://doi.org/10.1016/j.xplc.2020.100032

Chang TG, Zhu XG (2017) Source–sink interaction: a century old concept under the light of modern molecular systems biology. J Exp Bot 68(16):4417–4431. https://doi.org/10.1093/jxb/erx002

Chen GX, Li YN, Jin KN et al (2025) Synthetic photorespiratory bypass improves rice productivity by enhancing photosynthesis and nitrogen uptake. Plant Cell 37(1):koaf015. https://doi.org/10.1093/plcell/koaf015

Crang R, Lyons-Sobaski S, Wise R (2020) In: Inglês E (ed) Plant anatomy: a concept-based approach to the structure of seed plants. Springer, Heidelberg

Crozet P, Margalha L, Confraria A et al (2014) Mechanisms of regulation of SNF1/AMPK/SnRK1 protein kinases. Front Plant Sci 5:1–17. https://doi.org/10.3389/fpls.2014.00190

De Souza AP, Burgess SJ, Doran L et al (2022) Soybean photosynthesis and crop yield are improved by accelerating recovery from photoprotection. Science 377(6608):851–854. https://doi.org/10.1126/science.adc9831

Engels C, Kirkby E, White P (2012) Mineral nutrition, yield and source–sink relationships. In: Marschner P (ed) Marschner's mineral nutrition of higher plants, 3rd edn. Academic Press, New York, pp 85–133. https://doi.org/10.1016/B978-0-12-384905-2.00005-4

Eom JS, Chen LQ, Sosso D et al (2015) SWEETs, transporters for intracellular and intercellular sugar translocation. Curr Opin Plant Biol 25:53–62. https://doi.org/10.1016/j.pbi.2015.04.005

Fernández V, Gil-Pelegrín E, Eichert T (2021) Foliar water and solute absorption: an update. Plant J 105(4):870–883. https://doi.org/10.1111/tpj.15090

Fernie AR, Bachem CWB, Helariutta Y et al (2020) Synchronization of developmental, molecular and metabolic aspects of source-sink interactions. Nat Plants 6(2):55–66. https://doi.org/10.1038/s41477-020-0590-x

Fichtner F (2025) Trehalose 6-phosphate–a central regulator at the crossroads of sugar signalling, metabolism, and development. New Phytol. https://doi.org/10.1111/nph.70533

Griffiths CA, Xue XC, Miret JA et al (2025) Membrane-permeable trehalose 6-phosphate precursor spray increases wheat yields in field trials. Nat Biotechnol. https://doi.org/10.1038/s41587-025-02611-1

Halford NG (2010) Photosynthate partitioning. In: Pua EC, Davey MR (eds) Plant developmental biology - biotechnological perspectives, vol 2. Springer, Heidelberg, pp 67–82

Koley S, Chu KL, Mukherjee T et al (2022) Metabolic synergy in Camelina reproductive tissues for seed development. Sci Adv 8(43):eabo7683. https://www.science.org/doi/10.1126/sciadv.abo7683

Lalonde S, Wipf D, Frommer WB (2004) Transport mechanisms for organic forms of carbon and nitrogen between source and sink. Annu Rev Plant Biol 55:341–372. https://doi.org/10.1146/annurev.arplant.55.031903.141758

Lawlor DW, Paul MJ (2014) Source/sink interactions underpin crop yield: the case for trehalose 6-phosphate/SnRK1 in improvement of wheat. Front Plant Sci 5:418. https://doi.org/10.3389/fpls.2014.00418

Lawson T, Milliken AL (2023) Photosynthesis–beyond the leaf. New Phytol 238(1):55–61. https://doi.org/10.1111/nph.18671

Li B, Liu H, Zhang Y et al (2013a) Constitutive expression of cell wall invertase genes increases grain yield and starch content in maize. Plant Biotechnol J 11(9):1080–1091. https://doi.org/10.1111/pbi.12102

Li J, Baroja-Fernández E, Bahaji A et al (2013b) Enhancing sucrose synthase activity results in increased levels of starch and ADP-glucose in maize (*Zea mays* L.) seed endosperms. Plant Cell Physiol 54(2):282–294. https://doi.org/10.1093/pcp/pcs180

Li YP, Li HB, Li YY (2017) Improving water-use efficiency by decreasing stomatal conductance and transpiration rate to maintain higher ear photosynthetic rate in drought-resistant wheat. Crop J 5(3):231–239. https://doi.org/10.1016/j.cj.2017.01.001

Li HF, Testerink C, Zhang Y (2021) How roots and shoots communicate through stressful times. Trends Plant Sci 26(9):940–952. https://doi.org/10.1016/j.tplants.2021.03.005

Liu J, Wu MW, Liu CM (2022) Cereal endosperms: development and storage product accumulation. Annu Rev Plant Biol 73(1):255–291. https://doi.org/10.1146/annurev-arplant-070221-024405

Liu XY, Nakajima KP, Adhikari PB et al (2025) Fertilization–dependent phloem end gate regulates seed size. Curr Biol 35(9):2049–2063.e3. https://doi.org/10.1016/j.cub.2025.03.033

Lou HC, Li SJ, Shi ZH et al (2024) Engineering source-sink relations by prime editing confers heat-stress resilience in tomato and rice. Cell 188(2):530–549.e20. https://doi.org/10.1016/j.cell.2024.11.005

Ma SC, Li FM, Xu BC et al (2010) Effect of lowering the root/shoot ratio by pruning roots on water use efficiency and grain yield of winter wheat. Field Crop Res 115(2):158–164. https://doi.org/10.1016/j.fcr.2009.10.017

MacNeill GJ, Mehrpouyan S, Minow MAA et al (2017) Starch as a source, starch as a sink: the bifunctional role of starch in carbon allocation. J Exp Bot 68(16):4433–4453. https://doi.org/10.1093/jxb/erx291

Mason TJ, Maskell EJ (1928) Studies on the transport of carbohydrates in the cotton plant II. The factors determining the rate and the direction of movement of sugars. Ann Bot 42(3):571–636. https://doi.org/10.1093/oxfordjournals.aob.a090131

Nada RM, Abogadallah GM (2016) Restricting the above ground sink corrects the root/shoot ratio and substantially boosts the yield potential per panicle in field-grown rice (*Oryza sativa* L.). Physiol Plant 156(4):371–386. https://doi.org/10.1111/ppl.12377

Paul MJ (2021) What are the regulatory targets for intervention in assimilate partitioning to improve crop yield and resilience? J Plant Physiol 266:153537. https://doi.org/10.1016/j.jplph.2021.153537

Paul MJ, Oszvald M, Jesus C et al (2017) Increasing crop yield and resilience with trehalose 6-phosphate: targeting a feast-famine mechanism in cereals for better source–sink optimization. J Exp Bot 68(16):4455–4462. https://doi.org/10.1093/jxb/erx083

Ral JP, Bowerman AF, Li Z et al (2012) Down-regulation of glucan, water-dikinase activity in wheat endosperm increases vegetative biomass and yield. Plant Biotechnol J 10(7):871–882. https://doi.org/10.1111/j.1467-7652.2012.00711.x

Ranathunge K, El-Kereamy A, Gidda S et al (2014) *AMT1;1* transgenic rice plants with enhanced NH_4^+ permeability show superior growth and higher yield under optimal and suboptimal NH_4^+ conditions. J Exp Bot 65(4):965–979. https://doi.org/10.1093/jxb/ert458

Rosado-Souza L, Yokoyama R, Sonnewald U et al (2023) Understanding source-sink interactions: Progress in model plants and translational research to crops. Mol Plant 16(1):96–121. https://doi.org/10.1016/j.molp.2022.11.015

Rossi M, Bermudez L, Carrari F (2015) Crop yield: challenges from a metabolic perspective. Curr Opin Plant Biol 25:79–89. https://doi.org/10.1016/j.pbi.2015.05.004

Sadras VO, Villalobos FJ, López-Bernal Á et al (2024) Radiation interception, radiation use efficiency, and crop productivity. In: Villalobos FJ, Fereres E (eds) Principles of agronomy for sustainable agriculture. Springer Nature Switzerland, pp 177–197

Sanchez-Bragado R, Vicente R, Molero G et al (2020) New avenues for increasing yield and stability in C3 cereals: exploring ear photosynthesis. Curr Opin Plant Biol 56:223–234. https://doi.org/10.1016/j.pbi.2020.01.001

Santiago JP, Tegeder M (2016) Connecting source with sink: the role of Arabidopsis AAP8 in phloem loading of amino acids. Plant Physiol 171(1):508–521. https://doi.org/10.1104/pp.16.00244

Savin R, Slafer GA (2019) Crop science. In: Meyers RA (ed) Encyclopedia of sustainability science and technology, 2nd edn. Springer, New York

Shen BR, Wang LM, Lin XL et al (2019) Engineering a new chloroplastic photorespiratory bypass to increase photosynthetic efficiency and productivity in rice. Mol Plant 12(2):199–214. https://doi.org/10.1016/j.molp.2018.11.013

Siddique K, Belford R, Tennant D (1990) Root: shoot ratios of old and modern, tall and semi-dwarf wheats in a Mediterranean environment. Plant Soil 121:89–98. https://doi.org/10.1007/BF00013101

Simkin AJ, Faralli M, Ramamoorthy S et al (2020) Photosynthesis in non-foliar tissues: implications for yield. Plant J 101(4):1001–1015. https://doi.org/10.1111/tpj.14633

Smith AM, Coupland G, Dolan L et al (2010) Plant biology. Garland Science, New York

Song XJ, Kuroha T, Ayano M et al (2015) Rare allele of a previously unidentified histone H4 acetyltransferase enhances grain weight, yield, and plant biomass in rice. Proc Natl Acad Sci USA 112(1):76–81. https://doi.org/10.1073/pnas.1421127112

South PF, Cavanagh AP, Li HW et al (2019) Synthetic glycolate metabolism pathways stimulate crop growth and productivity in the field. Science 363(6422):eaat9077. https://doi.org/10.1126/science.aat9077

Su J, Hu C, Yan X et al (2015) Expression of barley SUSIBA2 transcription factor yields high-starch low-methane rice. Nature 523:602–606. https://doi.org/10.1038/nature14673

Taiz L, Zeiger E, Møller IM et al (2015) Plant physiology and development, 6th edn. Sinauer Associates, Sunderland

Tambussi EA, Maydup ML, Carrión CA et al (2021) Ear photosynthesis in C3 cereals and its contribution to grain yield: methodologies, controversies, and perspectives. J Exp Bot 72(11):3956–3970. https://doi.org/10.1093/jxb/erab125

Tegeder M, Masclaux-Daubresse C (2018) Source and sink mechanisms of nitrogen transport and use. New Phytol 217(1):35–53. https://doi.org/10.1111/nph.14876

Thomas B (2017) Sources and sinks. In: Thomas B (ed) Encyclopedia of applied plant sciences, 2nd edn. Academic Press, New York, pp 119–127

Walker BJ, VanLoocke A, Bernacchi CJ et al (2016) The costs of photorespiration to food production now and in the future. Annu Rev Plant Biol 67(1):107–129. https://doi.org/10.1146/annurev-arplant-043015-111709

Wang L, Lu Q, Wen X et al (2015) Enhanced sucrose loading improves rice yield by increasing grain size. Plant Physiol 169(4):2848–2862. https://doi.org/10.1104/pp.15.01170

Wen S, Neuhaus HE, Cheng J et al (2022) Contributions of sugar transporters to crop yield and fruit quality. J Exp Bot 73(8):2275–2289. https://doi.org/10.1093/jxb/erac043

White AC, Rogers A, Rees M et al (2016) How can we make plants grow faster? A source–sink perspective on growth rate. J Exp Bot 67(1):31–45. https://doi.org/10.1093/jxb/erv447

Xia XD, Fan XR, Wei J, Feng HM et al (2015) Rice nitrate transporter OsNPF2.4 functions in low-affinity acquisition and long-distance transport. J Exp Bot 66(1):317–331. https://doi.org/10.1093/jxb/eru425

Xu HF, Wang ZX, Xiao F et al (2021) Dynamics of dry matter accumulation in internodes indicates source and sink relations during grain-filling stage of japonica rice. Field Crop Res 263:108009. https://doi.org/10.1016/j.fcr.2020.108009

Yang T, Wang H, Guo L et al (2022) ABA-induced phosphorylation of basic leucine zipper 29, ABSCISIC ACID INSENSITIVE 19, and Opaque2 by SnRK2.2 enhances gene transactivation for endosperm filling in maize. Plant Cell 34(5):1933–1956. https://doi.org/10.1093/plcell/koac044

Yu SM, Lo SF, Ho TD (2015) Source-sink communication: regulated by hormone, nutrient, and stress cross-signaling. Trends Plant Sci 20(12):844–857. https://doi.org/10.1016/j.tplants.2015.10.009

Zhai L, Yan A, Shao K et al (2023) *Large Vascular Bundle Phloem Area 4* enhances grain yield and quality in rice via source–sink–flow. Plant Physiol 191(1):317–334. https://doi.org/10.1093/plphys/kiac461

Zhang YP, Zhang YH, Wang ZM et al (2011) Characteristics of canopy structure and contributions of non-leaf organs to yield in winter wheat under different irrigated conditions. Field Crop Res 123(3):187–195. https://doi.org/10.1016/j.fcr.2011.04.014

Zhang W, Peng K, Cui F et al (2021) Cytokinin oxidase/dehydrogenase OsCKX11 coordinates source and sink relationship in rice by simultaneous regulation of leaf senescence and grain number. Plant Biotechnol J 19(2):335–350. https://doi.org/10.1111/pbi.13467

Zhao ZG, Wang CL, Yu XW et al (2022) Auxin regulates source–sink carbohydrate partitioning and reproductive organ development in rice. Proc Natl Acad Sci USA 119(36):e2121671119. https://doi.org/10.1073/pnas.2121671119

Canopy Structure and Function

6

Shaping a Healthy Canopy for a Highly Productive But Environmentally Benign Crop Industry

Contents

Z. Liu, *Crop Physiology*, https://doi.org/10.1007/978-981-92-2960-4_6

Abstract

The structure–function relationship delineates the interaction between the physical architecture and the functional roles of a biological system. Crop canopy refers to the aboveground organs within a population that actively exchange matter, energy, and information with the aerial environment. This chapter aims to clarify the structure–function relationship of the canopy, which is crucial for elucidating the mechanisms underlying crop yield and quality. The concepts of canopy structure and function are elaborately explained, together with the role of the microenvironment in mediating their relationship. The importance of cultivating a healthy canopy to promote a highly productive and environmentally sustainable crop industry is emphasized, and approaches, including genetic modification and chemical intervention, are outlined to achieve this goal. Finally, it underscores the necessity of crop modelling in bridging the gap between genomics and phenomics through the integration of structural and functional information across multiple levels.

Keywords

Canopy · Structure · Function · Microclimate · Microenvironment · Healthy canopy

Did You Know

- Leaf area index values range from less than 1 in sparsely vegetated areas such as deserts and tundra to 5–7 in croplands and 5–10 in forests. The highest value can reach 15–16 in certain oak species (e.g. *Quercus robur*).
- On average, about 340 watts of solar radiation reach every square meter of Earth's surface, equivalent to 1496 μmol m^{-2} s^{-1}, but only around 5% of this energy is ultimately converted into carbohydrates through photosynthesis.
- On a clear day under direct sunlight, the photosynthetic photon flux density (PPFD) at the top of a dense forest canopy can reach approximately 2000 μmol m^{-2} s^{-1} at a given site. However, due to absorption and scattering by the canopy, PPFD may drop sharply to as low as 10 μmol m^{-2} s^{-1} at ground level.
- There is evidence that grain yields decline by 1% to 10% per °C rise in night temperature, with the extent of reduction varying according to genotypes, growing conditions, and phenological stage.
- On a hot, sunny day, a single mature maple tree with approximately 100,000 leaves can evaporate around 200 litres of water per hour.
- Terrestrial vegetation stores approximately 610 Gt of carbon at any given time. Through photosynthesis, around 120 Gt of carbon is removed from the atmosphere annually. Meanwhile, plant respiration returns roughly half of this amount, about 60 Gt of carbon per year, back to the atmosphere.

Learning Objectives

After reading this chapter, students will be able to:

- Appreciate the emergent property of the canopy
- Comprehend the concepts of canopy structure and function
- Elaborate on the relationship between canopy structure and function
- Understand the heterogeneity of microclimates within the canopy
- Justify the concept of smart canopy
- List the genetic and agronomic approaches to modifying canopy structure

Introductory Remarks

A system is an organized assembly of interrelated components, elements, or parts that function collectively to achieve a shared purpose or function. These components interact through structured relationships, processes, and feedback mechanisms, resulting in emergent behaviours or outcomes that cannot be fully understood by examining the individual parts in isolation. Systems may be physical (e.g. a mechanical engine), biological (e.g. an ecosystem), social (e.g. an organization), or abstract (e.g. a mathematical model). They are commonly analysed in terms of inputs, transformation processes, outputs, and their surrounding environment, which influences and constrains system function.

The structure and function category reveals the system's constitution and its dynamic interaction with the environment. The *structure* is the constitution of the system and the relationships among its components. The *function* refers to the system's capacity for reciprocal exchange of substance, energy, and information with its surrounding environment. The structure and function are intricately interconnected and mutually influence each other. The function of a system generally depends on its structure, much like a machine's specific operation arises from its unique construction. However, in some cases, the function may also exert counterproductive effects on the structure, potentially leading to the emergence of a new structure. For instance, the principle of "use it or lose it" for human organs: regular activity helps maintain their health, while prolonged inactivity can lead to atrophy and a decline in function. This phenomenon highlights the critical role of physical exercise, mental stimulation, and consistent engagement of bodily systems in preserving optimal physiological function over time.

Structure–function relationship refers to the connection between the physical structure of a biological system and its biological function. It exists when the individual components of a system are arranged in a way that allows the system to perform its activities. An essential goal of basic science is to understand how things are organized at a smaller scale to provide functions that are typically performed on a larger scale. Investigations into the structure–function

(continued)

relationships in plant sciences have been conducted across multiple scales, spanning genes, cells, tissues, organs, whole plants, and populations.

The concept of a plant canopy was first developed in tropical rainforests and has since been applied to various situations, including crops, grasses, and forest plantations. A crop canopy is defined as the aboveground layer of a cultivated field, which consists of the green, photosynthetically active components, such as leaves, stems, and reproductive organs, which intercept sunlight and consequently use it to drive biomass production. The structural–functional analysis of the canopy enables us to gain insights into the interactions and mechanisms within the crop stand, which is instrumental in shaping a healthy canopy for a highly productive yet environmentally sustainable crop industry.

Concept Map

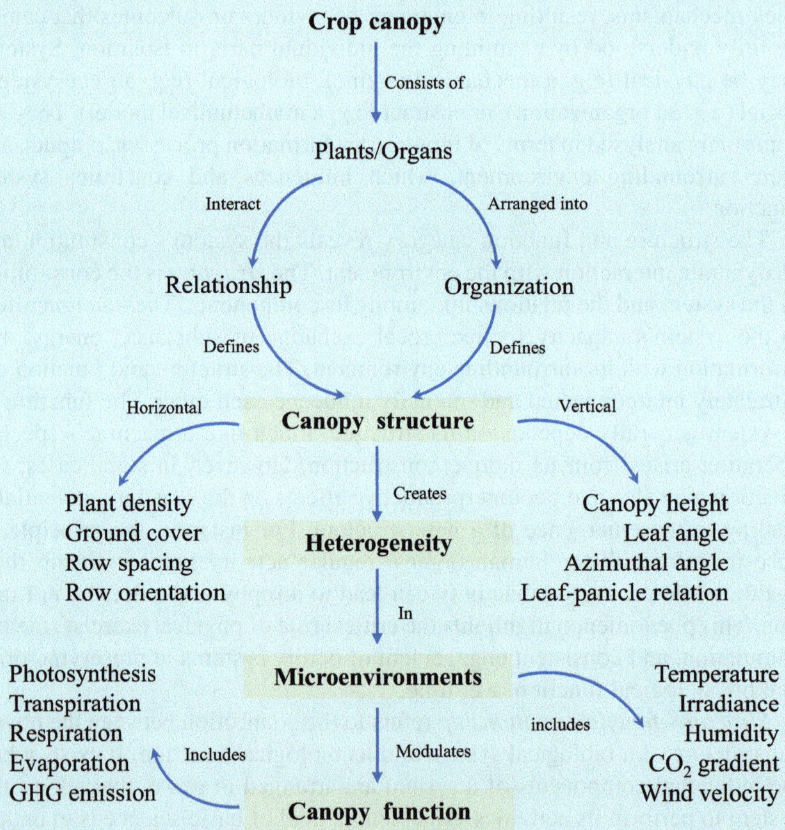

Annotation: The process of crop production, or the formation of crop yield, primarily occurs at the population or canopy level, which is composed of

(continued)

individual plants and their various organs. Canopy structure refers to the spatial organization of these components and the resulting competitive interactions among them. It can be described horizontally through parameters, such as plant density, ground cover, and row spacing, and quantified vertically by characteristics including canopy height, leaf angle, and azimuthal angle. Canopy structure predominantly influences canopy function by generating heterogeneous microenvironments within the canopy, particularly in terms of irradiance, temperature, and humidity distribution. Optimizing functional traits such as photosynthesis, evapotranspiration, and greenhouse gas (GHG) emissions by modifying canopy structure is a central research goal in crop science.

1 Emergent Property of the Canopy

Emergent properties are characteristics that arise from the interactions among components within a system, such that the whole system exhibits properties that are not evident in the individual parts. This concept is widely applicable across various disciplines, including ecology, where complex systems such as forests demonstrate emergent phenomena like biodiversity and ecosystem functioning. The crop canopy refers to the aboveground plant organs within a population composed of a single cultivar that has similar morphological and physiological traits. Like that of a forest, the crop canopy exhibits emergent properties that are not directly deducible from the attributes of individual constituents but depend on their interaction (Russell 2017).

In the crop stand, the growth of individual plants' aboveground parts is influenced by the proximity of neighbouring plants, which alters their aerial environment. Individual plants often vary in size, with larger individuals typically outcompeting smaller neighbours for resources. The canopy is structurally stratified, and the upper layer exerts a dominant influence on the lower one. Overlapping leaves generate complex shading patterns that reduce both the quantity and quality of light available to lower leaves and the soil surface.

As a physical barrier, the canopy also attenuates wind speed, moderates temperature fluctuations, and enhances moisture retention near the ground. These microclimatic conditions vary with the architectural properties of the canopy, such as leaf area index, the spatial arrangement of vegetative and reproductive organs, and canopy porosity. Microenvironments or microclimates within the canopy display considerable heterogeneity in both horizontal and vertical dimensions, resulting in distinct functional properties that differentiate the whole canopy from isolated plants.

The functioning of the canopy, like individual plants, can be analysed in terms of its integrated response to environmental factors. The structure determines the function; however, substantial changes in the structure of plants within the canopy may occur without significantly affecting overall functionality. For instance, the same canopy assimilate rate can be achieved through varying combination of leaf area and angles. Generally, quantitative descriptions of canopy traits are challenging due to spatial and temporal variability. The complexity increases from individual organs to plants, pure stands, and heterogeneous communities. Quantifying the structural and functional properties at the canopy level, where the crucial process of light harvest and transformation occurs, poses an exceedingly difficult task.

2 Canopy Structure

Canopy structure refers to the spatial arrangement of the aboveground plant organs within a crop stand, as well as the resultant competitive relations or interactions among individual plants. It is a central consideration in any description of plant–environment interactions. There exists a remarkable diversity in the structure of canopies among different crop species, wherein each plant community exhibits a distinct spatial arrangement of photosynthetic surfaces. Canopy structure determines the pattern of light interception, a key process in energy transformation within crop stands, thereby serving as a critical determinant of crop yield (Fig. 6.1).

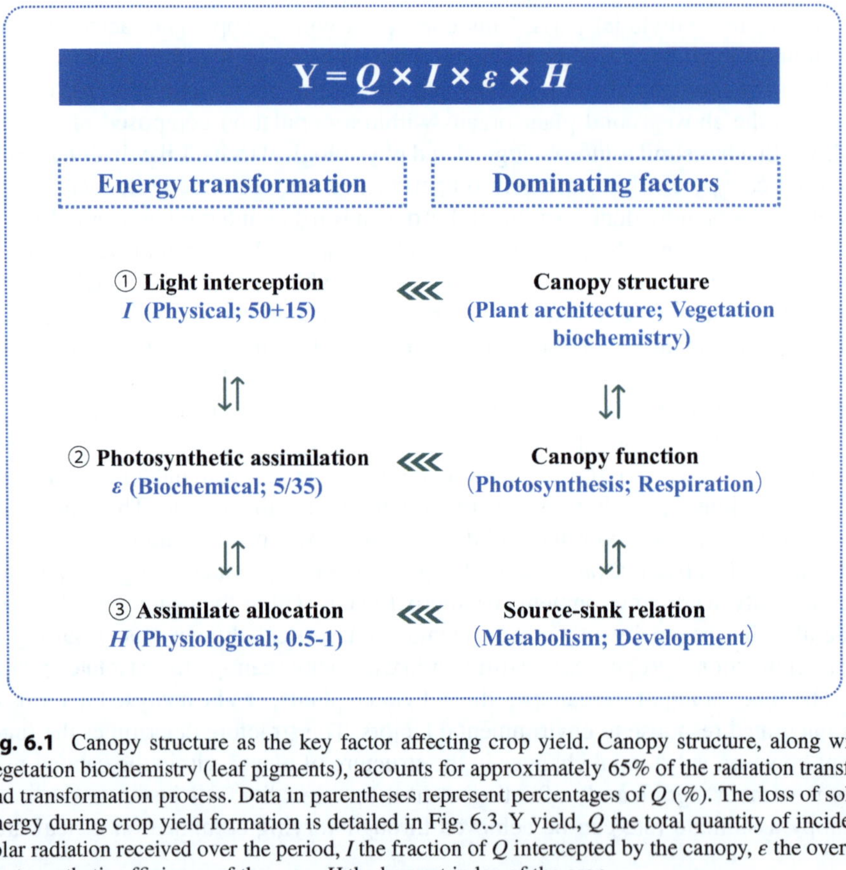

Fig. 6.1 Canopy structure as the key factor affecting crop yield. Canopy structure, along with vegetation biochemistry (leaf pigments), accounts for approximately 65% of the radiation transfer and transformation process. Data in parentheses represent percentages of Q (%). The loss of solar energy during crop yield formation is detailed in Fig. 6.3. Y yield, Q the total quantity of incident solar radiation received over the period, I the fraction of Q intercepted by the canopy, ε the overall photosynthetic efficiency of the crop, H the harvest index of the crop

Consequently, modifying canopy architectural traits has become one of the main objectives in crop breeding and management to enhance yield potential.

2.1 Spatial Arrangement of the Aboveground Organs

Spatial arrangement of the canopy refers to the size, shape, height, angle, and orientation of aboveground organs, such as leaves, stems, flowers, and fruits. It is a complex trait determined by the intricate interactions among genotype, environment, and management practices and can be quantified horizontally and vertically (Fig. 6.2).

Horizontally, spatial arrangement can be quantified by row orientation, row spacing, plant spacing, plant density, and ground cover such as LAI (*leaf area index*), GAI (green area index), and NDVI (*normalised difference vegetation index*). Among them, LAI is one of the most important biophysical parameters indicating the capacity of photosynthetic assimilation. Horizontal indices are primarily regulated by agronomic practices during seeding or transplanting stages when crop plants initiate their growth and development in the field. Precise control of plant density, as well as uniform spacing, is easily achieved with modern planters using uniformly sized seeds.

Vertically, spatial arrangement can be described through canopy height, leaf angle distribution, tiller angle, and *azimuthal canopy orientation*, which are mainly subjected to genetic control. As stated in Chap. 4, canopy height, the average plant height, is regulated by genes (*sd1* and *Rht1*) responsible for GA signalling, while leaf angle is determined by genes involved in BRs signalling, such as *lac1* (Tian et al. 2024). Shaping a highly efficient canopy by compacting plant architecture has been one of the main objectives of crop breeding.

With the emergence of robust phenotyping tools, several comprehensive indices for canopy structure have been proposed, including *canopy gap fraction* (CGF), leaf to panicle ratio (LPR), canopy roughness (CR), and canopy occupation volume (COV). CGF is defined as the probability of a light beam to pass through the canopy without any collision. Inverting gap fraction data is a powerful technique for estimating LAI without damaging the plants. LPR is a deep-learning based trait that indicates the light distribution between the leaf and panicle within the rice canopy (Yang et al. 2023). CR is a numerical value which characterizes the irregularities of the canopy surface measured by high-resolution 3D point clouds coming from RGB imagery acquired by UAV. It can discern aboveground biomass variations among different genotypes (Herrero-Huert et al. 2020). COV, an integrative index for leaf angle and leaf area index, explains more than 79% of the variation in canopy photosynthesis induced by changes in leaf angle and more than 84% of the variation induced by changes in leaf area (Liu et al. 2021).

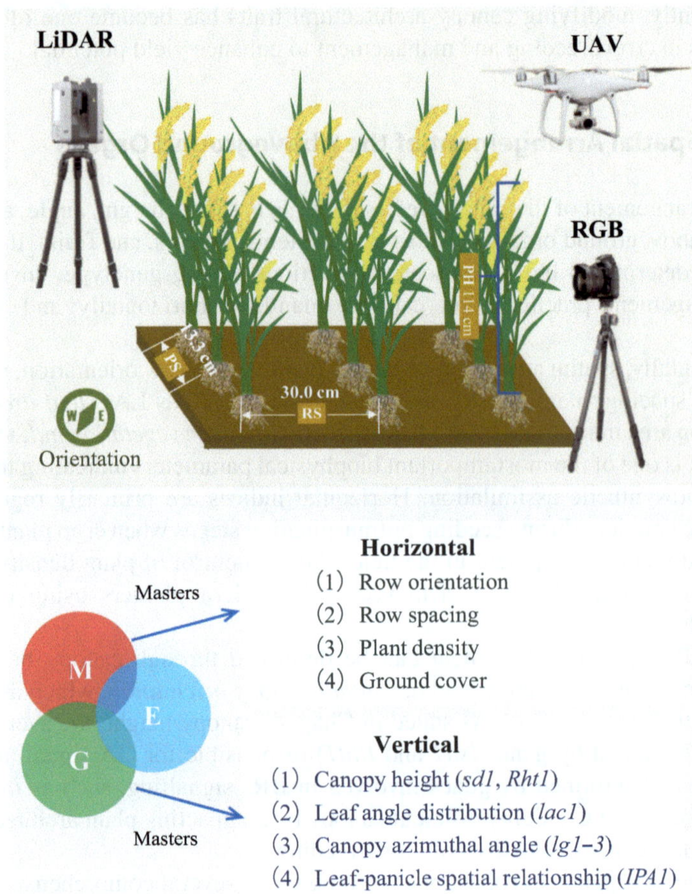

Fig. 6.2 Canopy structure as influenced by genotype (G), environment (E), and management practices (M). Horizontally, the traits of canopy structure such as row orientation, row spacing, plant density, and ground cover are primarily influenced by agronomic practices. Vertically, characteristics like canopy height, canopy azimuthal angle, and leaf angle distribution are predominantly determined by genetic factors. Genes such as *sd1*, *Rht1*, *lac1* (Tian et al. 2024), *lg1–3* (Zhou et al. 2024), and *IPA1* (Jiao et al. 2010) play crucial roles in regulating these architectural traits. High-throughput phenotyping platforms integrating UAV, RGB cameras, and LiDAR have revolutionized the research perspective on canopy structure. LiDAR light detection and ranging, PH plant height, PS plant spacing, RS row spacing, RGB red-green-blue, UAV unmanned aerial vehicle

2.2 Competition Among Individual Plants

Competitive interactions among plants arise when individuals growing at high densities compete for essential resources, such as light, water, and nutrients, resulting in mutual interference. In crop stands, increasing planting density reduces the average aboveground space available per plant and the amount of sunlight intercepted by each plant, thereby triggering adaptive changes in individual plant architecture (Connor et al. 2011). Plant competition occurs not only through direct physical contact but also indirectly by altering the shared environment. For example, when one plant depletes soil nitrogen and water, subsequent plants face reduced availability of these critical resources. Competition can be readily demonstrated when plants grown at low density attain a larger size than those grown at high density.

The crop population theoretically has an upper limit on the amount of biomass that can be produced from the available resources within a given area. As plants grow, they require a larger surface area to capture sunlight for producing sufficient sugars to support maintenance and growth respiration. This spatial expansion is achieved through the mortality of neighbouring plants, a process known as *self-thinning*, which is of great significance in natural ecosystems and forestry. In agriculture, it is avoided by precisely controlling planting density and spacing, as dead plants sequester minerals and photosynthates that could otherwise have benefited the surviving plants. Moreover, self-thinning typically occurs during the early growth stages in rice seedling nurseries with high seeding rates.

High plant densities create conditions favourable for disease and *lodging* of grain crops. The optimal density is influenced by various factors, such as resource availability, the growing duration of the crops, and their *morphological plasticity*. Tillering, or branching, is a crucial mechanism underlying morphological plasticity in cereal crops. Short-stalked species such as wheat and rice exhibit high plasticity, primarily due to their strong tillering capacity. Consequently, they can maintain consistently high grain yields per unit area across a wide range of planting densities. In contrast, tall-stalked crops such as maize and sorghum show limited morphological plasticity because they lack tillering and branching. As a result, their yield performance is highly sensitive to planting density, with only a narrow optimal range.

Increasing plant density represents one of the most effective strategies for enhancing the productivity of limited farmland. Nevertheless, dense planting exacerbates competition among plants for water, nutrients, and light. To tackle this challenge, maize breeders have modified plant architecture to accommodate dense planting conditions. A more upright leaf angle (i.e. a reduced leaf angle) minimizes mutual shading while ensuring optimal light interception and distribution for photosynthesis, even at higher plant densities. This adaptation also promotes the accumulation of leaf nitrogen for grain filling. Owing to these improvements, planting density in the United States has increased from approximately 30,000 plants per hectare in the 1930s to over 80,000 plants per hectare by the 2010s. Correspondingly, grain yield rose from 1287 kg/ha to 9595 kg/ha, representing a 7.5-fold increase over the same period (Mansfield and Mumm 2014).

2.3 Measurement of Canopy Structure

Measurements of canopy structure include the quantity and arrangement of aboveground plant organs, such as the position and orientation of each leaf. Canopy structure is dynamic, changing on timescales ranging from minutes to years (Norman and Campbell 1989). Leaves of *heliotropic* species follow the sun's movement across the sky during the day, maximizing light absorption for photosynthesis. In contrast, other plant species exhibit adaptive leaf behaviours, such as folding, drooping, or shedding, in response to environmental changes including water availability, light intensity, and temperature fluctuations. Due to the inherent complexity of canopy structure, a wide range of methodologies has been developed, exhibiting considerable variation in precision, accuracy, and technical complexity.

Direct methods, such as those involving area, shape, angle, or position, often disrupt the canopy structure and compromise the data quality (Welles 1990). These methods typically require significant field labour and involve straightforward data reduction. However, they prove to be arduous and labour-intensive in smaller canopies and are nearly impracticable for larger forest canopies. Indirect methods typically entail the measurement of radiation emitted from within or above the canopy (Welles 1990). Subsequently, the acquired data are processed by applying radiative transfer theory and an inversion procedure enabled by a computer program. These methods generally require straightforward and efficient field measurements, while complex algorithms are needed for data reduction.

Crop canopies can be digitized through diverse photographic and laser-based methods. However, the high density of crop stand presents practical challenges for such measurements. Nevertheless, continuous improvements are being made in terms of accuracy and realism. Recent advances in sensor technology, imaging techniques, and automation have facilitated the development of efficient phenotyping platforms at various scales. For example, the increasing accessibility of global measurements of sun-induced chlorophyll fluorescence (SIF) has been effectively utilized as a proxy to monitor vegetation physiology, biochemical composition, and structure.

Top-of-canopy (TOC) SIF results from three key processes: the absorption of incident PAR by chlorophylls, the emission of fluorescence by photosystems, and the scattering and reabsorption of emitted fluorescence (Yang et al. 2019). The magnitude of SIF emission is determined by physiological mechanisms that regulate the distribution of absorbed PAR into three de-excitation pathways: heat dissipation, photochemistry, and fluorescence. By contrast, the efficiency of incident PAR absorption and the extent of SIF scattering are primarily influenced by non-physiological factors, such as vegetation biochemistry (e.g. leaf pigments, water content) and canopy structure.

Yang et al. (2019) presented an approach that employs radiative transfer models (RTMs) and TOC reflectance to assess the influence of biochemical and structural factors on TOC SIF. This methodology was evaluated using airborne reflectance and SIF data collected at an agricultural experimental farm in Germany on 2 specific days, one before and the other during a heat event in the summer of 2015, when the maximum temperatures were 27 °C and 34 °C, respectively. The findings indicated that variations in vegetation biochemistry and structure accounted for over 76% of the observed SIF variability across different crops. Moreover, changes in these factors explained up to 73% of the variability in far-red SIF between the 2 days and 40% of the variability in red SIF. The remaining variability was primarily attributed to differences in physiological status. Thus, the integration of RTMs, reflectance, and SIF provides a novel means for detecting biochemical, structural, and physiological changes in the canopy. This study stands as a typical example illustrating the efficacy of advanced remote sensing techniques in monitoring the complex architectural and functional characteristics of the crop canopy. For more detailed information on indirect methods, please refer to plant phenomics in Sect. 4 in Chap. 8.

3 Canopy Function

The *canopy function* refers to the exchange of substances, energy, and information between the surrounding environments and the canopy. This can be quantified in terms of radiation capture, photosynthesis, respiration, evapotranspiration, and greenhouse gas emissions, all of which reflect the properties of the entire canopy.

3.1 Radiation Capture

3.1.1 Radiation Capture at Leaf Level

On average, about 340 W of radiant energy from the sun reach each square meter of Earth's surface.[1] The role of sunlight in plants is twofold: it serves as the primary source of energy for photosynthesis and as a signal that enables plants to adapt to the changing light conditions. The human eye perceives radiation within approximately 380 nm (violet) to 750 nm (red), which is known as visible light and encompasses the wavelengths necessary for photosynthesis. Additionally, plants can detect signals in wavelengths outside this visible range, including near-ultraviolet and far-red regions.

When evaluating the necessary light intensity for a specific plant process, it is more advantageous to conceptualize light as discrete particles termed photons rather than as a continuous waveform. Thus, it is appropriate to express light as photosynthetic photon flux density (PPFD), the flux of light (usually expressed as μmol m^{-2} s^{-1}) within the range of *photosynthetically active radiation* (400–700 nm; PAR). Approximately, PAR irradiance (W m^{-2}) multiplied by 4.4 is equivalent to PPFD (μmol m^{-2} s^{-1}).

The minimum light intensity required for plants to initiate net photosynthesis typically ranges from 1 to 20 μmol m^{-2} s^{-1}. Below this threshold, the light compensation point, plants experience a net release of CO_2 due to respiration surpassing photosynthetic carbon fixation. The saturation point of photosynthesis for most C3 plants occurs at approximately 500–1000 μmol m^{-2} s^{-1}, which is significantly lower

[1] Irradiance is the amount of energy that falls on a flat sensor of known area per unit time, expressed in watts per square meter (W m^{-2} or J m^{-2} s^{-1}). Quantum flux, or photon flux density (PFD), is the number of incident quanta striking the sensor, expressed in moles per square meter per second (mol m^{-2} s^{-1}), where moles refer to the number of photons (1 mol of light = 6.02×10^{23} photons, Avogadro's number)

than the full sunlight intensity of about 2000 μmol m^{-2} s^{-1}. For C4 plants, which require stronger solar radiation for photosynthesis, photosynthetic saturation is not achieved even under full sunlight.

When sunlight interacts with vegetation, only approximately 5% of the incident energy is converted into carbohydrates through the process of photosynthesis. The allocation of solar energy among the various components during the process of plant light utilization is depicted in Fig. 6.3. Firstly, approximately 50% of the light has wavelengths that are either too short or too long to be effectively absorbed by photosynthetic pigments, such as chlorophylls and carotenoids. For the sun, most of the energy is emitted over a waveband region of 200–3000 nm. However, crop growth is entirely dependent on the interception of PAR, which is absorbed by pigments and then converted into chemical energy stored in carbohydrates and other essential biological molecules. Of the PAR incident on a leaf, a portion is transmitted through the leaf and reflected from its surface (15%). Of the absorbed solar energy (35%), 10% is dissipated as heat or fluorescence, 20% is consumed by metabolic processes such as photorespiration in C3 plants and respiration, leaving only 5% to be stored as carbohydrates. The efficiency with which plants convert light energy into biomass is relatively low for both C3 and C4 photosynthetic pathways. According to Zhu et al. (2008), the theoretical maximum photosynthetic efficiency is estimated at 4.6% for C3 plants and 6.0% for C4 plants under standard conditions of 30 °C and atmospheric CO_2 concentration of 380 ppm.

3.1.2 Scaling Up Light Interception from Leaf to Canopy

Crop canopies must intercept or capture as much radiation in the 400–700 nm bandwidth as possible. Most of the light interception occurs through leaves; however, stems, petioles, leaf sheaths, and reproductive structures can also absorb significant amounts of radiation. To achieve high interception rates, plants must develop and present a canopy with a large leaf area. In many crops, the minimum leaf area index (LAI) is around 3. Typically, an LAI of 3–5 is generally required for 90% light interception. The amount of energy intercepted or captured by the canopy is determined by the organization of leaves into a 3D structure with a specific total surface area. The net photosynthesis and dry matter (DM) production during the early season are directly related to intercepted radiation, almost until anthesis. Thereafter, solar radiation becomes a less important factor as LAI declines and respiration increases. Both the amount of radiation captured by this structure and its allocation in time and space determine canopy photosynthesis and plant growth.

At the whole-plant level, leaves at the top of the canopy intercept the majority of sunlight, thereby reducing the amount of radiation reaching lower leaves within the canopy. The shaded leaves are exposed to reduced light intensity and altered light quality, and as a result, they exhibit significantly lower photosynthetic rates. This vertical heterogeneity in light distribution plays a critical role in shaping the emergent property of canopy photosynthesis (Hay and Porter 2006). For C3 species, which encompass most major cereals except maize, approximately 30% of solar energy is wasted as the upper canopy leaves are unable to utilize all the incident radiation. This occurs because the rate of photosynthesis in C3 species under

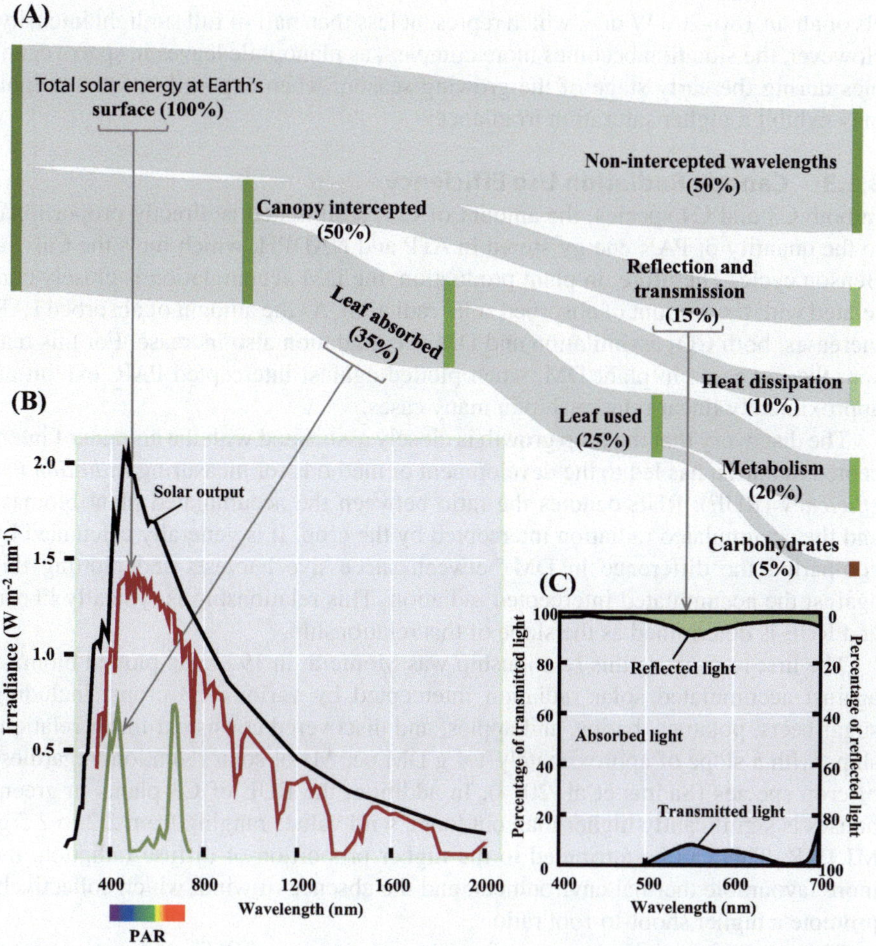

Fig. 6.3 The process of converting solar energy into carbohydrates by the leaf. Spectral characteristics determine the proportions of light that are absorbed, transmitted, or reflected, which in turn influence the efficiency of solar energy conversion into heat, metabolic activity, and stored carbohydrates. (**A**) The Sankey diagram depicts energy flow beginning with total incident solar radiation (100%). Approximately half of the incoming radiation (~50%) is not absorbed by the leaf due to the biochemical properties of photosynthetic pigments, as shown in (**B**). The remaining half (~50%) is absorbed. Of this absorbed portion, 15% is subsequently lost through reflection and transmission, as indicated in (**C**); the net energy retained by the leaf amounts to 35%, which is partitioned into heat dissipation (~10%) and biologically utilized energy (~25%). The utilized energy is further allocated to metabolic processes (~20%) and carbohydrate synthesis (~5%). (**B**) The solar spectrum and its relation to the absorption spectrum of chlorophyll. The black curve depicts the output of solar energy as a function of wavelength. The red curve represents the energy incident on the Earth's surface. The distinct valleys in the infrared region beyond 700 nm signify the absorption of solar energy by atmospheric molecules, predominantly water vapor. The green curve shows the absorption spectrum of chlorophylls, which exhibits strong absorption in the blue (~430 nm) and red (~660 nm) segments of the spectrum. (**C**) The percentages of light absorbed, reflected, and transmitted, as a function of wavelength. PAR, photosynthetically active radiation. (Redrawn after Taiz et al. (2015) and Goudriaan (2016))

ambient conditions either does not increase or becomes saturated at irradiance levels of about 180–200 W m^{-2}, which represent less than half of full sunlight intensity. However, the situation becomes more complex, as planophile leaves in sparse canopies during the early stage of the growing season, when exposed to full sunlight, may exhibit a higher saturation irradiance.

3.1.3 Canopy Radiation Use Efficiency

In both C3 and C4 species, the amount of CO_2 assimilated is directly proportional to the quantity of PAR energy stored in ATP and NADPH, which fuels the Calvin-Benson cycle. Therefore, in plant production, the DM accumulation is closely correlated with the amount of absorbed solar radiation. As the amount of absorbed PAR increases, both CO_2 assimilation and DM accumulation also increase. For this reason, the increases in plant DM, when plotted against intercepted PAR, exhibit an approximately linear relationship in many cases.

The discovery that the crop growth is closely associated with the amount of intercepted radiation has led to the development of methods for measuring *radiation use efficiency* (RUE). RUE denotes the ratio between the accumulated plant biomass and the accumulated radiation intercepted by the crop. It is generally calculated by comparing the difference in DM between successive harvests and plotting this against the accumulated intercepted radiation. This relationship is typically linear, and RUE is determined as the slope of this relationship.

The first to propose this relationship was Monteith in 1977. He plotted biomass against accumulated solar radiation intercepted by various C3 crops, including sugar beets, potatoes, barley, and apples, and discovered consistent linear relationships with a slope of approximately 1.4 g DM per MJ of solar radiation, regardless of crop species (Sadras et al. 2024). In addition, the RUE of C3 plants in greenhouses is significantly higher than outdoors, with values ranging from 2.2 to 2.5 g/MJ PAR. This can be attributed to the higher proportion of diffuse radiation, the more favourable thermal environment, and the absence of wind, which collectively promote a higher shoot-to-root ratio.

Timing is of crucial importance for the measurement of RUE, as it varies across different growth stages. To accurately determine RUE, it is essential to measure the incident PAR at the canopy surface, the amount of radiation intercepted by the canopy, and the biomass accumulated over a defined time interval. In many studies, the energy content of the DM is also required. Carbohydrates like starch and cellulose require less energy per unit weight, leading to higher RUE in crops such as wheat and rice. In contrast, proteins and oils demand more energy, resulting in lower RUE for crops like soybean and peanut.

For a given species, in the absence of biotic and abiotic stress, the amount of DM accumulated is linearly related to the intercepted PAR by the canopy, resulting in a relatively stable RUE value. However, significant variations exist across different crop species and cultivars due to the compound effects of photosynthetic capacity (C3 or C4) and biochemical compositions. For example, under non-stressed situations RUE based on solar radiation is about 1.7–1.8 g/MJ for maize in contrast to that of soybean (1.0 g/MJ). RUE on a PAR basis is twice these values,[2] being 3.4–3.6 g/MJ for maize and 2.0 g/MJ for soybean. It is now widely accepted that RUE is a fundamental measurement that underpins potential crop productivity. Given the current emphasis on global food security, there is currently significant interest in increasing the RUE of key crops in important agroecosystems.

3.2 Photosynthesis

3.2.1 Leaf Level Versus Canopy Level

The optimal design of crop canopy must consider not only the maximization of light interception but also the optimization of photosynthetic capacity across the entire canopy. In principle, a plant could achieve close to 100% interception with a single, planar chlorophyll-rich leaf and an LAI of 1.0. However, this does not result in optimum productivity, largely because photosynthesis saturates below full sunlight, and this is especially noticeable in C3 plants. At the leaf level, the light response curve exhibits a characteristic asymptotic shape of declining photosynthetic response with increasing *irradiance*, as depicted in Fig. 6.4. The measurement at leaf level is valuable for determining both the maximum rate of photosynthesis and the potential quantum yield of photosynthesis.

In contrast, at the canopy level, net photosynthesis shows a much more linear response to light (Fig. 6.4). Ideally, the most efficient canopy structure may be one in which all leaves are evenly illuminated at light intensities that saturate photosynthesis. However, this structure is rarely found in nature. More commonly, leaves positioned at the top of the canopy are exposed to irradiances that exceed those required for photosynthesis, while leaves at the bottom of the canopy receive very low levels of light. However, if the irradiance increases (e.g. moving from cloud cover to full sunlight), leaves lower in the canopy will be able to respond accordingly. This observation has led to a linear response being observed for canopies when plotting irradiance against photosynthesis, as opposed to the saturating response at the leaf level.

3.2.2 The Importance of Canopy Photosynthesis

Photosynthesis is a highly conserved biological process. Nevertheless, variations in the photosynthetic rate per unit leaf area can be observed. It remains unclear whether these variations in photosynthetic capacity drive higher yields or are instead a consequence of higher yields. Importantly, photosynthesis per unit leaf area is only one metric for assessing plant photosynthetic performance. Intuitively, canopy-level photosynthesis, not that of individual leaves, is widely regarded as the ultimate determinant of grain yield. This is because canopy photosynthesis integrates the photosynthetic activities of all photosynthetically active organs within the canopy

[2] PAR constitutes approximately fifty percent of the total incident solar energy

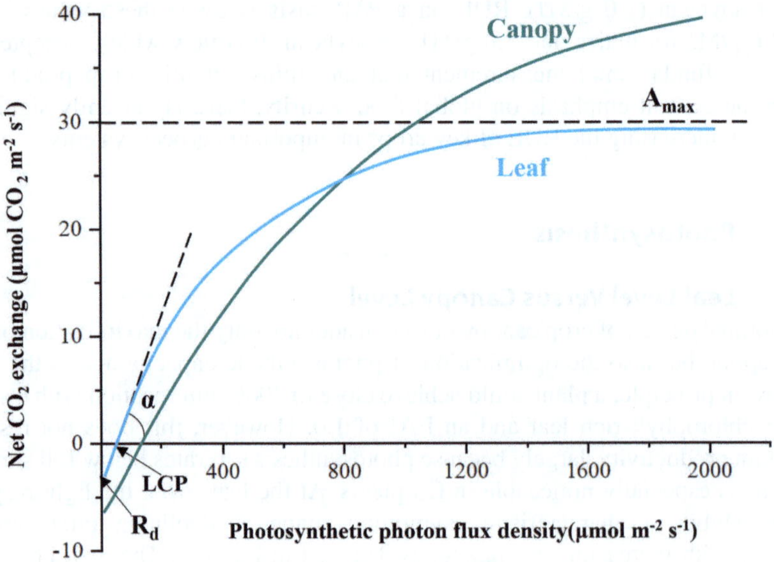

Fig. 6.4 Light response curve of leaf and canopy net photosynthesis at jointing stage of rice. This figure is constructed based on experimental data collected during the jointing stage of hybrid rice JYZK-6 in 2024. Notably, the leaf-level response curve to light intensity exhibits a typical rectangular hyperbolic pattern, whereas the canopy-level response demonstrates an approximately linear relationship. A_{max}, maximal rates of net photosynthesis; LCP light compensation point, Rd dark respiration, α the initial slope of the response curve, representing the quantum efficiency of photosynthesis

structure, including not only laminar leaves but also non-laminar components (e.g. stems, sheaths, and green bracts).

Using a retrospective approach, Lee and Tollenaar (2007) demonstrated that the significant genetic advances in hybrid maize over the past seven decades have led to increased canopy photosynthesis, both in terms of absolute rates and duration. The associated traits include the stay-green characteristic, which extends the photosynthetically active duration of the grain-filling phase; more erect leaves, which enhance the actual RUE of the whole canopy; and an increased capacity for planting density, which can be more than doubled, thereby further increasing canopy photosynthesis. However, the actual rates of leaf photosynthesis per unit leaf area, as measured at a given irradiance (since maize does not saturate at solar maximum), are likely unchanged or may have even decreased in modern hybrids. This could be attributed to the lower nitrogen concentration in leaf DM or an increased *specific leaf area* (the ratio of leaf area to biomass; SLA). This example illustrates how past breeding programs have significantly enhanced yield potential through improvements in canopy photosynthesis.

3.2.3 Measurement of Canopy Photosynthesis

Photosynthesis can be quantitatively measured through gas exchange, chlorophyll fluorescence, and reflectance spectroscopy at the field, canopy, and leaf levels (Song and Zhu 2024). Gas exchange methods involve the measurement of CO_2 and H_2O fluxes, including eddy covariance, canopy-level chambers, and organ-specific techniques. Chlorophyll fluorescence techniques, including solar-induced fluorescence (SIF), laser-induced fluorescence transients (LIFT), and pulse-amplitude modulation (PAM) fluorescence, are valuable for assessing photosynthetic efficiency and physiological responses to abiotic stresses. Hyperspectral reflectance is a widely used approach for estimating photosynthetic phenotypes. Moreover, deep learning-based segmentation of plant organs enables more accurate quantification of the individual contributions of leaf and non-leaf organs to overall canopy photosynthesis.

High-throughput phenotyping of photosynthesis can be achieved via the integration of the aforementioned techniques (Song and Zhu 2024). In terms of accuracy and efficiency, gas exchange techniques are regarded as the gold standard for measuring both photosynthesis and respiration; however, they pose challenges in achieving high throughput. Handheld non-imaging fluorescence devices offer a medium-throughput option suitable for field screening of photosynthetic traits, while fluorescence imaging provides a high-throughput alternative. The integration of fluorescence and hyperspectral imaging technologies into plant phenotyping platforms represents the emerging trend towards high-throughput field measurement.

3.3 Respiration

Assimilates obtained from photosynthesis serve as fundamental substrates for plant growth. *Respiration* (R) is a metabolic process through which plants obtain energy from carbohydrates to sustain cellular metabolism. Optimizing the balance between photosynthetic gain and respiratory loss is crucial for plant growth. Fast-growing crops exhibit higher rates of both photosynthetic assimilation and respiration, suggesting that respiration is an inevitable concomitant of rapid growth. Moreover, environmental factors, such as drought and nutrient scarcity, also exert a comparable stimulatory effect on respiratory processes.

3.3.1 Growth and Maintenance Respiration

Respiration evidently represents a significant depletion of the carbon metabolite pool, which could otherwise be allocated to plant growth (Connor et al. 2011). Multiple studies have demonstrated that crop respiration typically consumes a significant portion of the carbon fixed via photosynthesis. Under optimal growth conditions, the proportion of carbon lost to respiration generally ranges from 40% to 50%.

Respiration provides energy for both the construction of new biomass and the maintenance of existing structures. The fraction associated with growth is termed *growth respiration* (Rg), also known as synthetic or constructive respiration, which varies with the chemical composition of the newly formed biomass. *Maintenance respiration* (Rm), alternatively referred to as dormant or basal respiration, depends on tissue composition and takes priority over growth in terms of assimilate allocation. Rg is essential for the synthesis of new tissues, while Rm supplies the energy necessary to sustain the health and functionality of existing tissues.

Maintenance activities in higher plants encompass the turnover of proteins and lipids, as well as the maintenance of electrochemical gradients across membranes.[3] While some enzymes exhibit rapid turnover, the rates for the larger pools of structural proteins and lipids are relatively low. Consequently, a significant or even greater portion of Rm is attributed to the maintenance of electrochemical gradients. Specific Rm (Rm per g DM) increases with temperature but decreases with plant age. The exponential increase in Rm with temperature, up to an injury threshold between 40 °C and 50 °C, exhibits a Q_{10} value near 2, indicating that rates double for each 10 °C rise. As temperature rises and molecular activity intensifies, leakage and turnover rates are likely to increase. In contrast, mature tissues containing higher proportions of structural and storage compounds, such as cellulose, lignin, and starch, require less metabolic maintenance due to their reduced physiological activity. In addition, Rm rates increase linearly with N content. Nitrogen content not only indicates the amount of protein available for turnover but also reflects the total protoplasmic volume, thereby influencing membrane surface area and solute content. Reduction of maintenance costs represents an underexplored avenue for enhancing biological efficiency. For instance, ryegrass cultivars with lower Rm rates in mature leaf tissue exhibited higher forage yields compared to their parents.

Actual rates of growth respiration are indirectly influenced by temperature through its effects on growth rate and the balance between photosynthesis and Rm. Biochemical composition significantly impacts Rg. Among vegetative structures, leaves contain the highest levels of protein and lipid, resulting in higher construction costs compared to stems and roots. These compositional differences significantly influence Rg, as evidenced by the calculated growth yields for maize and soybean grains. Specifically, producing 100 g of DM requires 198.9 g of glucose for soybeans, compared to 139.1 g for maize (Connor et al. 2011). Given equal quantities of photosynthates allocated for grain production, it is evident that the DM yield of soybeans, which have higher contents of lipid and protein, will be lower than that of maize.

3.3.2 Quantification of Respiration Components

Respiration is a physiological process that can be quantified and modelled at various scales, such as the isolated cell, individual organ, entire plant, or canopy. Similar to photosynthesis, respiration is most readily quantified as a CO_2 flux in both closed

[3] Electrochemical gradients across cellular membranes are maintained through active transport mechanisms, including proton pumps and ion channels, which are responsible for generating and sustaining these gradients. Such gradients play a critical role in fundamental cellular processes, such as energy production and the transport of ions across the membrane

and open systems, because measuring the depletion of substrate or the release of O_2 presents greater technical challenges.

In measuring canopy respiration, the key challenges include the difficulty of accurately estimating daytime respiration and the lack of a clear distinction between Rg and Rm. Two primary methods, neither of which is entirely satisfactory, are used to distinguish Rg and Rm in growing plants (Connor et al. 2011). In the first method, plants are maintained in a controlled environment at the light compensation point, where gross photosynthesis equals dark respiration (Pn = 0). Total respiration (*R*) is measured during short intervals of darkness. Since photosynthesis is precisely balanced by respiration, it can be hypothesized that no assimilates are being allocated to new biomass formation; this implies that Rg = 0 and thus *R* = Rm. In the second method, the "starvation" approach, plants are placed in complete darkness, leading to the depletion of free assimilates and the cessation of growth. Total respiration declines to a minimum value (typically after 48 h), which is taken as an estimate of Rm. In both methods, the difference between normal (i.e. non-starved, light-grown) respiration and Rm provides an estimate of Rg. Nevertheless, these methods have limitations: the value of Rm may be higher in rapidly growing plants than in non-growing ones, and it may be substrate-dependent, occurring at a higher rate in non-starved plants.

Biscoe et al. (1975) carried out a field investigation in which barley crops were exposed to ambient air conditions, and CO_2 fluxes were estimated using aerodynamic techniques. Soil CO_2 flux, which represents the combined activity of roots and microorganisms, was measured using small enclosures. Daytime respiration was estimated from nighttime measurements, with adjustments made for temperature effects. As a result, they elucidated the temporal dynamics of the trade-off between growth and maintenance respiration costs, demonstrating how these factors interact to shape the seasonal pattern of respiration in barley crops.

Generally, canopy respiration accounted for 33% to 87% of gross photosynthesis (Pg) and increased over the season as biomass accumulated. During the heading stage, Rm increased significantly, rising from 11% to 40% of gross photosynthesis. This was accompanied by a decrease in net photosynthesis (Pn) due to the emergence of light green ears, which intercept a portion of sunlight but are unable to conduct photosynthesis at that time. Subsequently, Pn recovered to pre-heading levels due to the photosynthetic activities of green ears and peduncles. In the final week approaching maturity, respiration constituted 88% of Pg, with Rm accounting for 60% of the total R. Moreover, temperature-dependent respiratory losses became more prominent as the Rm increased. Thus, a 3 °C temperature increase during the vegetative phase reduced DM accumulation by only 3%. However, 2 months later, during the grain-filling phase, this same temperature rise caused a substantial 27% decline in DM production, owing to intensified Rm (Hay and Porter 2006).

3.3.3 Maintenance Respiration in Response to Elevated Night Temperature

Recent analyses of global climate records indicate that, over the past several decades, nighttime temperatures have risen at a faster rate than daytime temperatures (Serrago et al. 2025). On average, the rate of nighttime warming is approximately 1.4 times greater than that of daytime warming. In many terrestrial regions, the increase in nighttime temperatures relative to the 35-year baseline exceeds 0.25 °C. Accelerated nighttime warming has become a significant aspect of climate change and poses a direct threat to global food security.

Nocturnal temperatures exert an influence on crucial physiological processes, particularly the Rm. Even a moderate increase of 1–2 °C can lead to discernible reductions in the yields of staple crops. Katsura et al. (2009) quantified CO_2 assimilation dynamics in rice canopies from 35 days before anthesis to 40 days after anthesis, using high-temporal-resolution measurements taken every 30 min. Their results revealed that the mean daytime Pn and the nighttime respiration rate were 25 g CO_2 m^{-2} h^{-1} and 6 g CO_2 m^{-2} h^{-1}, respectively. Notably, when assuming a constant respiratory rate across day and night, daily respiratory carbon loss would account for approximately 40% of Pg, highlighting the substantial energetic cost of maintenance respiration in rice canopies. By employing a respiration model, Katsura et al. (2009) estimated that an elevation in Rm could result in a 3% reduction in grain yield per °C increase in temperature. Therefore, adaptation strategies, such as breeding heat-tolerant crop cultivars, adjusting planting schedules, and enhancing irrigation systems to alleviate the impacts of nocturnal heat stress, are becoming increasingly crucial for sustaining agricultural productivity in a warming global environment.

3.3.4 Optimum or Critical LAI

The initial hypothesis concerning the relationship between crop biomass and respiration posited that total respiration was a constant proportion of biomass, predominantly leaf DM. Consequently, as leaf area expanded, respiration would increase proportionally as well. The hypothesized linear relationship between R and DM or LAI implies that Pn initially rises but subsequently declines with increasing LAI until it reaches an optimum value that maximizes Pn (Fig. 6.5). This LAI value is optimal for canopy productivity because it is neither excessively high nor excessively low, making it ideal.

In fact, it is more common for Pn to increase up to a critical LAI value at which light interception becomes complete. Above this value, Pn remains constant at its maximal level. As crop growth advances, leaves lower in the canopy exhibit decreased respiration rates. Consequently, the overall canopy respiration rate decreases proportionally with increasing LAI and eventually becomes non-linear. Figure 6.5 shows that, as LAI increases, all three parameters initially ascend until the LAI reaches a threshold of approximately 3. Subsequently, their rate of increase decelerates, and they reach a plateau. This indicates that, beyond these thresholds, the difference between Pg and R remains relatively stable with further increases in LAI.

This insight enables an understanding of the rationale behind the relatively constant RUE, which represents the slope between the DM produced and the intercepted radiation (Hay and Porter 2006). For LAI values below 3 or 4, both the difference between Pg and R (i.e. net DM production) and light interception increase

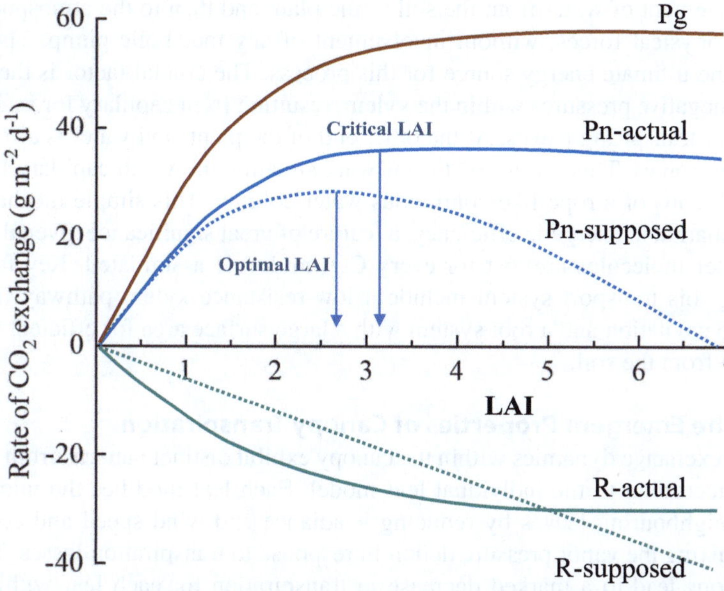

Fig. 6.5 Relationships between canopy CO_2 exchange rate and leaf area index (LAI). Dotted lines represent the hypothetical response curves of net photosynthesis (Pn-supposed) and respiration (R-supposed) to increasing LAI. Solid lines depict the observed response curves of net photosynthesis (Pn-actual) and respiration (R-actual) to increasing LAI. The critical and optimal LAIs are indicated by the arrows. Pg gross photosynthesis. (Redrawn from Hay and Porter (2006))

linearly. Once LAI exceeds approximately 4, the canopy stops intercepting additional radiation, and DM production no longer increases proportionally; in other words, Pg−R remains nearly constant. Essentially, when canopies stop intercepting more light, the rate of change in DM production becomes zero, resulting in a consistent RUE across all LAI values.

3.4 Evapotranspiration

3.4.1 Movement of Water from Soil to Plant

Living within Earth's atmosphere poses a significant challenge for terrestrial plants. On one hand, the atmosphere acts as a crucial source of CO_2, which is essential for photosynthesis. On the other hand, the generally arid atmospheric conditions lead to substantial water loss through evaporation. Since plants lack specialized surfaces that simultaneously promote CO_2 diffusion while preventing water loss, their uptake of CO_2 exposes them to the risk of dehydration. This issue is exacerbated by the fact that the concentration gradient driving CO_2 uptake is significantly smaller than that driving water loss. To address the conflicting requirements of maximizing CO_2 uptake while minimizing water loss, plants have evolved adaptations to control leaf water loss and replenish lost water from the soil.

The movement of water from the soil to the plant and then to the atmosphere is driven by physical forces, without involvement of any metabolic pump. The sun serves as the ultimate energy source for this process. The crucial factor is the generation of negative pressures within the xylem, resulting from capillary forces in the cell walls of transpiring leaves. At the other end of the plant, soil water is also held by capillary forces. This creates a "tug-of-war" situation, in which capillary forces act on both ends of a rope-like continuous water column. This simple mechanism confers remarkable energetic efficiency, a feature of great significance given that up to 400 water molecules are lost for every CO_2 molecule assimilated. Key factors facilitating this transport system include a low-resistance xylem pathway that is resistant to cavitation and a root system with a large surface area for efficient water absorption from the soil.

3.4.2 The Emergent Properties of Canopy Transpiration

The water exchange dynamics within the canopy exhibit distinct patterns from those expected according to the individual leaf model. Each leaf modifies the microclimate of neighbouring leaves by reducing irradiance and wind speed and concurrently adjusting the vapor pressure deficit in response to transpiration losses. These modifications lead to a marked decrease in transpiration for each leaf within the canopy compared to that predicted by the individual leaf model, which considers only the atmospheric conditions above the canopy. For instance, irradiance decreases exponentially with increasing LAI within the canopy, thereby reducing leaf energy absorption. Canopy-generated aerodynamic friction lowers wind speed near the canopy, and wind speed generally declines exponentially with canopy depth. Consequently, individual leaves within a canopy exhibit lower boundary-layer conductance than that predicted from the surrounding meteorological conditions. Finally, transpiration from each leaf leads to an increase in the water-vapor concentration inside the canopy, which is analogous to evaporation from a wet soil surface. The rising concentration of water vapor within the canopy weakens the vapor pressure gradient, the primary driving force for transpiration. Consequently, the rate of transpiration increases less than anticipated.

3.4.3 The Soil–Plant–Atmosphere Continuum

The soil–plant–atmosphere continuum (SPAC) represents a continuous pathway for water movement from the soil, through plants, to the atmosphere within a crop stand (Fig. 6.6). The term "continuum" underscores the continuous and interconnected nature of water transport along this pathway. Due to the difference in water potential between the atmosphere (highly negative) and the leaves (less negative), a diffusion gradient is established across the stomatal pores on the leaf surface, facilitating the loss of water vapor from plant tissues, the process of transpiration. Plants play a crucial role in this pathway, exhibiting diurnal and seasonal variations in water status in response to fluctuations in soil moisture availability and atmospheric evaporative demand. Moreover, water is lost directly to the atmosphere via soil evaporation. During the early vegetative phase, when LAI is too low to fully cover the ground, up

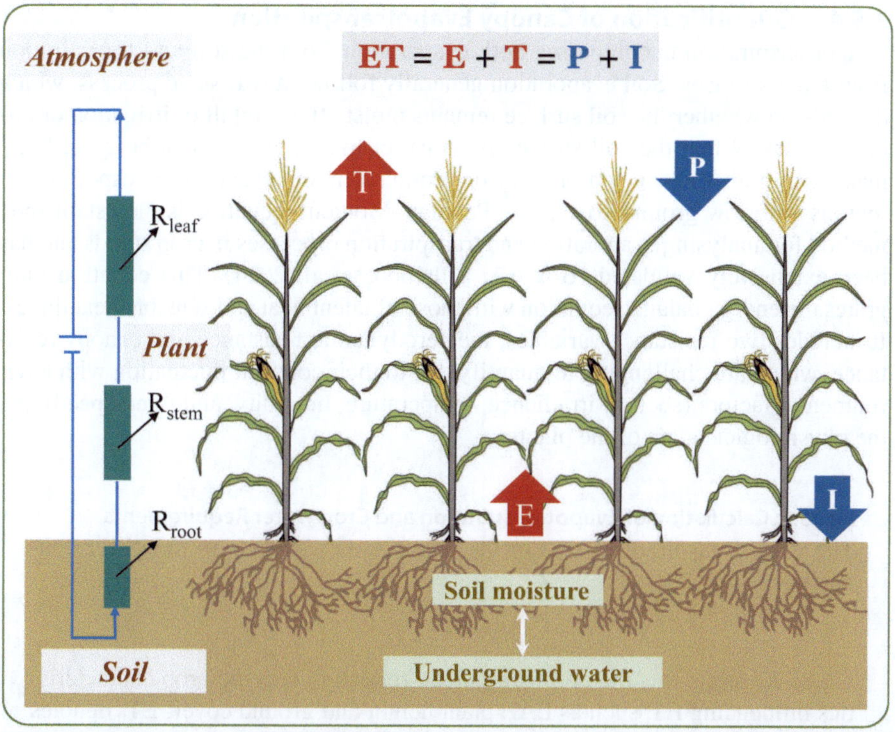

Fig. 6.6 The soil–plant–atmosphere continuum (SPAC) and the water balance equation in a maize field. (1) On the left, the diagram illustrates the theory of SPAC. Water flows more efficiently through certain components of the SPAC than others. For instance, water absorbed by roots must traverse multiple cell layers before entering the xylem vessels. These cell layers function as a filtration system within the root and exhibit significantly higher resistance to water flow compared to the xylem, where water transport occurs via open conduits. The relative ease with which water moves through different parts of the SPAC can be quantified using the equation: Flow = $\Delta\psi/R$. This equation parallels the description of electron flow in an electrical circuit according to Ohm's law: $i = V/R$, where R represents resistance, i denotes current or electron flow, and V signifies voltage. In the context of the SPAC system, V corresponds to the driving force of water flow, represented by the water potential difference ($\Delta\psi$), while i corresponds to the rate of water flow through or across a plant segment. By applying these analogies, Ohm's law can be utilized to calculate the hydraulic conductance (the inverse of hydraulic resistance R) of individual plant components (e.g. roots, stems, leaves) or the entire plant system (from soil to atmosphere). (2) On the right, the water balance equation for a maize field is presented, where the inputs consist of precipitation and irrigation, while the outputs include leaf transpiration and soil evaporation. This equation does not consider the contribution of groundwater. E evaporation from the soil, ET evapotranspiration of the field, I irrigation, P precipitation, R resistance, T transpiration from the leaves

to 50% of total water use may be lost through soil evaporation. In the studies of crop–water relationship and the irrigation management, the combined loss of water from the crop via transpiration and from the soil via evaporation is termed *evapotranspiration* (ET).

3.4.4 Quantification of Canopy Evapotranspiration

Evapotranspiration encompasses both evaporation from the soil and transpiration from plant surfaces. Soil evaporation generally follows a two-stage process, which depends on whether the soil surface remains moist after rainfall or irrigation or has become dry. When the soil surface is moist, evaporation rates can be quite high, making rainfall frequency a primary determinant of soil evaporation, especially in regions with low ground cover. The Penman–Monteith equation is the established method for analysing evaporation and transpiration processes in crop stands and has been extensively validated (Box 6.1; Villalobos et al. 2024). This equation integrates the energy balance equation with those of latent heat and sensible heat fluxes. It includes two resistance variables, the aerodynamic resistance and canopy resistance, which are challenging to quantify due to their constant fluctuation with environmental factors (such as irradiance, temperature, humidity, and wind speed) and the physiological state of the plants.

Box 6.1: Calculation of Evapotranspiration and Crop Water Requirements

The evapotranspiration (ET) in the absence of water stress may be calculated as:

$$ET = K_c\, ET_0$$

where K_c represents the crop coefficient, reflecting specific crop characteristics influencing ET, such as LAI, plant height and ground cover; ET_0 denotes the reference ET, defined as the ET of short (8–15 cm height) grass with complete ground cover and under non-limiting soil water and nutrient conditions. Ideally, ET_0 depends solely on meteorological variables, including irradiance, temperature, humidity, and wind speed. This equation remains valid unless water stress reduces ET, which typically occurs under the 65%–80% depletion of the available soil water. If depletion continues, K_c decreases linearly until it reaches zero at the permanent wilting point.

The application of this equation involves initially calculating ET_0 based on meteorological data, followed by applying a crop coefficient (K_c) that varies with crop developmental phases. The primary method for calculating ET_0 is through the FAO Penman–Monteith equation; however, when only air temperature data are available, the Hargreaves equation can serve as an alternative. K_c is determined using the FAO-recommended method, which employs linear functions to interpolate between the initial, peak, and harvest phases of the crop cycle. The initial Kc value is determined by the soil moisture status and the ET_0. For annual crops, the maximum Kc values generally range from 1.0 to 1.3.

Crop irrigation water requirement refers to the volume of water needed to maintain optimal ET. This can be determined using the water balance method, which calculates the difference between potential ET and the water supplied by precipitation or extracted from the soil during a specified period (Villalobos et al. 2024).

3.5 GHG Emission

Organic and inorganic materials, whether serving as inputs or outputs in agricultural systems, are predominantly decomposed through microbial processes, leading to the release of greenhouse gases (GHGs), such as CO_2, methane (CH_4), and nitrous oxide (N_2O), into the atmosphere. The agricultural sector is the leading global source of anthropogenic non-CO_2 greenhouse gases, contributing to the largest share of global CH_4 and N_2O emissions. For example, rice cultivation is considered as the major source of CH_4 emissions, while it also releases N_2O which primarily derives from the nitrogen fertilizers (Fig. 6.7). According to the IPCC (2023), with medium confidence, global agricultural activities contributed approximately 13% of

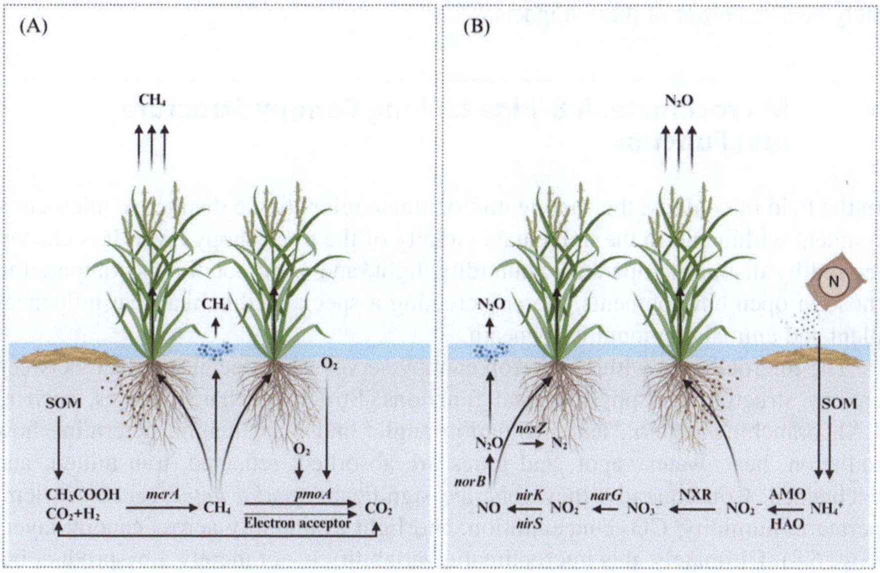

Fig. 6.7 Production and emission of CH_4 and N_2O in paddy fields. (**A**) Methane (CH_4) emissions are regulated by the interrelated processes of CH_4 production, oxidation, and transport. Methanogenesis is carried out by methanogenic archaea under anaerobic conditions, utilizing soil organic matter (SOM), including soil organic carbon, root exudates, and crop residues, as substrates. CH_4 oxidation is mediated by aerobic methanotrophic bacteria in surface soils and the rhizosphere, while anaerobic oxidation consumes 10%–20% of the methane produced through the reduction of nitrate, nitrite, iron, manganese, sulphate, and organic electron acceptors such as humic substances. During flooded conditions, plant-mediated transport of CH_4 through rice plants accounts for up to 90% of total emissions. *mcrA*, methyl coenzyme M reductase; *pmoA*, particulate methane monooxygenase. (**B**) Net nitrous oxide (N_2O) emissions are primarily driven by the microbial processes of nitrification and denitrification. Under flooded conditions, the majority of N_2O is emitted through the aerenchyma of rice plants; in contrast, under non-flooded conditions, N_2O is predominantly released via diffusion across the soil surface. AMO ammonia monooxygenase, HAO hydroxylamine oxidoreductase, NXR nitrite oxidoreductase, *narG* nitrate reductase, *nirK* copper-containing nitrite reductase, *nirS* haem-containing nitrite reductase, *norB* nitric oxide reductase, *nosZ* nitrous oxide reductase. (Redrawn from Qian et al. (2023))

CO_2 emissions, 44% of CH_4 emissions, and 82% of N_2O emissions during the period of 2007–2016. These activities accounted for around 23% (12.0 ± 3.0 Gt CO_2 equivalent yr^{-1}) of the total net anthropogenic GHG emissions. Paddy rice cultivation emerged as a significant source of global CH_4 emissions at approximately 11%, estimated to be between 493 and 723 Mt CO_2 equivalent yr^{-1} in 2010 (IPCC 2014).

Anthropogenic activities, primarily through the GHG emission, have unequivocally caused global warming, resulting in a global surface temperature increase of 1.1 °C above the average for 1850–1900 during the period from 2011 to 2020 (IPCC 2023). This has resulted in significant and observable changes in weather and climate extremes worldwide, triggering widespread adverse effects on food and water security, human health, economic stability, social systems, and biodiversity while simultaneously accelerating environmental degradation. Moreover, vulnerable populations, who have contributed the least to current climate change, disproportionately bear the brunt of these impacts.

4 Microclimate: A Bridge Linking Canopy Structure and Function

In the field of ecology, the canopy microclimate refers to the distinctive microenvironment within and in the immediate vicinity of the tree canopy layer. It is characterized by distinct temperature, humidity, light, and wind conditions compared to those in open environments, thereby creating a specialized habitat that influences plant and animal communities beneath.

The microclimate within the crop canopy serves as a crucial interface between canopy structure and physiological functions. Structural characteristics, such as LAI, branching pattern, leaf inclination angle, and plant height, determine how radiation, heat, water vapor, and gases are absorbed, reflected, transmitted, and exchanged. Consequently, they generate significant *spatial heterogeneity* in temperature, humidity, CO_2 concentration, and light availability across canopy layers (Fig. 6.8). Ultimately, this microclimatic variability is not merely a by-product but a direct reflection of the dynamic exchanges of mass and energy between plants and their surrounding environment. By measuring microenvironmental conditions at multiple depths within the canopy, researchers can assess how specific architectural traits regulate resource distribution and use patterns. Thus, the microclimate serves as a mechanistic bridge that translates physical structure into functional outcomes and lies at the core of eco-physiological studies on crop canopies.

4.1 Canopy Structure Creates a Heterogeneous Microclimate

4.1.1 Attenuation of Light Intensity

At the canopy level, the upper leaves primarily capture solar radiation and simultaneously attenuate its transmission to the lower leaves. Shaded leaves experience a 20- to 50-fold reduction in light intensity, as well as an alteration in spectral

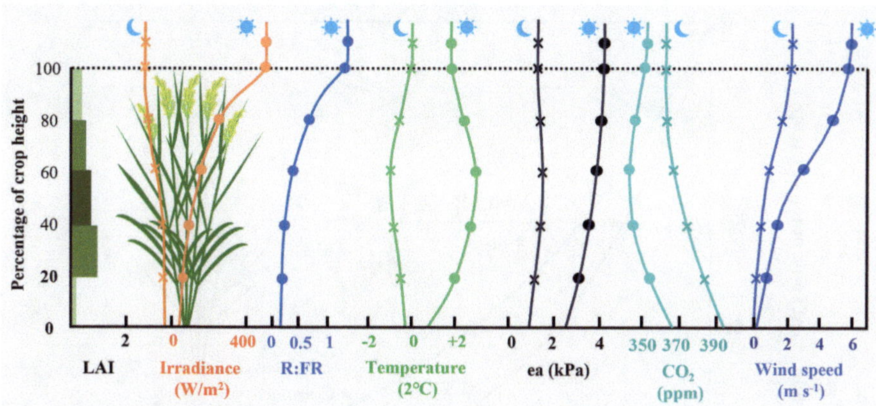

Fig. 6.8 Diurnal changes in microclimates within a rice canopy. (1) During the day, absorbed radiation causes an increase in temperature and water vapor pressure. Both light intensity (irradiance) and spectral composition (R:FR ratio) exhibit an exponential decrease from the top to the bottom of the canopy. CO_2 concentration reaches its minimum at the height of the assimilating leaf layer, particularly under conditions of reduced turbulence. Wind speed within the canopy is substantially reduced. (2) At night, a net loss of energy occurs, leading to the lowest temperatures at leaf level. CO_2 concentration peaks at the soil surface due to high rates of soil respiration and stable atmospheric conditions. The symbols of the sun and the moon represent daytime and nighttime, respectively. Microclimates of irradiance, R:FR, temperature, and actual water vapor pressure (ea) are depicted according to the field measurement of the canopy of a japonica rice cultivar NXJ-9 at heading stage; while those of wind speed and CO_2 are from Hay and Porter (2006) and Schulze et al. (2019). R:FR red to far-red, LAI leaf area index

composition, compared with their overlying counterparts. Precise characterization of the radiation distribution pattern within a canopy is challenging due to the complex canopy architecture, the angular variability of incident radiation, and the spectral properties of leaves. Nevertheless, several well-established simplifications provide adequate accuracy for a wide range of applications. Under the idealized assumption of a statistically random 3D distribution of leaves within the canopy, the attenuation of radiation through the canopy follows an exponential decay, described by the Beer-Lambert-Bouguer law of absorption (Monsi and Saeki 2005):

$$I = I_0 \, e^{-KL}$$

where I is the irradiance beneath the canopy; I_0 is the irradiance at the top of the canopy; K is the extinction coefficient; and L is the leaf area index.

The *light extinction coefficient*, K, is determined by the angle and orientation of foliage as well as its transparency and varies with crop species and cultivar. The representative extinction coefficients of the major crops, measured at high solar elevations, show a general trend where canopies of crops with erect and narrow leaves, like cereals, have lower K values compared to those of crops with flat, broad, and horizontal leaves, such as soybean and cotton. A low extinction coefficient enables more efficient light penetrate through the canopy. Nevertheless, K values

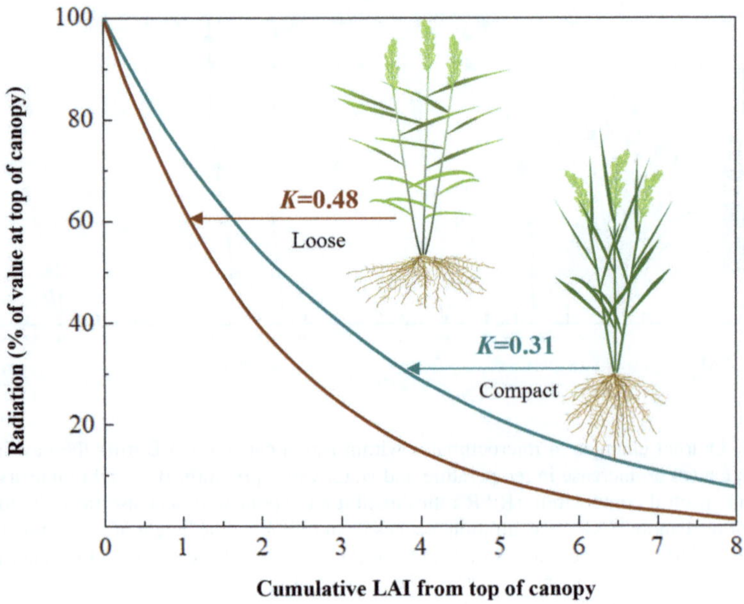

Fig. 6.9 The exponential attenuation of light within the canopy as affected by the light extinction coefficient *K*. The cultivar with a compact architecture exhibits a lower *K* value compared to that with a loose architecture, thereby enabling greater light penetration into the lower canopy strata at the same leaf area index (LAI). (The data utilized in drawing this figure were sourced from Li et al. (2024))

vary substantially among crop species, and this variation is influenced by developmental stage and genotypic structural features (Fig. 6.9).

4.1.2 Modification of Light Spectrum

Plant leaves exhibit unique spectral absorption and reflection characteristics, primarily determined by the absorption properties of chlorophyll. The leaves absorb blue (B) and red (R) light for photosynthesis in the chloroplasts, while reflecting unabsorbed green (G) light, which is why plants appear green to the human eye. In addition to green light, leaves reflect and transmit a significant portion of the incident far-red (FR) radiation, which is imperceptible to the human eye. Consequently, due to chlorophyll's selectivity for specific wavebands, the spectral composition of light within the canopy differs markedly from that of sunlight (Huber et al. 2021; Pierik and Ballaré 2021).

The initial light cue reflected from neighbouring plants manifests as a shift in the R:FR ratio. The sunlight generally shows an R:FR ratio of approximately 1.2. In contrast, under the deep shade of the canopy, neighbouring plants can decrease this ratio to as low as 0.1. Interestingly, R:FR can drop before true shading occurs due to neighbouring plants reflecting FR light, even before they overlap. The initial decrease in the R:FR ratio serves as an early warning signal of impending competition for light and is followed by a reduction in total light intensity and depletion of blue light as the canopy develops further and true shading occurs.

4.2 Microclimate Regulates the Functional Response of Crops

The canopy structure generates a heterogeneous microclimate within the crop stand, which significantly influences key functional traits including organ development and growth, physiological processes related to yield and quality formation, and resistance to both biotic and abiotic stresses.

4.2.1 The Attenuation of Light Intensity and Its Influence on RUE

Vertically oriented leaves exhibit a smaller projected surface area during midday hours, facilitating deeper light penetration into the lower canopy layers. Consequently, erectophile canopies enable more uniform light distribution across the canopy compared to planophile canopies. Long et al. (2006) carried out a simulation study to compare RUE between two plant types: Plant X, characterized by predominantly horizontal leaves, and Plant Y, featuring vertically oriented leaves at the top that gradually transition to a more horizontal orientation towards the base. The authors concluded that light distribution within the canopy of Plant Y is more optimal, resulting in higher canopy photosynthesis (Fig. 6.10). Older cultivars characterized by more horizontally positioned leaves, such as plant X, have been replaced by newer cultivars bred to possess predominantly vertical leaves in the upper layer, like plant Y. This change in canopy design offers maximum advantages under *direct radiation* when the sun is directly overhead; however, the benefits gradually diminish as the sun angles decrease or under *diffuse radiation*.

The influence of leaf angle distribution becomes increasingly significant in canopies with higher LAI. As solar angles changes throughout the day, the effect of leaf angle is minimal when LAI is below approximately 3. Previous studies have demonstrated that canopies with low LAI perform better under a more horizontal leaf posture, whereas those with high LAI achieve greater efficiency with a more upright leaf orientation. However, it has been observed that wheat lines characterized by upright leaves demonstrate higher grain yields, indicating that improvement in leaf erectness may confer additional advantages in certain cereal canopies. Modern high-yielding rice cultivars demonstrate a remarkable leaf erectness, a characteristic that is extensively acknowledged as a sign of superior yield potential. Moreover, this morphological characteristic enables higher planting densities, which contribute to more rapid canopy closure and improved light interception, particularly during the limited growing season in tropical environments.

4.2.2 Photoreceptors in Response to the Changes in Spectral Composition

The canopy acts as a spectral filter, attenuating blue and red wavelengths while transmitting a higher proportion of green and far-red light (Fig. 6.11). Higher irradiance is observed at the canopy surface across most wavelengths, particularly within

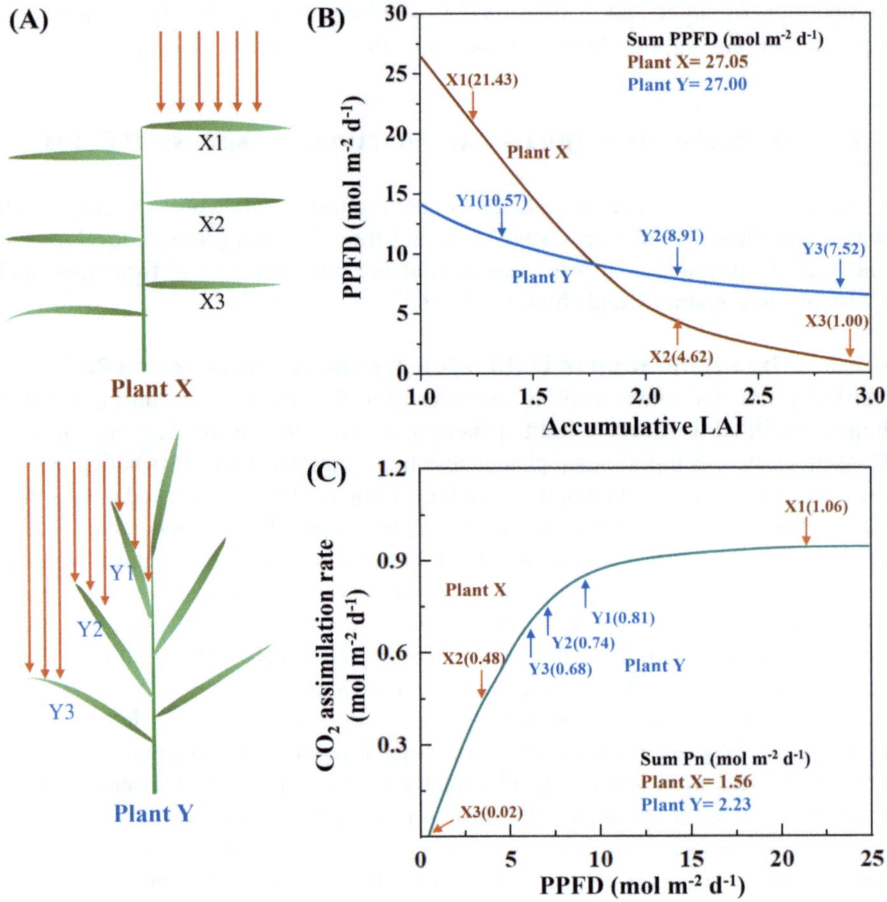

Fig. 6.10 Simulation of light use efficiency between plant types with contrasting leaf angles. The simulation was conducted using the Windows Intuitive Model of Vegetation response to Atmosphere and Climate Change (WIMOVAC) on the 190th day of the year at a latitude of 52° N, assuming an LAI of 3 and maintaining a constant canopy temperature of 25 °C. (**A**) Plant X exhibits its predominantly horizontal leaf orientation, resulting in the upper layer effectively intercepting a majority of incident solar radiation and providing shade to the lower layers. Conversely, the leaves of Plant Y display vertical orientation at the apex but gradually transition into a more horizontal arrangement towards the base, facilitating a more equitable distribution of light across different layers. (**B**) The predicted average photosynthetic photon flux density (PPFD) at noon at different canopy depths represented by accumulative LAI for plant X and plant Y. In plant X, the uppermost layer intercepts most of the light at midday, while about 10% may penetrate to the next layer and 1% to the layer below that. The PPFD intercepted per unit leaf area by an almost horizontal leaf at the top of a plant X is 1400 μmol m^{-2} s^{-1} or more (data not shown in the figure), which is approximately three times the amount required to saturate photosynthesis. Therefore, at least two-thirds of the energy intercepted by the upper leaves goes to waste. (**C**) The response of photosynthetic CO_2 uptake rate to PPFD. By evenly distributing the light energy among leaves, plant Y would achieve nearly 1.43 times the efficiency of RUE compared to plant X at midday under full sunlight. This example oversimplifies the situation, however, because the sun is only directly overhead within the tropics; at all locations, solar angle continually changes. LAI leaf area index, RUE radiation use efficiency. (Redrawn after Long et al. (2006))

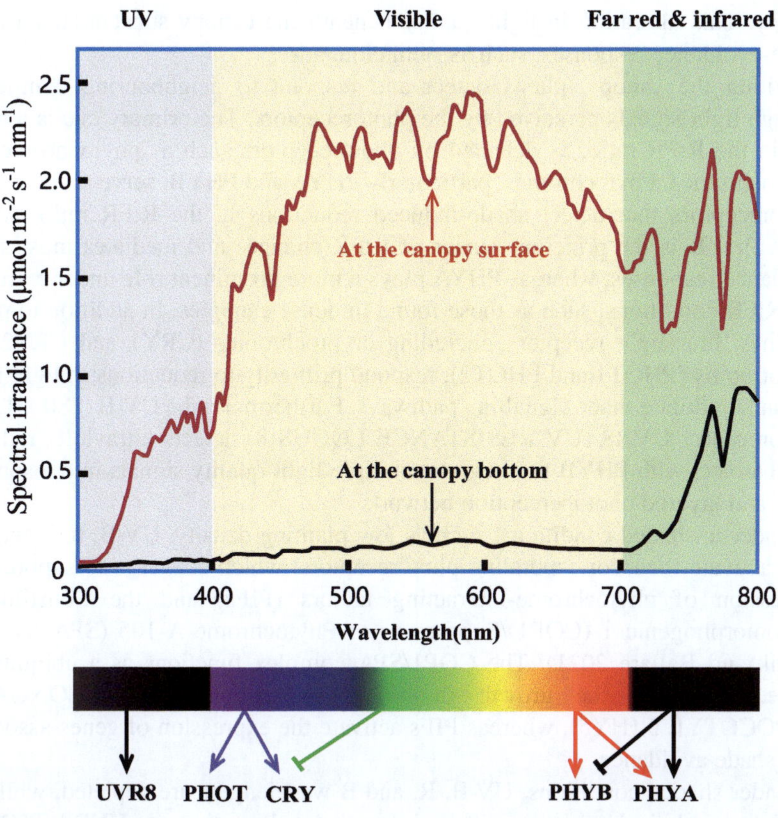

Fig. 6.11 Perception of shading signals by photosensory receptors. (1) The upper panel depicts the variation in light intensity as a function of wavelength for two distinct positions: at the canopy surface (red line) and beneath the canopy (black line). The red line shows the full solar spectrum with high irradiance across the visible range (400–700 nm) and a sharp decline in the ultra-violet (<400 nm) and far-red (700–800 nm). The black line illustrates the light environment under the canopy, where PAR is greatly reduced due to absorption by leaves. The relative proportion of far-red light is significantly increased in shaded conditions because chlorophyll absorbs less in this region. The measurement was conducted at 12:00 noon on September 12, 2025, at the Danyang Experimental Station (31°54′31″ N, 119°28′21″ E), using a spectral colour illuminometer (HP350UVE). The measured canopy consists of a japonica rice cultivar (NXJ-9) at the heading stage, with a LAI of 5.39. (2) The lower panel illustrates the effects of different spectral regions on the activation states of key photoreceptors, including phytochromes (PHYA and PHYB), cryptochromes (CRY), phototropins (PHOT), and UV RESISTANCE LOCUS 8 (UVR8). Lines terminating with an arrowhead indicate promotion, whereas those ending with a short line indicate suppression. (Redrawn from Casal (2013) and Taiz et al. (2015))

the visible spectrum (400–700 nm), which corresponds to PAR. In contrast, irradiance is substantially reduced beneath the canopy, with the most pronounced attenuation occurring in the blue (approximately 400–500 nm) and red (approximately 600–700 nm) regions. Light in the green (approximately 500–600 nm) and far-red (approximately 700–800 nm) ranges exhibits greater penetration through the canopy. This alteration in light quality beneath the canopy substantially triggers shade-avoidance responses, such as stem elongation.

Within the canopy, plants detect and respond to neighbouring competitors through light signals perceived by the photoreceptors. The primary cue is a reduction in the R:FR ratio, as detected by photoreceptors such as phytochromes and cryptochromes. Phytochromes, particularly PHYA and PHYB, serve as the primary photoreceptors that detect shade-induced reductions in the R:FR ratio. Among them, PHYB is the principal sensor of R:FR changes and mediates most shade-avoidance responses, whereas PHYA plays a more prominent role under extremely low R:FR conditions, such as those found in dense canopies. In addition to phytochromes, blue-light receptors, including cryptochromes (CRY1 and CRY2) and phototropins (PHOT1 and PHOT2), respond primarily to reductions in light intensity and modulate shade signalling pathways. Furthermore, the UV-B (280–315 nm) photoreceptor UVR8 (UV RESISTANCE LOCUS 8) detects ultraviolet radiation and interacts with PHYB to integrate multiple light-quality signals into a sophisticated and layered photoperception network.

Under unshaded conditions, such as low planting density, UV-B, red, and blue light activate their corresponding photoreceptors, which converge to suppress the expression of phytochrome-interacting factors (PIFs) and the Constitutively Photomorphogenic 1 (COP1)/Suppressor of Phytochrome A-105 (SPA) complex (Pierik and Ballaré 2021). The COP1/SPA complex functions as a ubiquitin E3 ligase, degrading the growth-inhibiting transcription factor ELONGATED HYPOCOTYL 5 (HY5), whereas PIFs activate the expression of genes associated with shade avoidance.

Under shaded conditions, UV-B, R, and B wavelengths are depleted, while far-red light is relatively abundant. This shift reduces the activity of UVR8, PHY, and CRY, thereby lifting their inhibition on the complex of COP1/SPA and PIFs. The progressive inactivation of all photoreceptors and the reduced light capture by chloroplasts synergistically enhance shade-avoidance responses, regulated by PIF-interacting proteins, such as DELLA and PHYTOCHROME RAPIDLY REGULATED (PAR).

Under high planting density, maize employs a sophisticated strategy to detect and transmit shading signals. Tian et al. (2024) identified a novel gene, *leaf angle and canopy architecture 1* (*lac1*), encoding a C-22 hydroxylase enzyme that catalyses a critical step in the biosynthesis of BRs. It is directly activated by RAVL1, a key transcription factor that regulates leaf angle. The *lac1* gene promotes BR accumulation in the leaf-sheath joint, thereby modulating leaf angle. Under shaded conditions, *lac1* expression is significantly downregulated. As planting density increases, the declining R:FR ratio triggers the accumulation of PHYA protein. PHYA then binds to RAVL1 and targets it for degradation in the 26S proteasome, thereby suppressing RAVL1-mediated activation of *lac1*. This leads to reduced BR levels and a smaller leaf angle. In *lac1* mutants, the PHYA–RAVL1–*lac1* signalling cascade is disrupted, resulting in diminished sensitivity to shade. Consequently, *lac1* plays a dual role: it optimizes canopy architecture for improved light distribution and attenuates shade-avoidance responses, coordinately enhancing both morphological and physiological traits to increase yield under high-density planting conditions.

4.2.3 Shade Avoidance Syndrome Under High Planting Density

High plant densities are prevalent in most farmlands. In such environments, intense competition for resources compels individual plants to adjust both their physiological processes and architectural traits to optimize resource acquisition. Light signals, particularly reductions in the R:FR beneath dense canopies, substantially influence intercalary meristem activity and internode elongation. Upon perceiving these signals, plants activate a suite of developmental responses collectively termed the *shade-avoidance syndrome* (SAS).

SAS denotes the coordinated morphological, physiological, and developmental changes plants undergo when they sense neighbouring vegetation and consequent light quality shifts. Key characteristics of the SAS include (1) stem and petiole elongation, plants increase in height or extend petioles to elevate leaves above neighbouring vegetation and reduce shading; (2) leaf *hyponasty*, leaves bend upward, adopting a more vertical orientation to enhance light interception; (3) suppressed branching and lateral growth, increased apical dominance limits the development of lower branches, minimizing self-shading; (4) accelerated flowering, plants hasten the transition to reproductive development to complete seed production before canopy closure intensifies shading; and (5) altered leaf morphology, leaves become thinner and larger in surface area, resulting in a higher SLA, while overall leaf expansion is reduced under shaded conditions.

At the population level, azimuthal canopy reorientation functions as an effective strategy for plants to optimize light distribution within the canopy, thereby minimizing mutual shading under high planting densities. This process entails dynamic adjustment of the azimuthal (horizontal) orientation of leaves around the stem relative to the planting row or axis. Specifically, plants can rotate or twist their leaves to direct them towards open inter-row spaces rather than maintaining parallel alignment with the rows. This adaptive behaviour is especially pronounced in densely planted crops such as maize. Girardin and Tollenaar (1994) observed that raising the planting density of from 70,000 to 100,000 plants ha^{-1} led to a marked increase in the proportion of upper canopy leaves oriented nearly perpendicular to the row directions. Maddonni et al. (2002) later demonstrated that this reorientation is driven by changes in light quality: at high densities, inter-row spaces exhibit a reduced R:FR ratio, which serves as a shade-avoidance signal. In response, maize leaves adjust their growth orientation away from neighbouring plants, increasing the leaf-to-row angle and minimizing mutual shading.

Zhou et al. (2024) explored how maize plants regulate the azimuthal orientation of their leaves, through a GWAS conducted on a genetically diverse panel. They identified several candidate genes associated with this trait, including *liguleless 1* (*lg1*), along with *lg2* and *Lg3*, as well as SAS-related genes such as *phyC2*. Field experiments conducted under high planting densities revealed that mutants of both *liguleless* and *SAS* genes exhibit significantly reduced inter-row leaf development, resulting in markedly lower PAR interception compared to WT. The findings demonstrate that *liguleless* genes, previously known primarily for their role in ligule formation, are also involved in light-responsive signalling and canopy reconfiguration under high planting density. By establishing a genetic link between specific loci and canopy geometry, this study provides a mechanistic foundation for breeding maize cultivars with improved canopy structures, offering promising pathways to enhance yield stability in dense cropping systems.

5 Breeding and Cultivating a Healthy Crop Canopy

5.1 The Quest for a Healthy Canopy

The field of agriculture has experienced substantial transformations, evolving from the use of ox and horse carts and manual labour to the employment of tractors, planters, spreaders, sprayers, harvesters, and drones. Consequently, crop research has evolved into a more specialized, technologically advanced, and data-driven discipline. The purpose and significance of it, however, remain unchanged. It continues to provide an integrated systematic approach to crop production that meets the ever-growing food demands. The challenges, however, become significantly intricate as we endeavour to identify the most resilient, viable, and sustainable routes for delivering nutritious, wholesome, and prosperous food while concurrently enhancing and safeguarding the environment. For example, pests present a major constraint on global crop productivity. Roughly 2 billion kilograms of pesticides are applied to crops worldwide each year, yet pests still destroy 20% to 40% of global production. The widespread application of synthetic pesticides has contributed to the rapid development of resistance and has exerted adverse effects on soil quality, biodiversity, and human health.

The canopy structure generates heterogeneity in the microenvironments within the canopy, establishing diverse and variable conditions not only for the growth, development, and functioning of the plant itself but also for numerous other organisms, including pathogens, insects, and wildlife. Shaping a *healthy canopy* may be an effective way to meet the challenges facing the crop industry. A healthy canopy can be defined as a high-quality crop canopy that exhibits vigorous growth, vitality, and resilience. It is characterized by high resource use efficiency, low incidence of pests and diseases, strong lodging resistance, and good balance between quantity and quality. These characteristics can be achieved through optimizing the canopy structure to improve microenvironments that favour the functioning of crop plants.

While achieving high yield and quality is the top priority for growers, these goals are often challenged by a wide variety of pests and diseases. However, there has been limited attention given to using ideotype design to create plants that can naturally prevent or limit the spread of diseases or pests. Canopy structure alters the susceptibility of organs or promotes mechanisms that facilitate infection escape or enhance tolerance. It influences microclimatic conditions such as light, temperature, humidity, and wind velocity, which in turn impact the spread of plant diseases (Grumet et al. 2013). Morphological modifications, including a decrease in branching density, a reduction in tiller number, and a more upright leaf orientation, improve canopy porosity and facilitate greater light penetration through the canopy. This leads to cooler, less humid microenvironments that limit pathogen production, spread, and effectiveness, slowing the progression of disease outbreaks. Therefore, cultivating a healthy canopy can be a valuable part of a broader disease control strategy that also includes genetic resistance and pesticide controls (Ando et al. 2007).

Fusarium head blight (FHB), caused by a complex of *Fusarium* species, results in substantial yield losses in wheat under prolonged moist conditions. It is widely recognized for its production of mycotoxins, such as deoxynivalenol (DON, also known as vomitoxin), which are highly toxic to both humans and animals. The *Fusarium* pathogens are disseminated by wind or rain and land on wheat ears, the most vulnerable organ, particularly during the flowering and early grain-filling stages. To date, no cultivars have been identified that exhibit complete immunity to FHB. QTL analysis indicates that resistance to the disease is governed by multiple genes. Interestingly, taller winter wheat lines exhibited lower FHB severity compared to shorter lines, presumably due to the greater distance between the ears of tall lines and the soil surface, where the disease-causing agents are predominantly located (Hilton et al. 1999). Near-isogenic lines harbouring the dwarfing gene *Rht1* exhibited a high degree of sensitivity to FHB compared with those lacking the gene, regardless of genetic background. In addition, no significant differences in relative humidity at ear height were observed between tall and short lines. These findings collectively suggest that the lower disease incidence in taller lines may be more directly attributable to the greater distance from the soil surface rather than to alterations in canopy microclimate conditions. The FHB case illustrates that modifying structural traits, such as plant height and ear position within the canopy, can effectively enhance crop resistance to disease, thus shaping a heathy canopy.

5.2 Genetic Approaches

Radiation capture is a highly dynamic process influenced not only by solar angle and variations in radiation intensity but also by the morphological structures that develop above ground throughout a plant's life cycle. In crop genetics and breeding, there is considerable interest in genetically modifying canopy structural traits that optimize light interception and distribution to improve RUE and grain yield.

5.2.1 Erect Canopy

Theoretical considerations indicate that crops with an erectophile canopy, in which the upper leaves are in an upright position, present advantages compared to those with floppy upper leaves when the LAI prior to flowering surpasses 3. The key advantage of an erectophile canopy lies in its ability to facilitate greater light penetration. Specifically, the upright orientation of the upper leaves allows more light to reach the lower leaves while also deflecting light deeper into the canopy. Therefore, under moderately favourable conditions and when row spacing is less than 30 cm, gross canopy photosynthesis and biomass production are expected to be higher in canopies with erect leaves.

Sadras et al. (2012) carried out a retrospective analysis of wheat cultivars covering five decades of breeding in Australia. They found that the fractions of incident radiation at different depths within the canopy increased over time, corresponding to the year of cultivar release. They also observed a continuous increase in RUE during the period from stem elongation to anthesis, the critical period determining grain number and potential grain size. In this case, the photosynthetic capacity and respiration activities of the flag leaf at anthesis did not differ significantly among the cultivars released in contrasting decades. Therefore, the observed changes could not be ascribed to variations in the photosynthetic rate of individual leaves; rather, they resulted from changes in canopy-level photosynthesis.

Similarly, Richards et al. (2019) investigated the canopy architecture and yield of an irrigated experiment involving 1000 wheat lines from a 4-way MAGIC population and another experiment with almost 300 lines from a more diverse MAGIC population formed by eight parents. After aggregating lines with similar canopy architecture, they discovered a tight association between grain yield and canopy architecture, with erect canopies typically exhibiting higher yields. In both populations, the most erectophile lines exhibited an average grain yield 24% higher than that of the most planophile lines. This notable increase can be mainly attributed to greater aboveground biomass and a larger grain number.

Leaf angle, which is defined as the angle between the culm and the leaf midrib, serves as a key determinant of an erect architecture. The ligular region, situated between the maize leaf blade and sheath, consists of the ligule and auricle and plays a pivotal role in regulating leaf angle. Liguleless 1 (LG1) and Liguleless 2 (LG2) are critical regulators involved in lamina joint formation. LG1 has been identified as a transcription factor belonging to the Squamosa Promoter-Binding Protein (SBP) family, whereas LG2 functions as a basic leucine zipper (bZIP) transcription factor. Genetic analyses of natural variations suggest that *LG1* and *LG2* have the potential to significantly improve the field productivity of maize cultivars. Furthermore, it has been demonstrated that the functions of *LG1* and *LG2* are conserved across various Gramineae crops, including maize, rice, wheat, sorghum, and sugarcane.

Mutant studies in maize have identified genes that are essential for the development of the ligular region. For instance, *lg1* and *lg2* mutants exhibit an erect leaf phenotype due to defects in ligule and auricle development. This erect leaf structure enhances light penetration into the middle and lower canopy layers of the maize, thereby facilitating denser planting arrangements. Comparisons of hybrids released in the United States from 1930 to 1990 reveal that more hybrids with erect leaves were introduced after 1970. Similar trends have been observed in Chinese and European hybrids (Perez et al. 2019). These leaf angle changes are thought to be an indirect response to higher plant densities, which reduce light penetration due to increased shading (Wei et al. 2018). This hypothesis is supported by studies that introgressed single-gene mutants and regulatory elements, such as *lg2*, *upright plant architecture 2* (*upa2*), and *rav-like 1* (*ravl1*), into inbreds or hybrids to generate more upright leaf architectures. At high plant densities, these introgression lines exhibited significantly higher grain yields compared to the WT (Tian et al. 2019).

In comparison to the substantial progress made in improving the plant architecture of major cereal crops like maize and rice, relatively less attention has been devoted to the architectural characteristics of foxtail millet. Zhang et al. (2025) employed genome-editing technology to target *SiLG1* and *SiLG2*, genes that regulate the leaf angle in foxtail millet, generating two mutants, *silg1* and *silg2*, with compact plant architectures. The leaf angles of these mutants were decreased by 83% and 80%, respectively, compared to the WT, while light penetration enhanced by 10.5‰ and 7.1‰, respectively. Field experiments demonstrated that the yields of the two mutants were increased by 20% and 9.6%, respectively. This study demonstrates the feasibility of enhancing the yield of foxtail millet by shaping an erect canopy.

In addition to leaf angle, inflorescence morphology also contributes to the erectness of crop canopy. In japonica rice, the erect panicle architecture is conferred by a heterotrimeric G protein gene named *DENSE AND ERECT PANICLE 1 (DEP1)*, which plays a role in the improvement of yield. Fei et al. (2019) compared the canopy structure of a curved panicle variety Sasanishiki (WT) and an erect panicle mutant derived from CRISPR/Cas9 gene editing at the *DEP1* locus. The results showed that more light could reach to the leaves under the panicle in the *CRISPR-dep1* population compared with the WT. The canopy of the *CRISPR-dep1* population exhibited higher temperature and lower humidity and CO_2 concentration and achieved higher yield through increased transplanting density.

5.2.2 Smart Canopy

While erect canopies facilitate dense plantings and consequently higher yields, a specific arrangement of leaves within the canopy, the smart canopy, has been proposed as the optimal architecture to improve RUE (Ort et al. 2015). The smart canopy ideotype features leaves that are more upright at the top and progressively more horizontal towards the bottom of the plant, the same as plant type Y in the proceeding section (Fig. 6.10). It envisions an aggregation of plants that participate in cooperative interactions at the canopy level, rather than competitive ones, to optimize light interception, transformation, and biomass production per unit land area. Additional components of the smart canopy include enhanced metabolic characteristics, such as differential catalytic efficiencies of Rubisco, as well as spatially varying antenna sizes and reaction centre densities across the canopy.

To date, genetic control of leaf angle has been studied primarily using one or two representative leaves, or by aggregating data from multiple leaves to calculate average values. This approach has limited the understanding of intraplant variation in leaf angles across canopy layers and the underlying genetic regulation. Mantilla-Perez et al. (2020) conducted GWAS and QTL mapping in sorghum to identify genes associated with the smart canopy architecture. They utilized layer-specific angle data obtained through both manual measurements and high-throughput phenotyping methods. Several genomic regions were associated with leaf angle within layers or across the canopy. The expression of the brassinosteroid-related transcription factor *BZR1/BES1* and the auxin-transporter *Dwarf3* were identified as being highly correlated with leaf angle distribution across the canopy.

Increasing planting density is a critical strategy for enhancing maize yields. An ideal plant type for high-density planting requires an optimized canopy structure. Tian et al. (2024) identified the *leaf angle architecture of smart canopy 1* (*lac1*), a natural mutant with upright upper leaves, less erect middle leaves and relatively flat lower leaves. *lac1* exhibits improved photosynthetic capacity and attenuated responses to shade under dense planting conditions. Accordingly, Tian et al. (2024) proposed an accelerated strategy for developing high-density-tolerant maize cultivars by leveraging *lac1* as a genetic chassis to facilitate the engineering of a smart canopy.

Another strategy for shaping a smart canopy involves modifying the light-harvesting apparatus (Smith et al. 2023). In densely cultivated fields, the majority of light is absorbed by the uppermost leaves in the canopy, significantly reducing the amount of light available to lower layers. In this scenario, plants with lighter-coloured leaves can facilitate a more uniform distribution of light throughout the canopy, which has been demonstrated to enhance overall crop productivity (Cutolo et al. 2023). Furthermore, excess nitrogen that is no longer required for the construction of the pigment-binding machinery in photosynthesis could be allocated to Rubisco and rate-limiting proteins of the photosynthetic machinery, thereby enhancing leaf photosynthetic capacity (Niinemets 2023).

5.3 Chemical Intervention

The ability to resist lodging is a critical trait for maintaining a healthy canopy structure. Lodging, defined as the permanent displacement of cereal stems from their vertical position, typically occurs after ear or panicle emergence. It can result in yield losses of up to 80%, along with several adverse effects such as diminished grain quality, increased drying costs, and prolonged harvesting times, posing a significant challenge to cereal production worldwide. Controlling lodging is a multifaceted process influenced by various factors, including wind, rainfall, soil characteristics, crop management practices, and disease prevalence (Savin and Slafer 2019). The increased availability of nitrogen, whether from organic matter mineralization or inorganic fertilizers, frequently exacerbates lodging issues.

Chemical intervention involves the application of *plant growth regulators* (PGRs) to regulate the overall morphology of plants, including plant height, internode length, branching angle, and leaf orientation. This regulation enables crops to more effectively adapt to high-density, mechanized production systems and enhances their resistance to lodging. In the United Kingdom, PGRs are applied to ~90% of winter wheat crops. These regulators are primarily categorized into two groups: inhibitors of GAs biosynthesis and ethylene-releasing compounds. Common GA biosynthesis inhibitors used in cereal production include chlormequat chloride, mepiquat chloride, trinexapac-ethyl, and prohexadione-calcium, while ethephon is the most widely used ethylene-releasing compound.

Increasing the planting density of maize crop is one of the most effective strategies to enhance grain yield. However, exceeding the optimal plant density can lead to increased lodging, which in turn alters the canopy structure and negatively impacts crop performance. Spraying a chemical called DHEAP[4] at the seven-expanded-leaf stage can effectively reduce maize lodging rate and improve photosynthetic performance of the maize ear leaf, thereby increasing grain yield under high densities. This is achieved by reducing plant and ear heights, increasing mean tilt angle, and improving light penetration in the middle leaf strata of the canopy (Huang et al. 2021).

6 Concluding Remarks and Perspectives

In the context of a changing climate and a growing global population, the demand for more productive and resilient crops is escalating. A multitude of studies have demonstrated that modifying one or a few genes can lead to yield increases ranging from 10% to 68% in rice, maize, tobacco, and soybean. Nevertheless, only a small number of these advancements have resulted in quantifiable yield improvements under field conditions (Khaipho-Burch et al. 2023). This disparity arises because yield evaluations in these studies were conducted at the organ or individual-plant level within controlled environments, such as greenhouses or small experimental plots, rather than at the crop level under real-world farming conditions. To bridge this gap, crop geneticists should improve experimental designs to minimize edge effects and other environmental and biological factors that may bias yield measurements. More importantly, it is essential to maintain standardized planting densities comparable to those used in commercial farmland. In this regard, a comprehensive understanding of canopy structure and function is indispensable for addressing the challenges facing the contemporary molecular breeding.

Progress in crop improvement can be significantly accelerated if the phenotypic outcomes of genetic modifications can be accurately forecasted. Nevertheless, the development of a scalable predictive framework from genotype to phenotype continues to pose challenges, owing to the biological complexity of genetic regulation, environmental influences, and the interactions among the fundamental processes of growth and development. Plant modelling offers a promising approach for dealing with this complexity (Hammer et al. 2006). Functional–structural plant models (FSPMs) examine and integrate the relationships between plant structure and the physiological and developmental processes underlying growth and development (Louarn and Song 2020). FSPMs are expanding research themes from microscopic cellular processes to macroscopic canopy-scale phenomena, thereby facilitating the translation of genetic variation into field performance (Soualiou et al. 2021).

[4] N, N-Diethyl-2-hexanoyl oxygen radicals-ethyl amine (2-ethyl chloride) phosphonic acid salt

Moreover, rapid progress in phenomics enables precise, high-throughput quantification of canopy structure and function, as well as structured integration of genetic and phenotypic information. With this support, crop modelling can focus on analysing the emergent properties of the canopy that stem from the interactions of individual plants or organs, thus bridging the knowledge gap between gene function and crop performance. Close collaboration across multiple disciplines, such as crop modelling, genetics, physiology, breeding, and agronomy, is essential to accelerate progress towards this goal.

Summary

1. The crop canopy demonstrates emergent properties that cannot be directly inferred from the characteristics of individual plants, but instead arise from their interactions.
2. Canopy structure refers to the spatial arrangement of aboveground plant organs within a crop stand, as well as the competitive relationships and interactions that arise among individual plants.
3. Canopy function is defined as the exchange of matter, energy, and information between the canopy and its surrounding environment, including radiation capture, photosynthesis, respiration, evapotranspiration, and greenhouse gas emission.
4. The canopy structure creates a heterogeneous microenvironment within the crop stand, significantly influencing key functions such as yield and quality formation, as well as resilience to both biotic and abiotic stresses.
5. A healthy canopy refers to a high-quality crop canopy characterized by vigorous growth, vitality, and resilience. Shaping a healthy canopy can serve as an effective alternative strategy for addressing the challenges facing the crop industry.

Important Terms

Canopy; Canopy function; Canopy structure; Emergent property; ET; Function; GHG; Growth respiration; Healthy canopy; Heterogeneity; Maintenance respiration; Microclimate; Microenvironment; RUE; Smart canopy; Spatial arrangement; SPAC

Review Questions

Multiple Choice

1. Within the canopy, (   ) shows a descending trend from the top to bottom.
   (a) CO_2
   (b) Temperature
   (c) Actual water vapour pressure
   (d) Solar radiation
2. Regarding the canopy property, which of the statements below is NOT right (   )
   (a) The canopy is not single-layered; the upper canopy has a strong impact on the one below.
   (b) Microclimates within the canopy are largely heterogeneous, both horizontally and vertically.
   (c) The functioning of the canopy, unlike individual plants, cannot be analysed in terms of its integrated response to environmental factors.
   (d) Contrasting canopy structures shaped by different management approaches can achieve similar yield level (functional trait) in crop populations.
3. On the light environment within the canopy, which is NOT right (   )
   (a) Upper canopy leaves predominantly capture solar radiation while attenuating its transmission to lower canopy leaves, resulting in a substantial variation in light availability ranging from 20- to 50-fold.
   (b) Lower leaves experience reduced light intensity and altered spectral composition compared to their overlying counterparts, leading to considerably diminished rates of photosynthesis.
   (c) Although sunlight generally exhibits an R:FR ratio of approximately 0.1, the presence of neighbouring plants can elevate this ratio to as high as 1.2 under deep canopy shade.
   (d) R:FR drops before true shading occurs due to neighbouring plants reflecting FR light, even before they overlap, serving as an early warning sign of upcoming competition for light.

True or False

1. For most crops, in the absence of biotic and abiotic stress, the amount of dry matter accumulated is linearly related to the intercepted PAR by the canopy (   )
2. Typically, an LAI of 3–5 is generally required for 90% light interception (   )
3. The available evidence indicates that under optimal conditions, respiration accounts for an estimated 40–50% of carbon assimilates (   )
4. Maintenance respiration involves the turnover of proteins and lipids, as well as the maintenance of electrochemical gradients across membranes (   )

Blank Filling

1. The crop canopy exhibits emergent properties that are not directly deducible from the attributes of the constituent individuals but depend on their ___________.
2. Canopy structure refers to the ___________ of aboveground plant organs within a crop stand, as well as the resulting competitive relations or interactions among individual plants.
3. Canopy function encompasses the exchange of matter, energy, and information with the ___________, which can be quantified through radiation capture, photosynthesis, respiration, evapotranspiration, and greenhouse gas emissions.
4. The canopy structure creates a varied ___________ within the crop stand, influencing key functional properties like the physiological processes that affect yield and quality, as well as the plant's resilience to biotic and abiotic stresses.

Short Answer

1. Emergent properties of the crop canopy.
2. Comparison of direct and indirect methods for quantifying the structure of canopy.
3. The way in which crops create a unique micro-environment within the canopy.
4. Compare the light response curve at the leaf level with that at the canopy level.
5. Interpret the relation between canopy structure and function.
6. The concept of canopy structure and traits for describing it.
7. The concept of canopy function and its relevance to crop production.
8. Factors influencing crop maintenance respiration.

Essay Items

1. Explain the strengths and weaknesses of erect canopies in terms of light interception, and select a familiar crop species (e.g. rice, wheat, maize) to exemplify the use of an erect canopy in boosting crop yield.
2. The maximum biomass production is determined by a critical LAI rather than an optimum one. Elucidate the mechanism and apply it to interpret the relatively constant RUE for a given crop species, which represents the slope between dry matter produced and intercepted radiation.
3. Refer to the article titled "Redesigning photosynthesis to sustainably meet global food and bioenergy demand" published in PNAS (2015, 112:8529–36), and critically evaluate the concept of smart canopy.

Authentic Performance Task

Analyse the structure–function relationship of the population of one crop species at a specific location, identify the pivotal structural factors that govern its functionality, and devise a strategy to enhance its productivity.

References

Ando K, Grumet R, Terpstra K et al (2007) Manipulation of plant architecture to enhance crop disease control. CAB Rev. https://doi.org/10.1079/PAVSNNR20072026

Biscoe PV, Scott RK, Monteith JL (1975) Barley and its environment. III. Carbon budget of the stand. J Appl Ecol 12(1):269–293. https://doi.org/10.2307/2401733

Casal JJ (2013) Photoreceptor signaling networks in plant responses to shade. Annu Rev Plant Biol 64:403–427. https://doi.org/10.1146/annurev-arplant-050312-120221

Connor DJ, Loomis RS, Cassman KG (2011) Crop ecology: productivity and management in agricultural systems, 2nd edn. Cambridge University Press, Cambridge

Cutolo EA, Guardini Z, Dall'Osto L et al (2023) A paler shade of green: engineering cellular chlorophyll content to enhance photosynthesis in crowded environments. New Phytol 239(5):1567–1583. https://doi.org/10.1111/nph.19064

Fei C, Yu JH, Xu ZJ et al (2019) Erect panicle architecture contributes to increased rice production through the improvement of canopy structure. Mol Breed 39(9):128. https://doi.org/10.1007/s11032-019-1037-9

Girardin P, Tollenaar M (1994) Effects of intraspecific interference on maize leaf azimuth. Crop Sci 34(1):151–155. https://doi.org/10.2135/cropsci1994.0011183X003400010027x

Goudriaan J (2016) Light distribution. In: Hikosaka K, Niinemets Ü, Anten NPR (eds) Canopy photosynthesis: from basics to applications. Advances in photosynthesis and respiration, vol 42. Springer, Dordrecht, pp 3–22

Grumet R, Colle M, Ando K et al (2013) Modified plant architecture to enhance crop disease control: genetic control and possible value of upright fruit position in cucumber. Eur J Plant Pathol 135(3):545–560. https://doi.org/10.1007/s10658-012-0087-1

Hammer G, Cooper M, Tardieu F (2006) Models for navigating biological complexity in breeding improved crop plants. Trends Plant Sci 11(12):587–593. https://doi.org/10.1016/j.tplants.2006.10.006

Hay RKM, Porter JR (2006) The physiology of crop yield, 2nd edn. Wiley-Blackwell, New Jersey

Herrero-Huert M, Bucksch A, Puttonen E et al (2020) Canopy roughness: a new phenotypic trait to estimate aboveground biomass from unmanned aerial system. Plant Phenomics 2020:6735967. https://doi.org/10.34133/2020/6735967

Hilton AJ, Jenkinson P, Hollins TW et al (1999) Relationship between cultivar height and severity of Fusarium ear blight in wheat. Plant Pathol 48(2):202–208. https://doi.org/10.1046/j.1365-3059.1999.00339.x

Huang GM, Liu YR, Guo YL et al (2021) A novel plant growth regulator improves the grain yield of high-density maize crops by reducing stalk lodging and promoting a compact plant type. Field Crop Res 260:107982. https://doi.org/10.1016/j.fcr.2020.107982

Huber M, Nieuwendijk NM, Pantazopoulou CK et al (2021) Light signalling shapes plant-plant interactions in dense canopies. Plant Cell Environ 44(4):1014–1029. https://doi.org/10.1111/pce.13912

IPCC (2014) Climate change 2014 synthesis report. Contribution of working groups I, II and III to the fifth assessment report of the Intergovernmental Panel on Climate Change. In: Pachauri RK, Meyer LA (eds) IPCC, Geneva, p 151

IPCC (2023) Climate change 2023 synthesis report. Contribution of working groups I, II and III to the sixth assessment report of the Intergovernmental Panel on Climate Change. In: Lee H, Romero J (eds) IPCC, Geneva, pp 35–115

Jiao YQ, Wang YH, Xue DW et al (2010) Regulation of *OsSPL14* by OsmiR156 defines ideal plant architecture in rice. Nat Genet 42(6):541–544. https://doi.org/10.1038/ng.591

Katsura K, Maeda S, Horie T et al (2009) Estimation of respiratory parameters for rice based on long-term and intermittent measurement of canopy CO_2 exchange rates in the field. Field Crop Res 111:85–91. https://doi.org/10.1016/j.fcr.2008.11.003

Khaipho-Burch M, Cooper M, Crossa J et al (2023) Genetic modification can improve crop yields–but stop overselling it. Nature 621(7979):470–473. https://doi.org/10.1038/d41586-023-02895-w

Lee EA, Tollenaar M (2007) Physiological basis of successful breeding strategies for maize grain yield. Crop Sci 47:S202–S215. https://doi.org/10.2135/cropsci2007.04.0010IPBS

Li GH, Zhang Y, Zhou C et al (2024) Agronomic and physiological characteristics of high yield and nitrogen use efficient varieties of rice: comparison between two near-isogenic lines. Food Energy Secur 13(2):e539. https://doi.org/10.1002/fes3.539

Liu FS, Song QF, Zhao JK et al (2021) Canopy occupation volume as an indicator of canopy photosynthetic capacity. New Phytol 232(2):941–956. https://doi.org/10.1111/nph.17611

Long SP, Zhu XG, Naidu SL et al (2006) Can improvement in photosynthesis increase crop yields? Plant Cell Environ 29(3):315–330. https://doi.org/10.1111/j.1365-3040.2005.01493.x

Louarn G, Song YH (2020) Two decades of functional-structural plant modelling: now addressing fundamental questions in systems biology and predictive ecology. Ann Bot 126(4):501–509. https://doi.org/10.1093/aob/mcaa143

Maddonni GA, Otegui ME, Andrieu B et al (2002) Maize leaves turn away from neighbors. Plant Physiol 130(3):1181–1189. https://doi.org/10.1104/pp.009738

Mansfield BD, Mumm RH (2014) Survey of plant density tolerance in U.S. maize germplasm. Crop Sci 54:157–173. https://doi.org/10.2135/cropsci2013.04.0252

Mantilla-Perez MB, Bao Y, Tang L et al (2020) Toward "smart canopy" sorghum: discovery of the genetic control of leaf angle across layers. Plant Physiol 184(4):1927–1940. https://doi.org/10.1104/pp.20.00632

Monsi M, Saeki T (2005) On the factor light in plant communities and its importance for matter production. Ann Bot 95(3):549–567. https://doi.org/10.1093/aob/mci052

Niinemets U (2023) Variation in leaf photosynthetic capacity within plant canopies: optimization, structural, and physiological constraints and inefficiencies. Photosynth Res 158(2):131–149. https://doi.org/10.1007/s11120-023-01043-9

Norman JM, Campbell GS (1989) Canopy structure. In: Pearcy RW, Ehleringer JR, Mooney H et al (eds) Plant physiological ecology. Springer, Dordrecht, pp 301–325

Ort DR, Merchant SS, Alric J et al (2015) Redesigning photosynthesis to sustainably meet global food and bioenergy demand. Proc Natl Acad Sci USA 112(28):8529–8536. https://doi.org/10.1073/pnas.1424031112

Perez RPA, Fournier C, Cabrera-Bosquet L et al (2019) Changes in the vertical distribution of leaf area enhanced light interception efficiency in maize over generations of selection. Plant Cell Environ 42(7):2105–2119. https://doi.org/10.1111/pce.13539

Pierik R, Ballaré CL (2021) Control of plant growth and defense by photoreceptors: from mechanisms to opportunities in agriculture. Mol Plant 14(1):61–76. https://doi.org/10.1016/j.molp.2020.11.021

Qian HY, Zhu XC, Huang S et al (2023) Greenhouse gas emissions and mitigation in rice agriculture. Nat Rev Earth Env 4(10):716–732. https://doi.org/10.1038/s43017-023-00482-1

Richards RA, Cavanagh CR, Riffkin P (2019) Selection for erect canopy architecture can increase yield and biomass of spring wheat. Field Crop Res 244:107649. https://doi.org/10.1016/j.fcr.2019.107649

Russell G (2017) Canopy architecture. In: Thomas B, Murray BG, Murphy DJ (eds) Encyclopedia of applied plant sciences, vol 3, 2nd edn. Academic Press, New York, pp 13–17

Sadras VO, Lawson C, Montoro A (2012) Photosynthetic traits of Australian wheat varieties released between 1958 and 2007. Field Crop Res 134:19–29. https://doi.org/10.1016/j.fcr.2012.04.012

Sadras VO, Villalobos FJ, López-Bernal Á et al (2024) Radiation interception, radiation use efficiency, and crop productivity. In: Villalobos FJ, Fereres E (eds) Principles of agronomy for sustainable agriculture. Springer Nature Switzerland, pp 177–197

Savin R, Slafer GA (2019) Crop science. Springer, New York

Schulze E, Beck E, Buchmann N et al (2019) Plant ecology, 2nd edn. Springer, Berlin, Heidelberg

Serrago RA, Giménez VD, García GA et al (2025) Out of sight, out of mind: night-temperature impact on major field crops. Field Crop Res 333:110108. https://doi.org/10.1016/j.fcr.2025.110108

Smith EN, van Aalst M, Tosens T et al (2023) Improving photosynthetic efficiency toward food security: strategies, advances, and perspectives. Mol Plant 16(10):1547–1563. https://doi.org/10.1016/j.molp.2023.08.017

Song QF, Zhu XG (2024) Techniques for photosynthesis phenomics: gas exchange, fluorescence, and reflectance spectrums. Crop Environ 3:147–158. https://doi.org/10.1016/j.crope.2024.05.002

Soualiou S, Wang Z, Sun W et al (2021) Functional-structural plant models mission in advancing crop science: opportunities and prospects. Front Plant Sci 12:747142. https://doi.org/10.3389/fpls.2021.747142

Taiz L, Zeiger E, Møller IM et al (2015) Plant physiology and development, 6th edn. Sinauer Associates, Sunderland

Tian JG, Wang CL, Xia JL et al (2019) Teosinte ligule allele narrows plant architecture and enhances high-density maize yields. Science 365(6454):658–664. https://doi.org/10.1126/science.aax5482

Tian JG, Wang CL, Chen FY et al (2024) Maize smart-canopy architecture enhances yield at high densities. Nature 632:576–584. https://doi.org/10.1038/s41586-024-07669-6

Villalobos FJ, Testi L, Fereres E (2024) The components of evapotranspiration. In: Villalobos FJ, Fereres E (eds) Principles of agronomy for sustainable agriculture. Springer Nature Switzerland

Wei HB, Zhao YP, Xie YR et al (2018) Exploiting *SPL* genes to improve maize plant architecture tailored for high-density planting. J Exp Bot 69(20):4675–4688. https://doi.org/10.1093/jxb/ery258

Welles JM (1990) Some indirect methods of estimating canopy structure. Remote Sens Rev 5(1):31–43. https://doi.org/10.1080/02757259009532120

Yang P, van der Tol C, Verhoef W et al (2019) Using reflectance to explain vegetation biochemical and structural effects on sun-induced chlorophyll fluorescence. Remote Sens Environ 231:110996. https://doi.org/10.1016/j.rse.2018.11.039

Yang ZF, Qi XT, Dai Y et al (2023) Nitrogen fertilization produces divergent effects on canopy structure between indica and japonica rice reflected in leaf to panicle ratio based on deep learning. Field Crop Res 304:109184. https://doi.org/10.1016/j.fcr.2023.109184

Zhang RL, Guo RF, Zhi H et al (2025) *De novo* creation of narrowed plant architecture via CRISPR/Cas9-mediated mutagenesis of *SiLGs* in foxtail millet. Plant Biotechnol J 23(6):2400–2402. https://doi.org/10.1111/pbi.70037

Zhou Y, Kusmec A, Schnable PS (2024) Genetic regulation of self-organizing azimuthal canopy orientations and their impacts on light interception in maize. Plant Cell 36(5):1600–1621. https://doi.org/10.1093/plcell/koae007

Zhu XG, Long SP, Ort DR (2008) What is the maximum efficiency with which photosynthesis can convert solar energy into biomass? Curr Opin Biotechnol 19:153–159. https://doi.org/10.1016/j.copbio.2008.02.004

Crop Phenology

Timing is everything

7

Contents

Z. Liu, *Crop Physiology*, https://doi.org/10.1007/978-981-92-2960-4_7

Abstract

Crop phenology refers to the sequential progression of developmental phases, the durations of which are determined by the interplay between internal genetic factors and environmental cues. It is a crucial determinant of crop adaptation to seasonal changes and of their capacity to maximize the utilization of solar energy, precipitation, and heat resources. This chapter begins with the essential concepts of crop growth and development and stresses the imperative of adapting crop phenology to the environment by listing four key phenological phases or stages with agronomic importance. There follows developmental response to environmental conditions including temperature per se, photoperiod, and vernalization, with an emphasis on the molecular mechanism underlying flowering time. Finally, examples of genetic and agronomic modifications on flowering time are presented, and future research directions in crop phenology are discussed.

Keywords

Phenology · Photoperiodism · Vernalization · Flowering time · Growth · Development

Did You Know

- The Sun–Earth geometry undergoes changes over time due to cyclic variations in the tilt of Earth's rotational axis, which has a period of about 41,000 y and is currently at an angle of 23.5°. These changes occur at such slow rates that they do not significantly impact agricultural ecology within the span of a human lifetime.
- Higher plants are generally mesophiles, with optimum temperatures ranging from approximately 10–30 °C. In contrast, thermophiles thrive at temperatures ranging from 30 to 65 °C without any growth limitations. Notably, some studies have even reported cyanobacteria capable of growing at temperatures as high as 85 °C.
- In plants, as many as 60% of genes are regulated by the circadian clock, which plays a critical role in modulating various developmental and physiological processes, including stomatal regulation, leaf movement, transition to flowering, and flower opening.
- The blooming of roses is always a delightful sight in the summer, while chrysanthemums bring beauty to the autumn season. In fact, many plants bloom so predictably year after year that gardeners have incorporated them into their gardens as floral calendars, accurately marking the progress of each season.
- Flowering at the correct time of year is crucial for the reproductive fitness of the plant; plants that are cross-pollinated must flower in synchrony with other individuals of their species, as well as with their pollinators, during the optimal season for seed set.

Learning Objectives

After reading this chapter, students will be able to:

- Understand the fundamental concepts of crop growth and development
- Appreciate the necessity of adapting crop phenology to the local environment
- Explain how crop phenological phases or stages affect crop yield
- Describe the developmental response to temperature and photoperiod
- Elucidate the molecular mechanism underlying flowering time
- Name some cases that use genetic and agronomic approaches to modify flowering time

Introductory Remarks

Crop production can be defined as the conversion of solar energy into useful organic materials in the form of food, feed, and fibre. Essentially, this transformation is predominantly influenced by the quantity and quality of natural resources, particularly soil fertility and prevailing meteorological conditions such as solar irradiance, temperature, and precipitation. The cropping systems adopted by different regions are intricately linked to their specific environmental characteristics. In the face of the challenges posed by climate change and population growth, there is an urgent need to develop crops that can efficiently capture and utilize resources, thereby extending the growing season and expanding into stressful environments.

When designing the cropping system for a target area, timing is crucial. The efficient optimization of timing and duration in crop growth and development enables the effective utilization of light, heat, water, and nutrients, leading to enhanced yield and quality. Mounting evidence suggests that crop biomass production relies more on resource capture than on resource utilization. Dry matter production under intensive cultivation mainly depends on the amount of PAR intercepted by the canopy, while in arid or semi-arid areas, it relies on the quantity of water vapored and transpired by the crop stands. The situation is not as clear for crops limited by nitrogen (N), but many investigations have shown that total N uptake is more important than N use. Thus, it is crucial for the ontogeny of crop plants to temporally match the availability and quantity of natural resources, among which the solar energy is of the most importance.

Phenology refers to the timing of recurring biological events, such as leafing, flowering, migration, or hibernation, in response to climatic factors like temperature, precipitation, and day length. Because these events depend closely on environmental conditions, phenology strongly affects organismal fitness: individuals that align their life cycles with favourable conditions are more likely to survive and reproduce. For example, if plants flower before their pollinators emerge, pollination declines; similarly, predators missing peak prey activity face lower foraging success. On a broader scale, phenological shifts

(continued)

affect ecosystem processes such as nutrient cycling, because the timing of leaf fall, fruit maturation, and subsequent decomposition determines when nutrients are returned to the soil. Thus, phenology links climate variation to individual performance, species relationships, and ecosystem functioning.

Crop phenology refers to the predictable sequence of developmental stages throughout the crop's life cycle. Properly aligning these stages with favourable seasonal conditions is crucial for the efficient utilization of available heat, water, and solar energy. For example, in the case of rice and wheat, this entails ensuring that seeding, tillering, jointing, and flowering take place at optimal times when light, temperature, and precipitation are most conducive to the growth and development of plants. Among the phenological stages, flowering time plays a crucial role in determining the allocation of time and solar energy between vegetative growth and reproductive development, thereby directly influencing the yield potential in specific climatic environments. This chapter explores the significance of crop phenology, especially the flowering time, in yield formation and environmental adaptation, emphasizing the critical role of precise timing in boosting crop productivity.

Concept Map

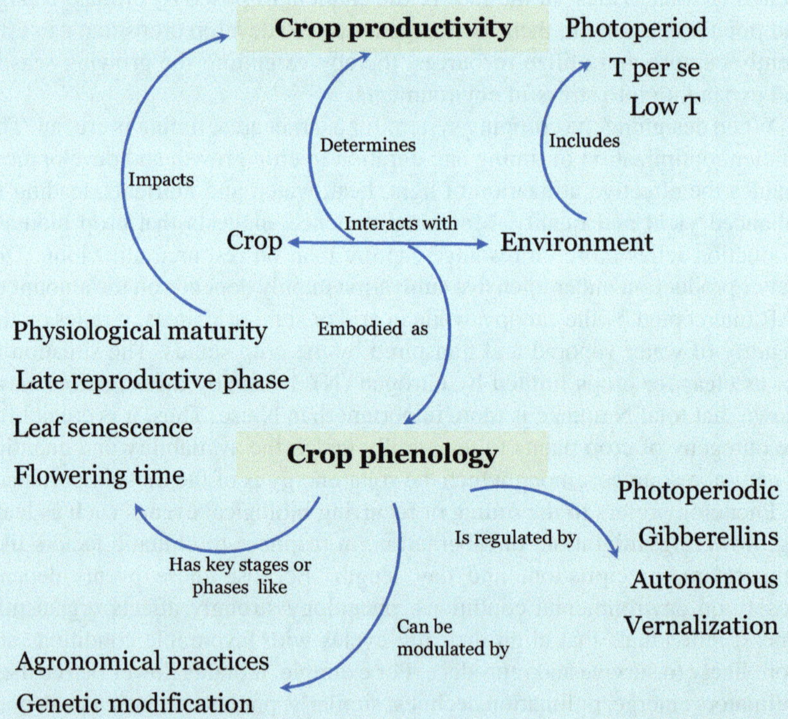

(continued)

Annotation: The interaction between crop plants and environmental conditions, such as temperature and photoperiod, is a critical determinant of crop productivity in specific regions and represents the central focus of crop science. Crop phenology, defined as the timing of plant developmental events, directly reflects the interaction between crops and their environment. Some key phenological phases/stages, including flowering time, the late reproductive phase, the onset of leaf senescence, and physiological maturity, are critical to yield formation and can be regulated through both agronomic practices and genetic manipulations. The photoperiodic pathway, autonomous pathway, gibberellin pathway, and vernalization pathway constitute the primary regulatory mechanisms underlying the phenological responses of crop plants.

1 Crop Phenology

The discipline of crop phenology focuses on the study of the timing of developmental stages of crops, such as emergence, leafing, tillering, flowering, grain filling, and maturation, within their life cycle. These phenological events are closely associated with weather, soil, and management conditions, which makes them crucial indicators for crop growth, water and nutrient requirements, and potential yield. By monitoring phenology, farmers can optimize seeding dates, plan irrigation and applications of fertilizer and pesticide, and anticipate harvest windows, thus enhancing productivity, resource use efficiency, and resilience to climate variability.

1.1 Fundamental Concepts of Crop Growth and Development

1.1.1 Growth and Development

The growth and development in crop science are related yet distinct processes. *Development* is the initiation and differentiation of organs and the progression of stages through which plants go during their life cycle, whereas *growth* is the change in size or mass of the initiated organs. The expansion of a leaf and the accumulation of crop biomass are typically considered growth processes, whereas the transition from a vegetative meristem that produces leaves to a reproductive meristem that produces flowers is characterized as a developmental process.

The *developmental stages* are the time points when distinct changes in morphology, dimensions, anatomy, and function occur during the progression of crop development. And the time between stages is referred to as *developmental phases*. It is common in the literature referring to these phases as "growth stages or phases". However, there is little growth in some developmental events such as anthesis. Both biotic (e.g. genetics, weeds, diseases) and abiotic factors (e.g. temperature, light, water, nutrients) impact plant organ initiation, growth, and thus developmental phases.

Generally, the growth of crop plant follows a sigmoid (S-shaped) curve, which consists of three distinct phases: the lag phase, the exponential phase, and the stationary phase. During the lag phase, growth is slow as seedlings develop roots and acclimate to their environment. Next, the exponential phase begins, characterized by rapid, nearly linear vegetative growth, driven by abundant resources and favourable environmental conditions. Finally, in the stationary phase, growth gradually slows and plateaus as the plant approaches physiological maturity. Understanding this logistic growth pattern is crucial for optimizing field management decisions throughout the crop life cycle.

The distinction between growth and development is crucial for two primary reasons (Sadras et al. 2024). Firstly, growth predominantly involves the acquisition and allocation of resources, while development is more closely associated with environmental factors such as low temperature, photoperiod, and light quality.[1] Second, the physiological processes involved are distinct: growth primarily depends on the metabolic activities associated with the biosynthesis of cellular components, whereas development is driven by the differentiation and functional specialization of the meristem cells. Although growth and development can be conceptually separated, in practice, organs, plants, and crops undergo these processes concurrently. For example, during the grain filling, a wheat grain not only increases in volume and accumulates reserves but also undergoes developmental changes, such as the differentiation of leaf and root primordia in the embryo.

1.1.2 Plastochron and Phyllochron

Plant development is a dynamic process that involves the differentiation, formation, growth, and senescence of phytomers and their components, resulting in a continuously changing architecture (*Refer to* Chap. 4). The rate of leaf primordium initiation is crucial for the formation of vegetative phytomers. The *plastochron* is defined as the interval between the formation of successive leaf primordia at the shoot apex. Similarly, the *phyllochron* is defined as the interval between the appearance of consecutive leaves on a shoot. If the phyllochron is short, tillers in cereals can be produced within a shorter period, resulting in greater uniformity in plant size. A short phyllochron also allows for synchronous development of inflorescences, which are crucial factors for yield and quality uniformity in cereals. The inverse of the phyllochron is referred to as the development rate, which can be computed as the reciprocal of the time interval between two developmental events. The relationship between plastochron and phyllochron varies depending on the species. In wheat, leaf primordia are produced at a higher rate than they emerge, indicating that the plastochron is shorter than the phyllochron. This implies the distinct mechanisms governing the two processes.

[1] In the field of ecology, the environment is defined as all external conditions that surround an organism, including both biotic and abiotic factors. In contrast, resources specifically refer to the usable substances or energy required by organisms for their survival, such as food, water, and nutrients. The environment represents the overall context, while resources are the specific elements within it that facilitate the survival and growth of organisms.

1.1.3 Crop Phenology

The phenology of crops refers to the sequential progression of their developmental stages and phases. The duration of these phases varies due to intrinsic factors, such as the genetic traits of a cultivar or species, and extrinsic biotic and abiotic factors, including diseases, temperature, water availability, light intensity, and nutrient levels. Achieving synchrony between each developmental phase and suitable seasonal conditions is essential for maximizing the utilization of available temperature and sunlight under specific ecological settings. For example, grain number, and consequently wheat yield, is primarily determined by growth rates during the critical period from the emergence of the penultimate leaf to early grain filling. From an agronomic standpoint, aligning this sensitive period with the most favourable seasonal conditions is therefore crucial (Hyles et al. 2020).

Crop development involves a continuous series of physiological and morphological changes, during which various organs are initiated and grow to complete the life cycle. Although this process is inherently continuous, ontogeny is commonly conceptualized as a sequence of discrete phenological phases influenced by environmental factors to enhance interpretability. Each phase is characterized by distinct morphological and/or functional transformations in specific organs. Thus, accurately identifying key developmental milestones that define the transitions between phenological phases is crucial for understanding crop growth dynamics.

In cereal crops such as rice and wheat, the principle of organ co-extension establishes a predictable relationship between *leaf age* and the concurrent development of tillers and roots. Leaf age pertains to the developmental stage of a crop plant, which is typically quantified by the number of fully emerged leaves on the main stem. According to the leaf-tiller co-extension model, new tiller initiation follows a consistent pattern synchronized with leaf emergence on the main stem. Specifically, when the main stem produces its N-th leaf (e.g., 4th leaf), a new tiller typically emerges from the axil of the (N–3)-th leaf (e.g. 1st leaf) node. In other words, tiller appearance lags behind leaf development by approximately three phytomer positions. Similarly, in the leaf-root co-extension model, root initiation at the (N–3)-th leaf node coincides with the emergence of the N-th leaf on the main stem. These models collectively offer a practical phenological framework for crop management. The Rice Leaf-age Model (RLM), developed by renowned agronomist Professor Qihong Ling (Ling et al. 2007), integrates leaf age with functional leaf and root dynamics and key developmental stages to guide decisions regarding transplanting density, fertilization, irrigation, and the optimization of yield components (Fig. 7.1). RLM has consistently demonstrated effectiveness across diverse rice-growing regions and cultivars throughout China.

1.2 Yield Components Formed at Various Developmental Phases

Crop yield is a complex characteristic that can be broken down into simpler, quantifiable components. When combined multiplicatively, these factors determine the

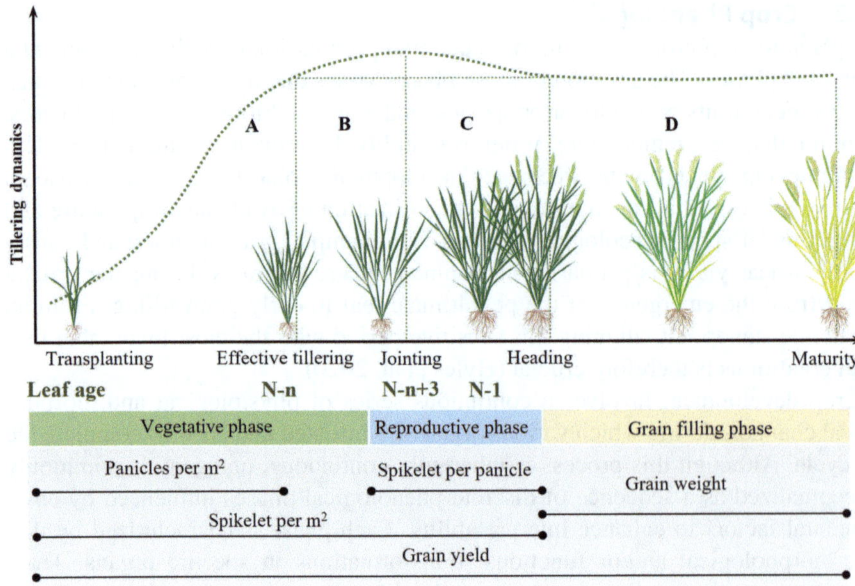

Fig. 7.1 The process of rice yield formation as indicated by the leaf age. According to the Rice Leaf-age Model (Ling et al. 2007), zone A represents the critical phase for effective tillering; zone B corresponds to the stage of ineffective tillering; zone C denotes the panicle differentiation; and zone D refers to the grain filling. The bottom boxes represent the formation periods of grain yield components. Note that the starting point of panicle differentiation (i.e. the reproductive phase) typically coincides with the stage at which the leaf age residual is 3.5, the second half of the fourth leaf from the top. It may occur earlier than, later than, or simultaneously with the jointing stage, depending on the genotype. N, the total number of leaves of the main stem; n, the number of elongated internodes of the main stem; N–n, the critical leaf age for effective tillering; N–n + 3, the leaf age at the jointing stage; N–1, the leaf age for spikelet initiation

total output per unit area. The fundamental equation for yield is: Yield = Plant density (plants/area) × Reproductive organs per plant (e.g. heads, pods, or ears) × Seeds per reproductive organ × Seed weight (or individual organ weight, e.g. grain, tuber, or fruit). For instance, wheat yield is commonly modelled as the product of productive tillers per square meter, grains per ear, and kernel weight. In legumes, key components include plants per square meter, pods per plant, seeds per pod, and seed weight. These yield components are interdependent (Carrera et al. 2024): an increase in one, such as more seeds per pod, may offset a reduction in another, like smaller seed size. Moreover, all components are shaped by genetic potential, environmental conditions, and agricultural management practices.

Although the negative relationship applies to the two main yield components, namely, the number of grains or seeds per unit land area (GN m^{-2}) and their average weight, it has been empirically demonstrated that GN m^{-2} is the critical component for achieving incremental changes in yield. Taking IR747B2-6, an early-maturing rice cultivar with grain yields ranging from 4.6 to 7.1 tons per hectare, as an example, grain number meter accounts for 74% of the total grain yield, whereas the combined contributions of filled-grain percentage and grain weight account for 26% (Yoshida 1981). A similar conclusion was obtained from IR8 crops grown at Los Baños. The grain yield increased linearly with an increasing GN m^{-2}, while the percentage of filled grains and grain weight remained constant regardless of the grain number. However, in specific locations and under particular weather conditions, the percentage of filled grains and grain weight may impose a greater limitation on yield than the GN m^{-2}. Nevertheless, this scenario rarely occurs under most conditions.

Crop yield components are developed across distinct growth phases, each contributing to final grain output through coordinated physiological processes (Fig. 7.1). The formation of these components follows a clear sequence: it begins with the vegetative phase, during which the crop produces a sufficient number of individual plants (e.g. tillers) to establish a solid foundation for a high-yielding population. This is followed by the reproductive phase, in which the number of potential harvested organs, such as ears in wheat, is determined. Finally, the grain-filling phase establishes grain size and weight. These developmental phases are highly interdependent: environmental conditions or management practices at any one stage can significantly affect the development of subsequent components, often triggering compensatory responses among them. Therefore, comprehending the dynamics of each yield component and the interactions among them is crucial for agronomists who aim to optimize interventions, such as fertilization and irrigation, at specific growth stages to maximize crop productivity.

1.3 The Imperative of Adapting Crop Phenology to the Specific Local Environment

Crop phenology is crucial in agriculture as it allows plants to withstand and adapt to both biotic and abiotic stresses. The key factor determining the success of a crop in a particular region depends on its phenology aligning with the prevailing environmental conditions. Successful crop cultivation at a specific site depends on two interrelated requirements. Firstly, the phenology of the cultivar must be precisely aligned with the local growing season. This alignment ensures that vegetative growth is adequately completed before the onset of reproductive stages and that sink organs not only fill efficiently but also reach full maturity prior to harvest. Secondly, the crop stand must efficiently capture and convert available meteorological resources, especially solar radiation, water, and temperature, into biomass and yield. Significantly, these two requirements are not independent: optimal phenology facilitates efficient resource utilization, and effective resource capture reinforces timely development. Collectively, they determine whether a cultivar can achieve reliable, high-yielding performance in a particular agroecological context.

Certain crops, like soybeans, have specific photoperiod requirements defined by light and dark durations, affecting their geographic distribution. The initiation of soybean flowering is determined by day length, which varies with latitude. Flowering must occur timely to allow sufficient time for seeds to develop and mature within the pod. Because soybeans are highly sensitive to photoperiod, distinct cultivars are planted across different latitudes from north to south to optimize the growing season. As the saying goes, "Wheat can be grown over a thousand miles, while soybeans only over a hundred miles". This adage highlights that every soybean cultivar has a limited adaptation range due to its high sensitivity to photoperiod, as explained at the end of this chapter.

2 Key Crop Phenology with Agronomic Importance for Cereal Crops

Cereal crops, such as rice, wheat, and maize, exhibit a similar pattern of development that can be broadly divided into the vegetative phase, reproductive phase, and grain-filling phase. The transition from the vegetative to reproductive phase occurs at the meristem level, where the production of vegetative phytomer ceases and the production of reproductive phytomer begins. A balance between (1) growth of vegetative source organs and (2) development of reproductive sink organs is crucial for achieving a good harvest. Overall, there are four critical phenological phases or stages that play a significant role in the formation of cereal yield.

2.1 The Late Reproductive Phase Is Critical for Balancing the Grain Number and Stem Growth

A commonly used conceptual model for crop growth analysis assumes that dry matter (DM) production is directly proportional to the radiation intercepted by the canopy over the entire growing season. Additionally, yield is determined by multiplying the shoot DM by the harvest index (HI). This model assumes that DM and HI are independent of each other and that biomass productivity remains constant across all growth stages. A deduction drawn from this model is that there exists a causal relationship between yield and the duration of the growing season. Slafer et al. (2023) contested this causal claim by proposing an alternative model centred on phenology, arguing that the late reproductive phase, rather than the length of the growing season, is the critical period for grain yield.

During the late reproductive phase of wheat, the intercalary meristems located above the nodes generate new cells, driving the rapid elongation and development of internodes. Each subsequent internode exhibits greater elongation than the preceding one, thereby promoting robust plant growth. As the stem elongates, spikelet differentiation and floret development proceed concurrently. This phase is characterized by the simultaneous determination of fertile floret number and vigorous growth of stems and spikes, leading to intensified competition for nutrients. Due to this heightened competition, nutrient supply becomes insufficient to sustain all emerged young tillers. Consequently, some tillers die, typically in reverse order of emergence, and the tiller density per square meter declines from its peak to the final number of effective tillers, which is generally fixed at anthesis. Therefore, the late reproductive phase, from terminal spikelet initiation to anthesis, is critical for determining potential grain yield (Reynolds et al. 2012).

There are distinct indications that decreases in crop growth before the onset of stem elongation have minimal effects on the ultimate grain yield. In contrast, decreases in crop growth during stem elongation show a direct relationship with grain number. The development of wheat spikes occurs only a few days before anthesis, simultaneously and in competition with stem elongation. If stem elongation were prolonged without compensatory adjustments in DM allocation, spike DM accumulated at anthesis would be higher, leading to an increased grain number. Consequently, prolonging the late reproductive phase would enhance DM accumulation during stem elongation, thereby potentially increasing the final grain number by reducing floret abortion and/or tiller death (Reynolds et al. 2012). Nevertheless, direct evidence substantiating this hypothesis remains limited, and its significance for crop breeding and management requires further investigation.

Prior research on the physiological basis of yield differences between elite and conventional rice cultivars has predominantly focused on the grain-filling phase, while the role of the pre-anthesis phase remains relatively underexplored. Cheng et al. (2024) adopted a comparative approach using representative indica and japonica cultivars, with each subgroup comprising one cultivar exhibiting ideal plant architecture (IPA) and one with conventional morphology. The results indicated that the IPA cultivars JYZK-6 and ZZY-1 consistently outyielded their conventional counterparts, NJ-5055 and HHZ. At the single-leaf level, no statistically significant differences in photosynthetic capacity were detected between IPA and conventional types. However, at the canopy scale, IPA cultivars exhibited markedly enhanced photosynthetic performance during the pre-anthesis phase, averaging 35.56% higher canopy photosynthesis than their counterparts. During the post-anthesis phase, no significant difference in canopy photosynthetic activity was observed between the two groups. Moreover, IPA cultivars accumulated substantially more DM in internodes and sheaths prior to anthesis and subsequently remobilized a larger proportion of these reserves for grain filling. In summary, enhanced canopy photosynthetic capacity before anthesis significantly contributes to the superior yield performance of high-yielding IPA cultivars, underscoring the critical importance of pre-anthesis growth and development in yield formation.

2.2 At Anthesis, the Storage Reserves in the Vegetative Organs Are Crucial for the Growth of Reproductive Organs

Crop phenology exerts a significant influence on yield through the modulation of the relative development and spatial arrangement of vegetative and reproductive organs. A crucial event that determines the balance between vegetative and reproductive development is the transition to anthesis. The modularity of plant ontogenesis necessitates the production of a specific quantity of vegetative phytomers to offer structural support for subsequent reproductive phytomers. Consequently, the genetic programming of a plant is initially configured in "vegetative mode" to prevent premature flowering, which could endanger the plant's survival and reproductive success.

The significance of anthesis timing can be further explained by quantifying the contribution of prestored *non-structural carbohydrates (NSCs)* in the stem and sheath to grain filling using carbon (C) isotope labelling analysis. The accumulation of NSCs, such as sugars and starch, sharply increases about 2 weeks before flowering and reaches its maximum concentration in the internodes and sheath around anthesis. As grain filling progresses, the stored NSCs begins to decline but may slightly rise again near maturity. The ^{14}C labelling experiment on rice revealed that 68% of the accumulated NSCs were translocated into the grain, 20% were respired during the ripening period, and 12% remained in the stems and sheaths. The translocated NSCs accounted for approximately 26% of the grain carbohydrate, which is equivalent to about 2 t/ha when the yield is 7.8 t/ha (Yoshida 1981).

In addition, the NSCs stored before anthesis support continuous grain filling under adverse weather conditions. When photosynthesis is restricted by 90% shading during the ripening stage, the accumulated NSCs in internodes and sheaths support grain growth at an almost normal rate for approximately 2 weeks after anthesis. This suggests that when the current supply of photosynthates is insufficient for grain filling under unfavourable conditions, such as nitrogen deficiency, low light intensity, or low temperature, the storage NSCs can be efficiently remobilized, thereby preventing unfilled spikelets and stabilizing grain yield.

2.3 The Onset of Senescence Signifies the Shift of Leaves from a Carbon-Based to a Nitrogen-Based Source

2.3.1 The Carbon–Nitrogen Transition in Leaf

Cell death is an inevitable yet essential component of plant life. Like animals, plants possess mechanisms to defend against pathogens and eliminate pathogen-infected cells through *programmed cell death (PCD)*. PCD entails localized cell death events that occur within a matrix of living cells. By contrast, *senescence* is a large-scale, organ-wide form of cell death that is characterized by a gradual process of nutrient redistribution to sink organs such as seeds and tubers. It is an active developmental program that requires the upregulation of numerous genes, distinguishing it from mere necrosis or passive cellular decline (Havé et al. 2017).

Many crop plants exhibit *monocarpic* behaviour, where they undergo a single reproductive cycle followed by senescence and death. In these species, the onset of reproduction triggers leaf senescence, facilitating the reallocation of nutrients from senescing leaves to developing seeds. As such, the removal of flowers and pods has been observed to significantly prolong the process of senescence in soybean. During the senescence process, cells undergo significant modifications in cellular metabolism, structure, and gene expression, subsequently leading to the degradation of chlorophyll as well as macromolecules such as proteins, lipids, nucleic acids, and mineral complexes. The elimination of inefficient and aging photosynthetic organs renders leaf senescence highly beneficial, as is clearly demonstrated during the autumnal senescence of tree leaves.

Crop leaf development follows distinct functional phases characterized by dynamic shifts in C and N assimilation and translocation (Thomas and Ougham 2014). During the initial phase, a newly expanded leaf acts as a strong sink for C and N, importing these nutrients to support cell differentiation and expansion. Upon full establishment of the photosynthetic machinery, the increased accumulation of Rubisco and chlorophyll enhances photosynthesis and, in turn, promotes the translocation of photosynthates to sink organs. As photosynthetic activity declines, chlorophyll degradation and Rubisco breakdown begin, releasing amino acids and other N-rich metabolites that are transported to developing sinks. This phase features an inverse pattern: while C export gradually diminishes, N export increases, peaking when 80%–90% of leaf N has been remobilized. In the final phase, the leaf becomes functionally inactive, ceasing all significant export of C and N. The initiation of leaf senescence can, therefore, be regarded as the crucial moment when the transition from a C-based to a N-based source takes place.

The pattern of C–N shift at the leaf level exhibits temporal variation, whereas that at the canopy level reveals clear spatial heterogeneity. Within the canopy, leaves display a vertical gradient across these developmental phases: upper leaves maintain photosynthetic activity longer, whereas lower, older leaves enter the N-remobilization phase earlier. This spatial heterogeneity enables the entire canopy to balance C acquisition with N supply to reproductive tissues. Understanding the spatiotemporal heterogeneity of the C–N transition within the crop stands helps to deepen the insights for RUE and yield formation of crops.

2.3.2 The Significance of Leaf Senescence in Nutrient Remobilization

Leaf senescence impacts on both carbon and nitrogen management. The decline in photosynthetic activity during leaf senescence restricts carbon fixation and reduces the translocation of sugars to developing seeds. Meanwhile, increased protein degradation promotes the remobilization of N from senescing leaves. These coordinated processes alter the C/N ratio of assimilates, thereby influencing nutrient redistribution from source tissues to sink organs, a process vital for improving seed quality and optimizing crop productivity (Zhang et al. 2021b).

The contribution of N remobilization from leaves to seeds varies with crop species, genotypes, and environmental conditions. For example, canola plants exhibit reduced expression and activity of root nitrate transporters during the reproductive phase, resulting in decreased nitrate uptake compared to the vegetative phase. Under abundant nitrate and water conditions, N uptake contributed up to 30% of silique N, while the majority (70%) originated from the translocation from vegetative organs. In maize, redistribution accounts for 62% of the kernel N content, with the remaining 38% acquired through root absorption after tasselling (Coque and Gallais 2007). In real-world farming operations, N fertilization is commonly applied during the early growth stage of crops, potentially leading to soil nitrate depletion during the late phase. Moreover, N uptake is closely dependent on soil moisture availability, which may be constrained in certain regions during the summer months. In such scenarios, crop plants primarily rely on N remobilization from proteins and storage pools in leaves to ensure adequate nutrition for reproductive sinks (Tegeder and Masclaux-Daubresse 2018).

2.3.3 Hormonal and Nutritional Regulation of Leaf Senescence

The regulation of senescence is modulated by the complex interplay of multiple phytohormones. CKs exhibit a remarkably potent anti-aging effect. However, large-scale increases in CKs production can also alter plant architecture and potentially reduce yields. An effective solution to this issue involves the use of senescence-inducible promoters to drive the expression of CKs biosynthesis genes (Zhong et al. 2012). This approach enables precise spatiotemporal regulation of CKs levels, ensuring their increase exactly where and when needed to delay senescence. Introducing a key CKs-synthetic gene, *isopentenyltransferase* (*IPT*), delays leaf senescence, maintains photosynthesis, boosts seed or fruit yield, and enhances tolerance to drought, salinity, and cold. In rice, expression of the bacterial *IPT* gene under the control of senescence- or stress-inducible promoters has been shown to elevate endogenous CKs levels, inhibit chlorophyll degradation, and confer a stay-green phenotype, resulting in increased grain yield. Unlike CKs, ethylene has been demonstrated to promote senescence. Plants with deficiencies in ethylene perception or signal transduction exhibit delayed leaf senescence. Additionally, ethylene-induced senescence is limited to leaves that have reached a specific developmental stage. Other hormones are implicated in accelerating senescence, but the precise mechanisms remain elusive. It is likely that defence and stress hormones such as JA, SA, and ABA are necessary to induce senescence responses.

Leaf senescence is significantly influenced by mineral nutrition (Engels et al. 2012). In *Arabidopsis*, K deficiency leads to the upregulation of genes encoding enzymes involved in JA biosynthesis, resulting in increased JA concentrations in leaves and subsequent leaf senescence. This phenotype can be mitigated or delayed by applying salicylate and aspirin, which inhibit JA synthesis. Limited N availability and high tissue C/N ratios are critical regulators of leaf senescence, leading to reductions in chlorophyll, protein, and starch concentrations. These reductions are associated with a significant decrease in the leaf concentration of isopentenyl adenosine (iP), an active form of cytokinin. Root application of the cytokinin 6-benzylaminopurine significantly increased the leaf iP concentration, thereby retarding leaf senescence. This suggests that N deficiency-induced leaf senescence may result from a reduced supply of root-derived CKs to the leaves.

2.3.4 The Meaning of Stay-Green for Grain Yield and Quality

The initiation and execution of the leaf senescence are orchestrated by a repertoire of senescence-associated genes (SAGs). Alterations in SAGs disrupt the normal process of senescence, often resulting in a typical chlorophyll retention, a characteristic known as *stay-green* (Thomas and Ougham 2014). The stay-green trait is merely one of numerous characteristics that synergistically contribute to the improved productivity of major grain crops. Empirical selection for the functional stay-green trait has played a role in enhancing crop yields, especially when integrated into a broader breeding strategy that also targets complementary traits, such as sink strength and resilience to abiotic stresses (Gregersen et al. 2013).

Research on rice, wheat, and barley indicates that postponing leaf senescence can increase grain yields, as leaves maintain photosynthetic activity for a longer duration. However, in modern high-yielding cultivars, the correlation between stay-green traits and yield is relatively weak. This phenomenon can be partly attributed to a scaling effect: gains in productivity achieved by selecting for specific phenotypic traits at the organ level, e.g. leaf photosynthetic rate or longevity, often fail to translate into improved whole-plant or field-level performance (Pedró et al. 2012). Besides, the stay-green trait may increase plant susceptibility to weather-related stresses associated with delayed harvests, such as heat or drought stress in summer crops and cold stress in autumn crops.

In addition, stay-green may confer limited or even negative effects on certain species, particularly grain legumes such as soybean (Kumudini 2002). Although it is well known that delaying leaf senescence can increase seed yield by maintaining C fixation, it also leads to a decline in grain protein content due to sustained carbohydrate fluxes and delayed N translocation from source leaves. A dilemma therefore emerges for legume crops grown for grain protein content, as they must balance yield performance with seed quality. However, this trade-off is not straightforward, as it depends critically on the timing of senescence onset and offset, as well as the duration of senescence relative to seed development. A more in-depth understanding of the molecular mechanisms triggering senescence and regulating nutrient remobilization from leaves to seeds could help alleviate the trade-off between increased grain yield and reduced nutritional quality in legumes.

2.4 The Precise Estimation of Physiological Maturity Is a Prerequisite for Determining the Optimal Harvest Time

Accurately estimating the timing of physiological maturity enables farmers to mitigate yield losses stemming from lodging, pre-harvest sprouting, hail damage, and other harvest-related risks. Moreover, timely maturity and harvest of the first crop play a vital role in maximizing the growth window and yield potential of the second crop in double cropping systems. For instance, harvesting early rice at physiological maturity allows timely sowing of wheat, ensuring optimal tillering and grain filling, thereby improving annual grain productivity in rice–wheat rotations.

Using a shared time frame, the dynamics of grain weight gain and grain water content exhibit a weak correlation, as they are substantially influenced by both genotype and environmental factors (Fig. 7.2A). However, there is a positive

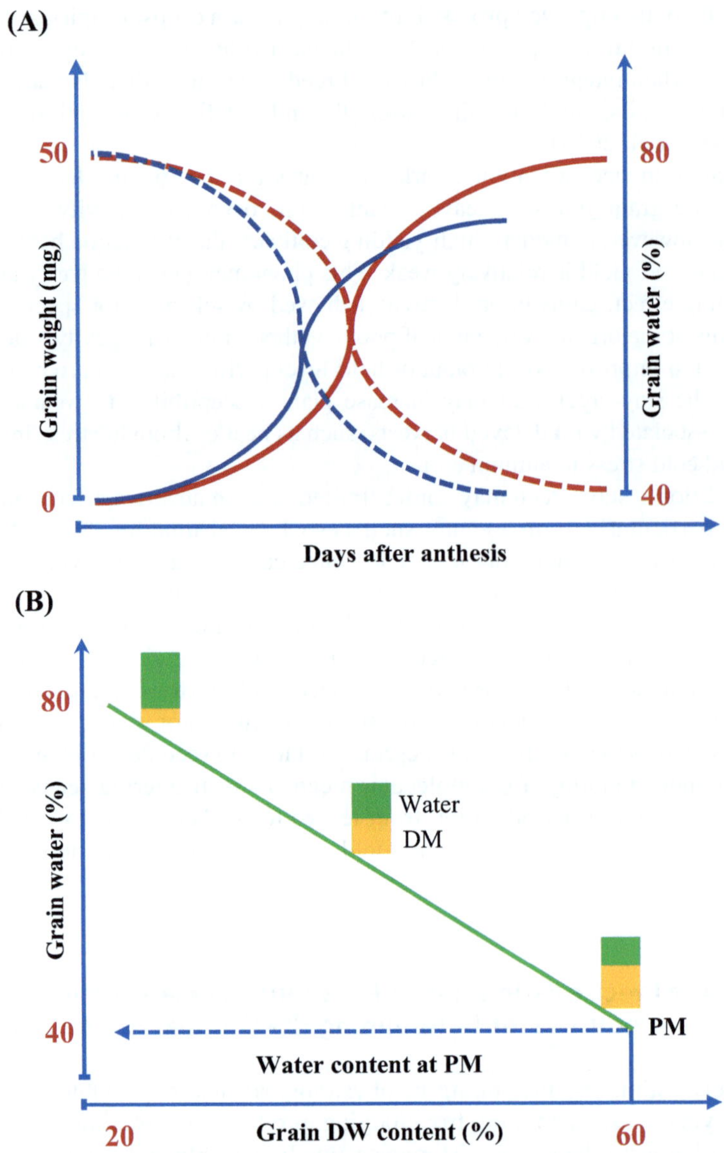

Fig. 7.2 Relationship between grain weight and its water content during ripening. (**A**) depicts the dynamics of grain weight (solid lines) and water content (dotted lines) during grain filling for two cultivars, shown in different colours. Note that the growth dynamics of grain are similar to those of the whole plant, both exhibiting a classic sigmoidal (S-shaped) pattern. (**B**) shows the relationship between grain water content and grain dry matter (DM) content. PM, physiological maturity. The data on grain weight and water content represent approximate estimates for the wheat samples. (Redrawn according to Calderini et al. (2000))

correlation between the rate of grain growth and the rate of water loss. When grain weight is normalized as a percentage of the final grain weight, a consistent and significant negative relationship is observed between grain water content and grain dry matter content (Fig. 7.2B). This finding reveals a close association between dry matter accumulation in developing grains and the simultaneous loss of water (Calderini et al. 2000). During grain-filling process, storage carbohydrates, proteins, and lipids accumulate as water is progressively lost, until the grain reaches a critical minimum moisture content. Studies on maize, wheat, sorghum, soybean, and sunflower show that final grain weight is reached at a specific moisture content, regardless of grain size variations due to genetic or environmental factors. These robust patterns suggest that grain moisture content serves as a reliable, intrinsic indicator of developmental progress towards maturity. For wheat, the water content is approximately 80% just after anthesis and around 40% at physiological maturity (Calderini et al. 2000). Therefore, we can directly determine the proportion of the grain-filling period that has elapsed at any given time by measuring grain moisture content in the field. This quantitative assessment of the extent to which grain filling has already been completed may be crucial in management decisions, such as determining when to apply a desiccant to the crop to advance harvest without compromising yield.

3 Developmental Response to Environmental Conditions

The adaptation of crop development to specific environmental conditions depends on the perception of temperature and day length, which are governed by two critical mechanisms: vernalization and photoperiod sensitivity. Moreover, temperature per se can directly influence flowering time independently of vernalization and photoperiod.

3.1 Temperature per se

Earth's thermal conditions create an ideal environment for the emergence and sustenance of life, primarily due to its stable aqueous state and temperature range that maintains liquid water across much of its surface. Biological processes are constrained at the lower end by the freezing point of water and at the upper end by irreversible protein denaturation. Within this range, the relationship between growth and temperature for individual organisms typically follows an asymmetric bell-shaped curve, a pattern also observed in individual and multiple enzymatic reactions. An organism's overall growth temperature profile is effectively a composite of the thermal responses of key physiological processes such as photosynthesis, respiration, and other essential metabolisms. Notably, these growth curves exhibit three critical temperature points, minimum, optimum, and maximum, consistent with those seen in individual enzymatic reactions.

The development of plants is highly influenced by temperature, which regulates the pace of growth and the shift from vegetative to reproductive stages. Temperature plays a crucial role in crop development through two distinct mechanisms. First, within a broad range of thermal conditions, higher temperatures accelerate developmental rates, a fundamental biological response to temperature. Because the relationship between development rate and temperature is linear under suboptimal conditions, the progression towards flowering can be quantified using *thermal time* units, such as degree-days. Second, many temperate crops, including wheat, require prolonged exposure to low temperatures to initiate the transition from vegetative to reproductive growth. This physiological process, known as vernalization, depends on specific cold treatments that promote flowering.

The period from seeding to flowering is mainly regulated by temperature. It is well established that the time to flowering decreases in a curvilinear manner as temperature increases. Within a specific temperature range, the rate of progress towards flowering increases linearly with temperature, beginning at a base temperature and extending up to the optimum temperature, where the rate reaches its maximum and flowering time is minimized. Beyond this optimum temperature, elevated temperatures often impede developmental progress, resulting in a longer time to flowering. The base and optimum temperatures delineate the temperature interval within which development progresses linearly. Due to this consistent linearity, a single slope is sufficient to represent the entire temperature segment. The reciprocal of this slope corresponds to the thermal time (in degree-days) required to reach flowering. For a given genotype, the thermal time to flowering remains constant across temperatures, provided environmental conditions stay within the base-to-optimum range and the requirements for photoperiod and vernalization are satisfied.

3.2 Vernalization

Vernalization refers to the acceleration of plant development by exposing sensitive cultivars to cool temperatures during the early stages of crop ontogeny. A large number of plant species growing at higher latitudes require vernalization, which usually involves a 4- to 12-week exposure to the cold temperatures typically encountered in winter. Without cold treatment, plants that need vernalization will exhibit delayed flowering or remain in a vegetative state and will be unable to respond properly to floral induction signals, such as suitable photoperiods.

The vernalizing stimulus can be detected by seeds in the soil, juvenile green plants, and even developing grains within the maternal plant when exposed to sufficiently low temperatures. Devernalization may occur if the low-temperature period is interrupted, thereby reversing the vernalization process. Vernalization requirement is a characteristic feature of crops originating from temperate regions, such as wheat, barley, canola, and peas. There are "winter" and "spring" cultivars, classified by the requirement for the duration and intensity of vernalization. Winter wheat requires a prolonged cold period to flower and is typically sown in autumn and overwinters and resumes growth in spring after vernalization. Spring wheat can flower without cold exposure and is therefore planted in spring and matures in early summer.

The temperatures that induce this process are known as vernalizing temperatures. Rather than being defined by a specific numerical range, they are characterized by their functional outcome, which is the promotion of flowering after cold exposure. Different crop species and cultivars have distinct requirements for temperature ranges and durations to meet the vernalization need. Any temperature regime that consistently induces flowering after chilling can be classified as a vernalizing temperature. There is variability in the literature regarding the optimal temperature ranges for vernalization. Most studies characterizing vernalization have been conducted on seedlings. The highest vernalization efficiency occurs between 1 and 4 °C, whereas temperatures from 6 to 10 °C also induce vernalization, but with reduced effectiveness. Vernalization can still occur at temperatures as high as 18 °C, albeit less efficiently.

3.3 Photoperiodism

The Earth's rotation on its tilted axis gives rise to diurnal cycles and seasonal changes in day length. Plants use day-length variations to determine the optimal timing for flowering. *Photoperiod* is defined as the duration of daylight, which varies across seasons and with latitude. The photoperiod increases after the winter solstice and decreases after the summer solstice (Fig. 7.3). Garner and Allard (1920) discovered that many plants flower regardless of their sowing date, a phenomenon they termed *photoperiodism.* Subsequently, photoperiod has been recognized as a crucial factor affecting flowering time.

Photoperiodism, defined as an organism's ability to detect and respond to changes in day length, enables organisms to synchronize key life events with specific times of the year, thereby facilitating seasonal adaptations. This phenomenon occurs in both animals and plants. In animals, photoperiodism regulates a wide range of seasonal activities, including hibernation, moulting for seasonal coats, and reproductive cycles. In plants, photoperiodic control governs diverse physiological responses, such as flowering induction, vegetative propagation, storage organ formation, and dormancy onset.

Flowering plants perceive day length and night length as seasonal cues to initiate flowering. They exhibit varying degrees of photoperiod sensitivity and require specific minimum photoperiods to transition from the vegetative to the reproductive phase. Photoperiod sensitivity is generally quantitative rather than qualitative: when the photoperiod is inappropriate, plant development is delayed rather than arrested. Plants are classified into three groups based on their photoperiodic flowering response. *Long-day plants* (LDPs) flower when exposed to photoperiods longer than a critical threshold; *short-day plants* (SDPs) flower only when the dark period exceeds a critical duration. *Day-neutral plants,* such as tomato and cucumber, flower regardless of day length. Photoperiodic flowering behaviour is strongly influenced by evolutionary origin: LDPs typically originate from temperate regions (e.g.

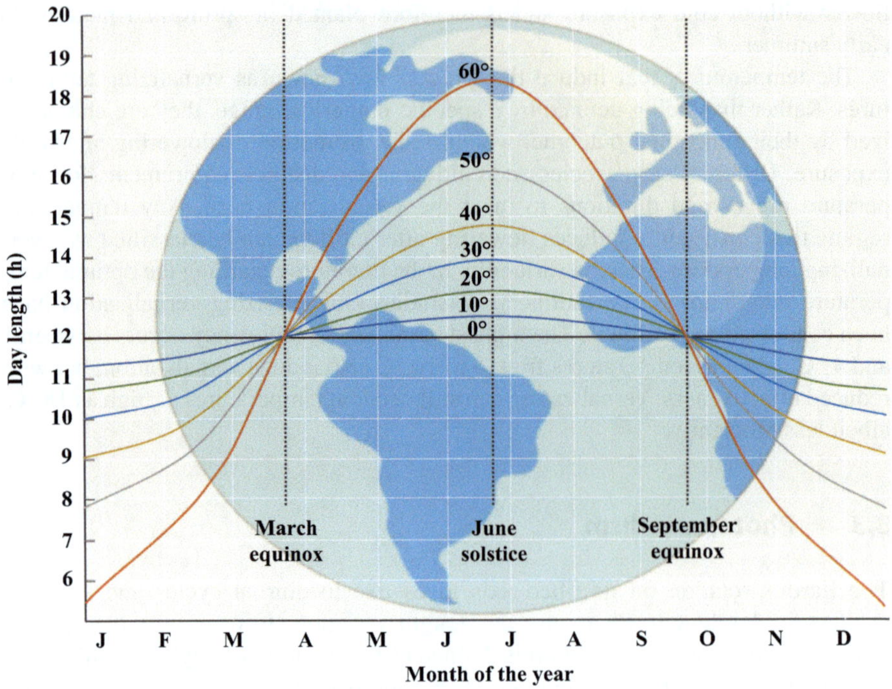

Fig. 7.3 Day length variation by latitude in the Northern Hemisphere throughout the year. The annual variation in day length depends on the geographic location on Earth. This phenomenon is a consequence of the planet's axial tilt relative to its orbital plane, which causes the poles to alternate between being tilted towards and away from the Sun throughout the year. At the December solstice, which occurs approximately on December 22, the North Pole is tilted away from the Sun. In the Northern Hemisphere, this event marks the winter solstice, characterized by shorter days and longer nights. Conversely, in the Southern Hemisphere, it is the summer solstice, featuring longer days and shorter nights. Six months later, around June 21, the situation is reversed during the June solstice. The equinoxes, which occur around March 20 and September 22, indicate the times when Earth's axis is neither tilted towards nor away from the Sun, resulting in nearly equal day and night lengths at all latitudes. Under theoretical, cloud-free conditions, the Earth's symmetrical axial tilt ensures that, over a complete solar year, every location on Earth receives approximately equal total durations of daylight and darkness, about 6 months each

wheat, barley, oats, canola, peas), whereas SDPs are native to tropical or near-equatorial regions (e.g. maize, sorghum, rice, soybean).

SDPs exhibit a distinct mechanism for measuring day length compared to LDPs. In the context of photoperiodic flowering, the critical day length (CDL) serves as the photoperiodic cue that determines the successful transition to flowering. LDPs, such as *Arabidopsis*, do not exhibit a discernible CDL and initiate flowering progressively earlier as day length increases. Conversely, SDPs, like rice, possess a CDL of approximately 13.5 hours and can detect daily photoperiod changes as small as 15 minutes to time flowering optimally. Moreover, SDPs respond to the duration of the dark period, not the length of the light period, making the dark period the decisive photoperiodic signal. This conclusion is supported by the night-break response phenomenon: a brief light pulse during the night strongly inhibits flowering in most SDPs but has little or no effect on LDPs. Consequently, the critical dark period length and the night-break response are two key criteria for distinguishing photoperiodic responses between SDPs and LDPs (Sun et al. 2022). Nevertheless, the molecular and physiological basis underlying this fundamental divergence remains unclear.

4 Molecular Mechanism Underlying Flowering Time

Flowering time refers to the transition from the vegetative to the reproductive stage or from the juvenile to the adult phase, representing a critical developmental milestone in the plant life cycle. Prior to flowering, plants must accumulate sufficient nutrients in their vegetative organs to support seed development and ensure that environmental conditions are conducive to successful reproduction. In crop species, flowering time is a key determinant of regional and seasonal adaptation, facilitating timely maturation under favourable climatic conditions and thereby realizing yield potential.

The developmental transition to flowering in plants encompasses a highly complex regulatory network that responds to a variety of flowering cues. These cues consist of endogenous factors, such as leaf age and nutritional status, as well as exogenous factors, such as seasonal fluctuations in light quantity, light quality, and temperature. Genetic studies in the long-day plant *Arabidopsis* have identified four major pathways regulating flowering: the photoperiod pathway, the autonomous pathway, the vernalization pathway, and the gibberellin pathway. Furthermore, the core components of the genetic network governing flowering time, particularly those associated with the photoperiodic and autonomous pathways, are evolutionarily conserved among *Arabidopsis*, rice, and temperate grasses, even though these species belong to divergent evolutionary lineages (Cao et al. 2021). This conservation provides a solid theoretical foundation and practical potential for cross-species gene transfer and precise genome editing through the exploitation of natural allelic variation.

4.1 The Photoperiodic Pathway

Although floral evocation takes place at the shoot apical meristem (SAM), the inductive photoperiod is detected by the leaves. This suggests that a long-distance signal must be transmitted from the leaves to the shoot apex, a hypothesis strongly supported by grafting experiments conducted across a wide range of plant species. Molecular genetic studies have identified the floral stimulus as a protein named florigen (FT), which functions as the systemic flowering signal. Moreover, several other biochemical signals have been found to serve as either promoters or inhibitors of flowering, collectively contributing to the precise regulation of the floral transition.

4.1.1 Phytochromes and Cryptochromes Detect Photoperiod

Light quality (wavelength) and quantity (irradiance) govern multiple aspects of plant growth and development, ensuring optimal exposure of photosynthetic tissues to light and maximizing photosynthetic activities. Plants are highly responsive to variations in light intensity, spectral composition, and incident angle, dynamically modulating both their structural features and physiological processes accordingly. It has been discovered that floral induction in SDPs can be effectively inhibited by brief red-light interruptions during the dark period and reversed by subsequent exposure to far-red light. This discovery prompted researchers to investigate how plants perceive specific wavelengths of light, leading to the identification of specialized photoreceptors.

Plants possess a variety of *photoreceptors*, which are light-sensitive molecules that experience structural, energetic, or conformational changes upon the absorption of light energy. These photoreceptors detect light and transform its physical energy into chemical signals, thereby initiating a downstream signalling cascade. Different photoreceptor families in plants exhibit sensitivity to specific wavelengths of light, as determined by their unique molecular structures.

Phytochromes, a family of photoreceptors highly sensitive to red and far-red light, play a crucial role in regulating flowering time. The *Arabidopsis* genome encodes five phytochromes (PHYA to PHYE), each exhibiting distinct yet partially overlapping functions. Cryptochromes detect blue and ultraviolet light, playing essential roles in regulating flowering time and mediating phototropism. *Arabidopsis* possesses two cryptochromes, CRY1 and CRY2. In *Arabidopsis* and closely related species, both PHYA and CRY2 promote flowering. However, in other plant species, this function is primarily mediated by PHYA alone. For example, in the LDP pea, *phyA* mutants display photoperiod insensitivity and flower at the same time with the WT regardless of day length. Furthermore, FLAVIN-BINDING, KELCH REPEAT, F-BOX 1 (FKF1) acts as an additional blue-light receptor involved in photoperiodic flowering under long-day conditions (Song et al. 2012).

4.1.2 Circadian Clock Regulates the Physiological Sensitivity to Photoperiod

Numerous physiological and biochemical processes exhibit *circadian rhythms*, as evidenced by the rhythmic opening and closing of leaves in mimosa plants at specific times on a daily basis. The most prominent characteristic of circadian rhythms is their ability to persist in the absence of daily light-dark cycles, such as under conditions of constant darkness or constant light. These self-sustained oscillations indicate that circadian rhythms are not merely passive responses to external environmental changes; rather, they reflect an internal timekeeping mechanism capable of tracking 24-hour cycles and autonomously generating rhythmic outputs, commonly known as the *circadian clock* or circadian oscillator.

This endogenous clock regulates daily oscillations in gene expression, protein phosphorylation, and cytosolic calcium levels, enabling plants to temporally synchronize essential physiological processes with specific phases of the day or night. These processes include photoperiodic flowering, petal opening, leaf movement, stomatal regulation, and intracellular chloroplast repositioning. Harmer et al. (2000) utilized high-density oligonucleotide microarrays to investigate gene expression in *Arabidopsis*. They discovered that 6% of the over 8000 genes on the array demonstrated circadian fluctuations in steady-state mRNA levels. Notably, these rhythmically expressed genes exhibit distinct peak times throughout the day. Certain mRNAs reach maximal levels at dawn, whereas others peak at midday or in the evening.

Plants utilize the circadian clock to regulate physiological processes in response to photoperiodic stimuli. In SDPs, exposure to light during the night can hinder the flowering process, and the intensity of this inhibitory effect varies depending on the timing of the light interruption. Red goosefoot (*Chenopodium rubrum*) has been widely used as a model species for studying the physiological mechanisms underlying photoperiodism. This short-day plant flowers when grown under exposed to a single 72-hour dark period. To investigate the influence of light on flowering, plants were given 4-minute pulses of red light at different intervals throughout the dark phase. Flowering occurred only when the light pulse was delivered approximately 24, 48, or 72 hours after the onset of darkness; light pulses at other times suppressed flowering. These results strongly support the hypothesis that photoperiodic flowering is governed by a circadian clock with rhythmic sensitivity to light at specific phases of its cycle.

4.1.3 CONSTANS Is Regulated by the Circadian Clock and Functions as a Central Regulator of Photoperiodic Flowering

The circadian clock offers an internal timing mechanism, which allows plants to sense day length and commence flowering solely when the seasonal photoperiod is conducive. *CONSTANS* (CO) serves as a crucial regulator of flowering time, integrating photoperiodic and circadian signals to govern the floral transition in plants. As a nuclear transcription factor, CO is characterized by two conserved B-box zinc-finger domains and a C-terminal CCT domain, both of which are essential for its light-dependent stability and transcriptional activity.

Under short-day conditions, the *CO* mRNA levels begin to rise at dusk and peak during the night, with minimal overlap with daylight hours. CO protein is unstable and fails to accumulate to functional levels, thereby maintaining the plant in a vegetative state. In contrast, under long-day conditions, the peak of *CO* mRNA abundance (between hours 12 and 16) coincides with daylight (Fig. 7.4). Exposure to blue and far-red light activates CRY2, PHYA, and PKF1 to stabilize the CO protein, enabling its accumulation and promoting the activation of its target gene, *FT* (Fig. 7.4). *CO* expression and the subsequent induction of *FT* occur in phloem companion cells, where the FT protein is synthesized and subsequently transported to the SAM to induce flowering. As a result, *Arabidopsis* flowers only when *CO* transcription and translation coincide with light exposure, a condition met under

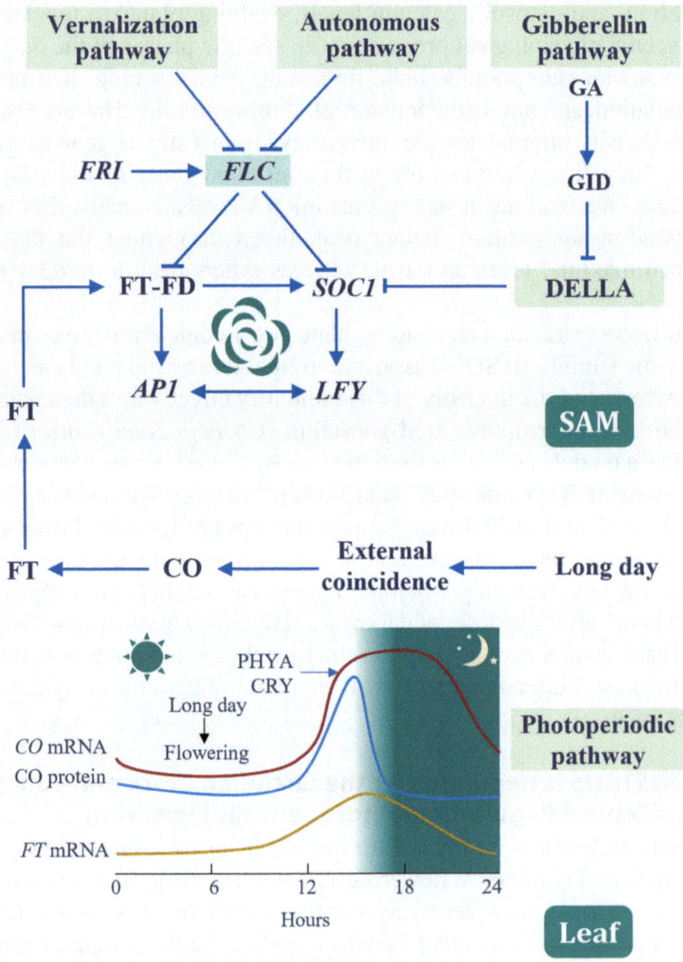

Fig. 7.4 Integration of multiple signalling pathways for flowering. The transition from vegetative to reproductive growth is regulated by four major pathways: photoperiod, autonomous, vernalization, and gibberellin. These pathways converge through the regulation of *FT* and *SOC1*, which subsequently activate the floral meristem identity genes *AP1* and *LFY*. According to the external coincidence model, in LDPs such as *Arabidopsis*, *FT* mRNA is synthesized in the companion cell of leaves in response to CO protein accumulation under long-day conditions. Subsequently, FT protein undergoes translocation through sieve tubes to reach the shoot apical meristem (SAM), where it interacts with FD to trigger the differentiation of inflorescence meristem. *AP1*, *APETALA 1*; CO, CONSTANS; CRY, cryptochrome; *FLC*, *FLOWERING LOCUS C*; FD, FLOWERING LOCUS D; FT, FLOWERING LOCUS T; *FRI*, *FRIGIDA*; GA, gibberellic acid; GID1, GIBBERELLIN INSENSITIVE DWARF 1; *LFY*, *LEAFY*; PHY, phytochrome; *SOC1*, *SUPPRESSOR OF OVEREXPRESSION OF CONSTANS 1*. (Redrawn after Taiz et al. (2015) and Buchanan et al. (2015))

long-day photoperiods. In other words, flowering is triggered when the internal circadian phase, represented by *CO* expression, aligns with external light signals, a mechanism referred to as the external coincidence model.

In rice, a short-day plant, *Heading date 1* (*Hd1*) serves as the ortholog of *CO* in *Arabidopsis*, while *Hd3a* is functionally equivalent to *FT*. The key distinction between these two systems lies in their regulatory mechanisms: Hd1 suppresses *Hd3a* transcription under long-day conditions, but this repression is lifted under short-day conditions. Thus, a minor adaptation of the *Arabidopsis* flowering model adequately explains the flowering mechanism in rice. Under long-day conditions, the peak of *Hd1* mRNA expression occurs during the daytime, promoting the accumulation of the Hd1 protein, which in turn represses *Hd3a* expression. As a result, *Hd3a* mRNA levels are suppressed, and the plant remains in the vegetative phase. Under short-day conditions, the temporal misalignment between *Hd1* mRNA expression and daylight prevents effective accumulation of the Hd1 protein. In the absence of Hd1-mediated repression, *Hd3a* mRNA is upregulated, leading to the synthesis and transport of the Hd3a protein to the SAM, where it promotes flowering through a mechanism analogous to that in *Arabidopsis*.

4.1.4 Florigen Is Produced in the CCs and Translocated to the SAM

Pioneering grafting experiments conducted in the 1930s demonstrated that a mobile flowering signal, subsequently named "*florigen*", is transmitted from leaves to the SAM to initiate the floral transition. Nevertheless, for more than six decades, despite extensive efforts, florigen remained resistant to biochemical characterization. A significant breakthrough occurred with the genetic identification of FLOWERING LOCUS T (FT) in *Arabidopsis*, which was later shown to encode the long-awaited florigen protein. Biochemically, FT is a small globular protein belonging to a conserved family of regulatory proteins found in organisms ranging from budding yeast to vertebrates. *FT* gene expression is induced across diverse plant species during floral-inductive photoperiods. Notably, *ectopic expression* of FT in species with photoperiod-insensitive flowering pathways is sufficient to induce flowering independently of day length. Furthermore, the FT protein moves from leaves to the shoot apical meristem, fulfilling all key criteria expected of the elusive florigen.

There are two critical steps in the process of FT-induced flowering (Fig. 7.4): (i) the export of FT protein from companion cells (CCs) to sieve tube elements (STEs) and (ii) the activation of FT target genes at the SAM, which triggers flowering. First, *FT* mRNA is transcribed in leaf CCs in response to multiple environmental cues, including photoperiod, light quality, and temperature. The endoplasmic reticulum serves as a major route for FT protein trafficking from CCs to STEs, mediated by the ER-localized protein FT INTERACTING PROTEIN 1 (FTIP1). Upon arrival at the SAM, the FT protein moves into the nucleus and forms a functional complex with FLOWERING LOCUS D (FD), a bZIP transcription factor specifically expressed in the meristem. This FT–FD complex then activates key floral meristem identity genes or *homeotic genes*, such as *APETALA 1* (*AP1*) and *LEAFY* (*LFY*). AP1, a member of the MADS-box family of transcription factors, determines the identity of the outermost (first) whorl of the flower, thereby promoting sepal formation and initiating early stages of floral organ development. LFY, which functions at a later stage in the floral developmental program, directly activates numerous downstream genes essential for flower morphogenesis and maintains the floral meristem in a stable, indeterminate, and organized state.

The movement of the flowering signal (FT) from companion cells to shoot apical meristems occurs at a rate similar to that of sugar translocation in the phloem. For example, in SDPs such as *Chenopodium*, FT export from mature leaves is completed within 22.5 hours after the onset of the long night. In contrast, in LDPs like *Sinapis*, FT can be transported out of the leaf as early as 16 hours after the initiation of long-day conditions. The difference in phloem translocation rates suggests that sunlight may influence the mass flow in the sieve tube elements. In LDPs, translocation is faster during daylight hours when solar radiation is abundant, whereas in SDPs, it slows down at night without light. Since FT is translocated through the phloem along with sugars, its distribution is governed by source–sink relationship. Consequently, an induced leaf that is located closer to the SAM has a higher probability of promoting flowering compared to a leaf at the base of the stem, which usually allocates sugars towards root development.

4.2 The Vernalization Pathway

Unlike photoperiod, which is perceived in the leaves, vernalization relies on molecular events occurring at the SAM. FLOWERING LOCUS C (FLC), a MADS-box transcription factor, acts as the central repressor of flowering and is responsible for mediating the vernalization requirement in *Arabidopsis* (Fig. 7.4). In wheat, flowering time is regulated by a set of vernalization genes, namely, *VRN1*, *VRN2*, and *VRN3*. *VRN1* encodes a MADS-box transcription factor, whose expression is strongly upregulated following prolonged cold exposure. *VRN3* corresponds to the ortholog of *Arabidopsis FT* gene. *VRN2* produces a protein containing both zinc-finger and CCT domains; it is predominantly expressed in non-vernalized seedlings and functions as a floral repressor to downregulate both *VRN3* and *VRN1*, thereby delaying flowering onset.

The alterations in gene expression within the SAM following vernalization persist throughout the remainder of the plant's life cycle. This regulatory mechanism relies on stable, epigenetically mediated remodelling of chromatin structure. Vernalization initiates a transition in the chromatin state of FLC, which is characterized by the loss of histone modifications in the transcriptionally active euchromatin and the acquisition of methylation at specific lysine residues in the transcriptionally silent heterochromatin. This cold-induced shift from an active to a repressed chromatin state results in the stable silencing of the *FLC* gene. Upon completion of vernalization, the *FLC* gene is epigenetically silenced, enabling flowering in response to long-day conditions. In the subsequent generation, however, FLC is reactivated, thereby restoring the requirement for vernalization.

The response of *FLC* to cold exposure occurs gradually. Specifically, the longer the duration of cold treatment, the more significant the reduction in *FLC* expression.

This dose-dependent repression of *FLC* elucidates why vernalization is a quantitative process: shorter cold exposures promote flowering to a lesser degree than prolonged ones. Nevertheless, the downregulation of *FLC* is not solely mediated by vernalization; the autonomous pathway also contributes to the regulation of *FLC* expression.

4.3 The Autonomous Pathway

The autonomous pathway promotes flowering independently of photoperiod (Fig. 7.4). In *Arabidopsis*, this pathway initiates flowering in response to endogenous developmental cues, particularly the attainment of a specific leaf number. It involves multiple factors associated with RNA processing and epigenetic regulation that collectively repress the floral inhibitor FLC, which is positively controlled by *FRIGIDA* (*FRI*). All known components of this pathway are expressed in the SAM. As plants reach maturity, the expression of the floral repressor FLC gradually declines through age-dependent regulatory pathways. Once the level of FLC drops below a critical threshold, its inhibitory effect on FT and SUPPRESSOR OF CONSTANS 1 (SOC1) is alleviated, allowing the plant to shift from vegetative growth to flowering.

4.4 The Gibberellin Pathway

The gibberellin pathway promotes flowering by degrading the DELLA protein, the key negative regulator of GA signalling, thereby enabling expression of floral identity genes and integrating hormonal signals with environmental cues (Fig. 7.4). GAs can induce flowering under non-inductive short-day conditions. GA is perceived by the soluble receptor GIBBERELLIN INSENSITIVE DWARF 1 (GID1). Upon binding GA, GID1 interacts with DELLA proteins. The GA-GID1-DELLA complex facilitates ubiquitination and subsequent proteasomal degradation of DELLA proteins. Removal of DELLA-mediated repression enables activation of key flowering integrators such as LFY, SOC1, and AP1.

Exogenous GAs can induce flowering when applied either to LDPs like *Arabidopsis* or to day-neutral plants such as *Bryophyllum* grown under short-day conditions. Moreover, the application of GAs can trigger flowering in a small number of SDPs under non-inductive conditions and in cold-requiring plants that have not undergone vernalization. Thus, in certain plant species, exogenous GAs can bypass both the age-dependent autonomous pathway and the primary environmental signals such as photoperiod and temperature. Besides GAs, other plant hormones also regulate flowering. A commercially important example is the robust induction of flowering in pineapple plants following exposure to ethylene or ethylene-releasing compounds, a response predominantly observed within the Bromeliaceae family.

4.5 Integration of Flowering Signals

The phenological response of plant is regulated by multiple pathways, each of which responds to distinct internal or external cues and promotes flowering. Nevertheless, all these pathways converge on a shared downstream module reprograming the SAM to transition from vegetative growth to reproductive development. In *Arabidopsis*, the major four flowering pathways converge on a small set of floral integrator genes, primarily *FT* and *SOC1*, which in turn activate floral identity genes such as *AP1* and *LFY* (Fig. 7.4).

1. In the photoperiod pathway, long-day or short-day information is sensed by the circadian clock in conjunction with photoreceptors like PHYA, CRY2, and FKF1. The CO protein accumulates under inductive photoperiods and directly activates the transcription of *FT* in the leaf vasculature. Subsequently, the FT protein is transported via the phloem to the SAM, where it upregulates *SOC1* and its downstream genes.
2. In the autonomous pathway, internal developmental signals are received by age-dependent RNA-processing factors, which mediate chromatin-level repression of the floral repressor FLC. This epigenetic silencing of *FLC* alleviates its repression on *FT* and *SOC1*, allowing the plant to initiate flowering at the appropriate time.
3. In the vernalization pathway, prolonged exposure to cold induces epigenetic silencing of *FLC*, thereby releasing the repression of *FT* and *SOC1*. Thus, the vernalization signal converges on the same floral integrators as the autonomous pathway.
4. In the gibberellin pathway, GAs promote flowering primarily under non-inductive photoperiods or when other flowering pathways are compromised. GAs perception triggers the degradation of DELLA repressors, thereby enabling the activation of *SOC1* (and in some cases *FT*) and the upregulation of *LFY* expression.

All four pathways ultimately converge to modulate the transcription of *FT* and *SOC1*. Once FT and SOC1 accumulate to threshold levels, they form protein complexes with other MADS-box transcription factors, such as FD, and directly activate the floral identity genes *AP1* and *LFY*. The synergistic action of *AP1* and *LFY* reprograms the SAM from a vegetative state producing leaves to a reproductive state generating floral primordia, thereby completing the flowering program.

5 Genetic Modification

Given the escalating frequency of extreme weather events, adjusting flowering time has emerged as a crucial strategy to desynchronize critical phenological phases from periods of severe abiotic stress (Burgess et al. 2022). Crop cultivars developed through current breeding techniques are closely adapted to local climatic conditions, which are increasingly unpredictable due to the impacts of climate change. Therefore, attaining control over phenological events via a straightforward chemical treatment has emerged as a compelling objective in plant breeding.

Okada et al. (2017) produced transgenic rice plants overexpressing the floral repressor gene *Grain-number, plant height and heading date 7* (*Ghd7*). The *Ghd7* overexpressing rice (*Ghd7ox*) did not flower, even after several years of planting. Using a chemical genomics approach, the researchers isolated a promoter that responds to agrochemical spraying. Pesticides such as Routine or Oryzemate are commonly applied as defence activators to control rice blast. *Hd3a* under the control of the agrochemical-responsive promoter was introduced into the *Ghd7ox* rice. The resultant plants exhibited a strong response to chemical spraying, initiating early floral transition (Fig. 7.5). Moreover, yield-related traits were enhanced due to

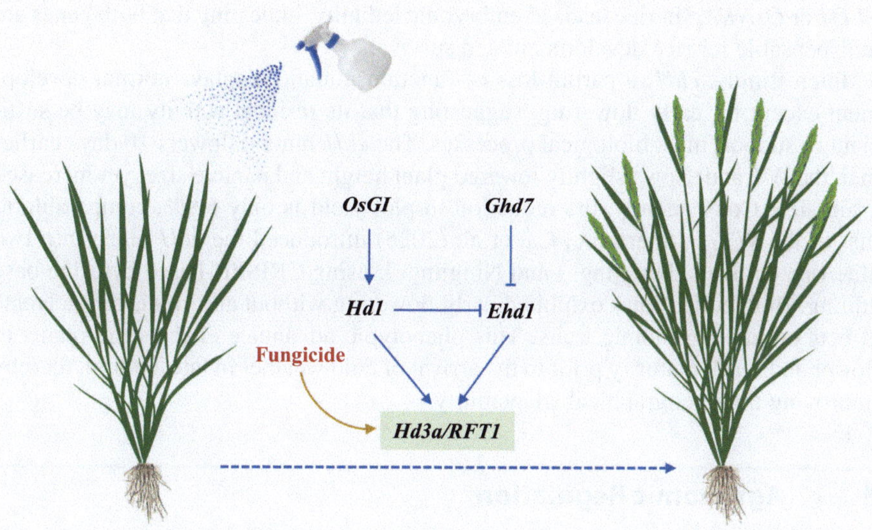

Fig. 7.5 A simplified scheme of rice floral transition under the control of agrochemical-responsive promoter. There are two major signalling pathways that regulate flowering in rice. The first is the evolutionarily conserved photoperiodic pathway, *OsGI-Hd1-Hd3a/RFT1*, which is analogous to the *GI-CO-FT* pathway in *Arabidopsis*. *OsGI* is the *GIGANTEA* orthologue. *Heading date 1* (*Hd1*), the ortholog of *Arabidopsis CONSTANS* (*CO*), functions as a central regulator of photoperiodic flowering in rice. *Hd1* generally promotes flowering under short-day conditions by activating the florigen genes *Heading date 3a* (*Hd3a*) and *RICE FLOWERING LOCUS T 1* (*RFT1*), but suppresses flowering under long-day conditions. In addition to this conserved pathway, rice possesses another monocot-specific flowering pathway in which *Early heading date* 1 (*Ehd1*) acts as a central regulatory hub. *Ehd1* is regulated by multiple flowering genes, including *Ghd7*, *MADS50*, *Ehd2*, *Ehd3*, and *Ehd4*. The red text indicates that the gene *Hd3a* is under the control of the fungicide-responsive promoter. (Redrawn after Jung (2017) and Cai et al. (2025))

both an extended vegetative growth period and an increased grain number per panicle under field trials. Desirably, this innovative technology enables farmers to precisely control the flowering period in rice cultivation through regular agrochemical applications. Meanwhile, it facilitates the transfer of genes from elite germplasms originating from diverse ecological regions, as flowering time can be flexibly adjusted to ensure successful crossbreeding (Jung 2017).

Essential genes are indispensable for the growth and survival of organisms and are typically involved in core metabolic and cellular processes, such as DNA replication, transcription, and translation. In crop species, loss-of-function mutations in these genes frequently lead to lethality or severely compromised agronomic traits, thereby limiting their direct application in breeding programs. Cai et al. (2025) identified ELD1, a CCHC-type zinc finger protein essential for rice embryo survival. Partial loss of ELD1 function results in early flowering under long-day conditions. Further investigation reveals that ELD1 regulates the splicing profile of *OsCCA1*, a key component of the circadian clock oscillator. Knockout of either *ELD1* or *OsNKAP* in rice leads to embryonic lethality, indicating that both genes are indispensable for rice development and survival.

Interestingly, *eld1*, a partial loss-of-function mutant, displays normal development except for early flowering, suggesting that its residual activity may be sufficient to support most biological processes. The *eld1* mutant flowers 10 days earlier than the WT and shows slightly lowered plant height and panicle size, yet increased grain size. Consequently, the reduction in plot yield is only 5.9%, comparable to that of the WT. Furthermore, Cai et al. (2025) introduced the *eld1* allele into two elite rice cultivars, Ningjing-4 and Ningjing-7, using CRISPR-based cytosine base editing. Both edited lines exhibited early flowering without any significant adverse effects on key agronomic traits. This phenotypic advantage enables the plants to flower and reach maturity prior to the arrival of cold weather in late autumn, thereby improving their geographical adaptability.

6 Agronomic Regulation

The adoption of suitable cultivars and appropriate fertilization methods is the two primary agronomic approaches that substantially affect crop phenology. However, due to the complex regulatory mechanisms for flowering time, the effectiveness of agronomic interventions has to be evaluated on a case-by-case basis.

6.1 Adaptation of Soybean Cultivars to Different Latitudes

Soybean is a typical short-day plant characterized by high photoperiod sensitivity. Its yield at a specific latitude depends on appropriate photoperiod responsiveness, as demonstrated by the restricted latitudinal range within which each traditional cultivar is adapted. Domesticated originally in China between 30° N and 45° N, soybean is now cultivated globally across a broad latitudinal range from 53° N to 35° S. This extensive ecological adaptability is largely attributable to genetic variation in major *gene loci* and *quantitative trait loci (QTLs)* that regulate flowering time and maturity. Forward genetic approaches have identified key gene loci including *E1* to *E11* and *J* (long-juvenile), as well as several QTLs such as *Tof11/Gp11*, *Tof12/Gp1/qFT12-1*, and *qDTF-J* (Lin et al. 2021). Specifically, dominant alleles at *E1*, *E2*, *E3*, *E4*, *E7*, *E8*, and *E10* confer late flowering, while dominant alleles at *E6*, *E9*, *E11*, and *J* promote early flowering.

Multiple naturally occurring variants at these loci have been subjected to human selection, endowing soybeans with the capacity to thrive in environments with diverse photoperiods (Duan et al. 2026). In high-latitude (long-day) environments, genotypes carrying recessive *e1*, *e2*, and *e3* alleles in combination with the dominant *E4* allele exhibit significantly reduced photoperiod sensitivity, enabling timely flowering and widespread agronomic utilization. Conversely, adaptation to low-latitude (short-day) conditions depends on the long-juvenile (LJ) trait, which is characterized by delayed flowering and prolonged vegetative growth under short-day conditions. This trait enhances biomass accumulation and yield potential in tropical and subtropical regions. The LJ phenotype is primarily controlled by two major recessive loci: *j* (at the *J* locus) and *e6* (at the *E6* locus). Homozygosity at both loci (*j/j* and *e6/e6*) is required for full expression of this adaptive trait.

Selection within the intricate genetic background of the American soybeans has resulted in the emergence of two distinct cultivar types: (1) indeterminate, photoperiod-insensitive cultivars adapted to northern environments with long photoperiods, where late frosts may damage early flowers, yet subsequent flowering at higher nodes can compensate for this loss, and (2) determinate cultivars carrying alleles that prolong the inductive phase by several weeks, rendering them better suited to southern and subtropical regions. Moreover, there is a need for indeterminate, short-duration cultivars in drought environments. The classification of these environmentally responsive growth patterns and their corresponding agronomic traits has promoted the development of the maturity group system, which guides the selection of cultivars and the planting of soybean crops across North and Central America (Hey and Porter 2006).

6.2 Does Nitrogen Fertilization Delay the Flowering Time of Cereal Crops?

Nitrogen (N) is an essential macronutrient for plants and a key structural component of numerous macromolecules. Its availability represents a primary limiting factor for plant growth and crop productivity. Low N levels typically accelerate flowering, whereas high N levels delay it. Excessive N application not only raises production costs and environmental risks but also frequently results in "late maturity due to excessive greenness", thereby disrupting the planting schedule of subsequent crops.

Hall et al. (2014) conducted a comprehensive review of 1130 studies, critically evaluating the effects of N fertilizer on time to flowering (TOF) in wheat and barley. They used relative yield responses, compared to unfertilized controls, as an indicator of plant N status and measured phenological differences through relative changes in TOF. Their dataset, comprising 180 cases, covered a wide range of unfertilized crop yields (1–8 t ha^{-1}) and TOF durations (47–168 days). Notably, only a small fraction of cases (19 out of 118) showed a relative TOF difference exceeding 5% between fertilized and unfertilized treatments. Accordingly, Hall et al. (2014) concluded that there is limited support for a consistent responsive pattern of flowering time to N availability in wheat and barley. In rice, Zhang et al. (2022) examined flowering time across 100 accessions grown under four N rates and identified four distinct patterns: flowering induced by high N, inhibited by high N, a U-shaped response (delayed under both low and high nitrogen), and insensitivity to N supply. These findings collectively indicate that the variation in flowering time in response to N fluctuations is genotype-dependent.

The underlying molecular and genetic mechanisms by which N regulates flowering are now being progressively elucidated. Nitrogenous compounds, including nitrate and ammonium, can modulate both local and systemic signalling pathways. Zhang et al. (2021a, b) demonstrated that *nitrogen-mediated heading date 1* (*Nhd1*) directly activates the florigen gene *OsHd3a* in rice. Loss-of-function mutations in either *Nhd1* or *OsHd3a* lead to delayed flowering and reduced responsiveness to N supply. Further, Zhang et al. (2025) identified a reciprocal feedback loop between two transcription factors, Nhd1 and Ghd7, that orchestrates the U-shaped flowering response in rice under long-day conditions. Under low N, flowering is delayed through Ghd7-mediated repression of Nhd1. In contrast, high N delays flowering by activating Ghd7 via glutamine-induced Nhd1. Genetic and molecular evidence further shows that Hd3a is essential for the U-shaped response mediated by the Nhd1–Ghd7 module. Notably, an analysis of natural variation indicates that antagonistic combinations of *Nhd1* and *Ghd7* alleles were selected during domestication and are associated with geographic patterns of soil nitrogen deposition. Collectively, these findings offer mechanistic insights and potential targets for the breeding of high-NUE rice cultivars.

7 Concluding Remarks and Perspectives

The economic success of crop production heavily depends on the precise timing of economic inputs and agronomic interventions. To get the most out of a single piece of land within a limited period, farmers must align the growth stages of their crops with the prevailing local climate conditions. This synchronization ensures that each phenological event occurs when the environment is most favourable, thereby maximizing yield and profitability. It is particularly critical under the scenario of global change, as the frequency of heatwaves, droughts, and floods is increasing. However, our understanding remains inadequate for this grand mission, as several key issues remain unresolved.

In crop physiology, a central challenge is understanding how key phenological phases are precisely coordinated to maximize grain yield. Since grain yield is the result of the cumulative integration of multiple physiological processes throughout the entire plant life cycle, it is unlikely that any single phase alone determines the final yield. Instead, yield formation is shaped collectively by all developmental phases. However, the interactions among specific phases, such as early and late reproductive phases, the onset of leaf senescence, and grain filling, in regulating the formation of yield components, remain poorly understood. It is still unclear how these phases interact synergistically or antagonistically. Accurately identifying these critical developmental windows and clarifying how they interact would allow researchers and agricultural practitioners to design integrated management strategies, such as optimizing seeding dates and planting densities and fine-tuning the timing and amounts of irrigation and fertilization.

Adjusting flowering time remains a primary objective in crop breeding, especially in the development of new cultivars with enhanced adaptation to local environmental conditions and shifting climate patterns. Significant progress has been made in understanding the molecular mechanisms that regulate flowering time, and a substantial quantity of evolutionarily conserved yet functionally diverse genes have been identified. Nevertheless, the perception of light and temperature signals in major crop plants seems to be more complex, as is evidenced by the significant variation in flowering time across a wide range of growing environments. Consequently, further research is needed to translate gene discovery into improved field performance, particularly in the following areas (Wang et al. 2023): (1) comprehensive screening of germplasm for diverse genetic loci or alleles; (2) identification of all genomic regions associated with phenology and their underlying genes; (3) development of diagnostic markers that fully capture the spectrum of allelic variation; (4) high-throughput field phenotyping for phenological and agronomic traits across controlled and field environments; and (5) advancement of crop models that link genotype to phenotype under diverse growing conditions.

Summary

1. Crop phenology refers to the sequential progression of developmental stages. The success of a crop in a specific region largely depends on how well its phenology aligns with the local climate.
2. Senescence is normal, actively controlled processes characterized by the remobilization of nutrients into seeds. Delayed senescence confers yield advantages in certain crops, but only under certain circumstances.
3. The adaptation of crop development to ambient environment relies on the perception of low temperature and day length via vernalization and photoperiod sensitivities. Moreover, the ambient temperature per se can directly affect flowering time.
4. The initiation of flowering involves a highly intricate regulatory network that responds to internal and external factors. Genetic studies have established four distinct developmental pathways that control flowering in the long-day plant *Arabidopsis*: the photoperiod, vernalization, autonomous, and gibberellin pathways.
5. In general, flowering is accelerated under low nitrogen conditions but delayed under high nitrogen conditions. However, a comprehensive review of the

literature indicates that the timing of flowering in wheat, barley, and rice does not show a consistent pattern in response to nitrogen availability.

Important Terms

CO; Crop phenology; Development; FLC; Florigen; FT; Growth; Late reproductive phase; Phenology; Photoperiodism; Physiological maturity; Senescence; Vernalization

Review Questions

Multiple Choice

1. In the ( ) pathway of *Arabidopsis*, flowering occurs in response to internal signals, specifically the production of a fixed number of leaves.

   (a) Gibberellin
   (b) Autonomous
   (c) Photoperiodic
   (d) Vernalization

2. The gene ( ) is exclusively found in the photoperiodic pathway and is absent from the autonomous, vernalization, and gibberellin pathways.

   (a) *CO*
   (b) *FT*
   (c) *FLC*
   (d) *AP1*

3. The four flowering-inducing pathways, namely, photoperiod, autonomous, vernalization, and gibberellin pathways, act as parallel regulatory inputs that converge on a central regulatory hub ( ), which subsequently activates the master regulators AP1 and LFY to initiate the floral transition.
   (a) FT–FD
   (b) FLC–FD
   (c) FLC–SOC1
   (d) FT–SOC1

4. Regarding the photoperiodic pathway, which of the following statements is NOT correct? ( )

   (a) Although floral induction occurs at the shoot apical meristem, the inductive photoperiod is perceived by the roots.
   (b) Plants employ the circadian clock to regulate physiological processes in response to photoperiodic signals.

(c) CONSTANS, which is regulated by the circadian clock, acts as a central regulator of photoperiodic flowering.
(d) The movement of the flowering signal, FT, from companion cells to the shoot apical meristem occurs at a rate comparable to that of sugar translocation through the phloem.

True or False

1. There is mounting evidence suggesting that the production of crop biomass relies more on utilizing resources rather than harvesting them, with a few exceptions ( )
2. Leaf senescence is an inevitable yet passive and negative process in the plant life cycle ( )
3. Plant development can broadly be divided into vegetative and reproductive phases, which often overlap ( )
4. Sensitivity to photoperiod is generally qualitative rather than quantitative ( )
5. Development is delayed instead of being prevented when the photoperiod is inappropriate ( )

Blank Filling

1. Across a wide range of environmental and genotypic combinations, yield variations have been consistently far more strongly related to ____________ than to the average weight of grains or seeds.
2. The adaptation of flowering time to environmental conditions depends on the perception of temperature and day length. Additionally, flowering time is further regulated by ____________.

Short Answer

1. Describe the hormonal control of leaf senescence.
2. Illustrate how flowering signals are converged to control the expression of meristem identity genes.
3. Compare the photoperiod pathway between rice and *Arabidopsis*.
4. Name some key phenological stages or phases that critically influence cereal crop yield formation and explain their physiological basis.
5. Provide a few examples of cases that utilize genetic and agronomic methods to alter flowering time.

Essay Items

1. Senescence involves nutrient remobilization. How can you measure the amount of leaf nitrogen transferred to seeds during senescence? Discuss different methods to investigate this, and identify the most feasible and accurate options.
2. Justify the strategy of manipulating crop phenology to realize the yield potential, such as lengthening the late reproductive phase in the loss of the early phase.
3. Discuss the dual effects of leaf senescence on seed formation, illustrating with a positive and negative example from a crop species you are familiar with.
4. Efforts to either delay or accelerate leaf senescence can significantly impact both seed yield and the nutritional quality of seeds. What factors can accelerate or delay senescence? How would delaying leaf senescence in a crop plant be expected to influence both yield and quality?

References

Buchanan BB, Gruissem W, Jones RL (2015) Biochemistry & molecular biology of plants, 2nd edn. Wiley Blackwell, New Jersey

Burgess AJ, Masclaux-Daubresse C, Strittmatter G et al (2022) Improving crop yield potential: underlying biological processes and future prospects. Food Energy Secur 12(1):e435. https://doi.org/10.1002/fes3.435

Cai L, Hao B, Xu Z et al (2025) ELD1 mediates photoperiodic flowering via *OsCCA1* alternative splicing and interacts with phytochrome signaling in rice. Nat Commun 16:5329. https://doi.org/10.1038/s41467-025-60839-6

Calderini DF, Abeledo LG, Slafer GA (2000) Physiological maturity in wheat based on kernel water and dry matter. Agron J 92:895–901. https://doi.org/10.2134/agronj2000.925895x

Cao S, Luo X, Xu D et al (2021) Genetic architecture underlying light and temperature mediated flowering in Arabidopsis, rice and temperate cereals. New Phytol 230(5):1731–1745. https://doi.org/10.1111/nph.17276

Carrera CS, Savin R, Slafer GA (2024) Critical period for yield determination across grain crops. Trends Plant Sci 29:329–342. https://doi.org/10.1016/j.tplants.2023.08.012

Cheng YX, Xiao F, Huang D et al (2024) High canopy photosynthesis before anthesis explains the outstanding yield performance of rice cultivars with ideal plant architecture. Field Crop Res 306:109223. https://doi.org/10.1016/j.fcr.2023.109223

Coque M, Gallais A (2007) Genetic variation for nitrogen remobilization and post silking nitrogen uptake in maize recombinant inbred lines: heritabilities and correlations among trait. Crop Sci 47:1787–1796. https://doi.org/10.2135/cropsci2007.02.0096

Duan Z, Fang C, Liu S et al (2026) Soybean. Curr Biol 36(9):pR370–pR374. https://doi.org/10.1016/j.cub.2026.03.039

Engels C, Kirkby E, White P (2012) Mineral nutrition, yield and source–sink relationships. In: Petra M (ed) Marschner's mineral nutrition of higher plants, 3rd edn. Academic Press, New York, pp 85–133

Garner WW, Allard HA (1920) Effect of the relative length of day and night and other factors of the environment on growth and reproduction in plants. Mon Weather Rev 48(7):415. https://doi.org/10.1175/1520-0493(1920)48<415b:EOTRLO>2.0.CO;2

Gregersen PL, Culetic A, Boschian L et al (2013) Plant senescence and crop productivity. Plant Mol Biol 82:603–622

Hall AJ, Savin R, Slafer GA (2014) Is time to flowering in wheat and barley influenced by nitrogen?: a critical appraisal of recent published reports. Eur J Agron 54:40–46. https://doi.org/10.1016/j.eja.2013.11.006

Harmer SL, Hogenesch JB, Straume M (2000) Orchestrated transcription of key pathways in Arabidopsis by the circadian clock. Science 290(5499):2110–2113. https://doi.org/10.1126/science.290.5499.2110

Havé M, Marmagne A, Chardon F et al (2017) Nitrogen remobilization during leaf senescence: lessons from Arabidopsis to crops. J Exp Bot 68(10):2513–2529. https://doi.org/10.1093/jxb/erw365

Hay RKM, Porter JR (2006) The physiology of crop yield, 2nd edn. Wiley-Blackwell, Hoboken

Hyles J, Bloomfield MT, Hunt JR et al (2020) Phenology and related traits for wheat adaptation. Heredity 125:417–430. https://doi.org/10.1038/s41437-020-0320-1

Jung C (2017) Flowering time regulation: agrochemical control of flowering. Nat Plants 3:1–2. https://doi.org/10.1038/nplants.2017.45

Kumudini S (2002) Trials and tribulations: a review of the role of assimilate supply in soybean genetic yield improvement. Field Crop Res 75:211–222. https://doi.org/10.1016/S0378-4290(02)00027-8

Lin X, Liu B, Weller JL et al (2021) Molecular mechanisms for the photoperiodic regulation of flowering in soybean. J Integr Plant Biol 63(6):981–994. https://doi.org/10.1111/jipb.13021

Ling Q, Zhang H, Ding Y et al (2007) The theory and technology of precise and quantitative cultivation in rice. China Agriculture Press, Beijing

Okada R, Nemoto Y, Endo-Higashi N et al (2017) Synthetic control of flowering in rice independent of the cultivation environment. Nat Plants 3:1–7. https://doi.org/10.1038/nplants.2017.39

Pedró A, Savin R, Slafer GA (2012) Crop productivity as related to single-plant traits at key phenological stages in durum wheat. Field Crop Res 138:42–51. https://doi.org/10.1016/j.fcr.2012.09.016

Reynolds M, Pask A, Mullan D (2012) Physiological breeding I: interdisciplinary approaches to improve crop adaptation. CIMMYT

Sadras VO, Villalobos FJ, Fereres E (2024) Crop development and growth. In: Villalobos FJ, Fereres E (eds) Principles of agronomy for sustainable agriculture. Springer

Slafer GA, Savin R, Sadras VO (2023) Wheat yield is not causally related to the duration of the growing season. Eur J Agron 148:126885. https://doi.org/10.1016/j.eja.2023.126885

Song YH, Smith RW, To BJ et al (2012) FKF1 conveys timing information for CONSTANS stabilization in photoperiodic flowering. Science 336(6084):1045–1049. https://doi.org/10.1126/science

Sun C, He C, Zhong C et al (2022) Bifunctional regulators of photoperiodic flowering in short day plant rice. Front Plant Sci 13:1044790. https://doi.org/10.3389/fpls.2022.1044790

Taiz L, Zeiger E, Møller IM et al (2015) Plant physiology and development, 6th edn. Sinauer Associates, Sunderland

Tegeder M, Masclaux-Daubresse C (2018) Source and sink mechanisms of nitrogen transport and use. New Phytol 217(1):35–53. https://doi.org/10.1111/nph.14876

Thomas H, Ougham H (2014) Senescence and crop performance. In: Sadras VO, Calderini DF (eds) Crop physiology. Applications for genetic improvement, agronomy and farming systems, 2nd edn. Academic Press, New York. https://doi.org/10.1016/B978-0-12-417104-6.00010-8

Wang F, Li S, Kong F et al (2023) Altered regulation of flowering expands growth ranges and maximizes yields in major crops. Front Plant Sci 14:1094411. https://doi.org/10.3389/fpls.2023.1094411

Yoshida S (1981) Fundamentals of rice crop science. International Rice Research Institute. https://doi.org/10.20710/dojo.58.6_757

Zhang W, Peng K, Cui F et al (2021a) Cytokinin oxidase/dehydrogenase OsCKX11 coordinates source and sink relationship in rice by simultaneous regulation of leaf senescence and grain number. Plant Biotechnol J 19(2):335–350. https://doi.org/10.1111/pbi.13467
Zhang S, Zhang Y, Li K et al (2021b) Nitrogen mediates flowering time and nitrogen use efficiency via floral regulators in rice. Curr Biol 31:671–683. https://doi.org/10.1016/j.cub.2020.10.095
Zhang S, Liu Y, Du M et al (2022) Nitrogen as a regulator for flowering time in plant. Plant Soil 480:1–29. https://doi.org/10.1007/s11104-022-05608-w
Zhang S, Shou G, Li X et al (2025) U-shaped response of flowering time to low and high nitrogen via a transcription feedback loop in rice. Adv Sci (Weinh):e08498. https://doi.org/10.1002/advs.202508498
Zhong Y, Taylor J, Williams M (2012) The end: senescence and cell death. Plant Cell 24(1):tpc.112.tt0112. https://doi.org/10.1105/tpc.112.tt0112

Physiological Traits

Mining Novel Traits for Crop Breeding and Management

8

Contents

Z. Liu, *Crop Physiology*, https://doi.org/10.1007/978-981-92-2960-4_8

Abstract

From an application-oriented perspective, crop science can be viewed as the systematic body of knowledge concerning the quantification, selection, and regulation of traits indicative of crop performance. The primary objective of crop physiology is to identify novel and applicable traits that can be utilized in crop breeding and management. As the concluding section of the textbook, this chapter begins with categorizing crop traits, followed by an elaborate exposition on a representative physiological trait, canopy temperature. Subsequently, it introduces the principles and applications of physiological breeding, along with phenotyping platforms for physiological traits. Ultimately, it underscores the paramount importance of prioritizing the exploration of novel traits for crop breeding and management through the integration of state-of-the-art AI technologies, using the leaf to panicle ratio (LPR) as an example.

Keywords

Physiological trait · Physiological breeding · Canopy temperature · Crop phenomics

Did You Know

- If a leaf had no means to dissipate solar energy absorbed, its temperature would reach 100 °C in less than 1 min.
- A common misconception is that plants regulate transpiration to control leaf temperature. In reality, transpiration cooling is an inevitable consequence of water loss through transpiration, a process essential for photosynthesis.
- Carbon stable-isotope ratios ($\delta^{13}C$) serve as a well-established proxy for estimating the composition of human diets derived from C4 or C3 plants. Due to reduced discrimination against ^{13}C during photosynthesis, tissues consuming a higher proportion of C4-derived foods exhibit relatively enriched $\delta^{13}C$ values (typically −10‰ to −12‰). In contrast, diets based on C3 plants result in more depleted $\delta^{13}C$ values, generally ranging from −20‰ to −30‰.
- Large-scale surveys of modern human hair and other tissues have revealed a distinct geographic pattern: populations in North America exhibit significantly higher $\delta^{13}C$ values than those in Europe, attributable to their greater consumption of maize and other C4-derived staple foods.
- Under water-deficit conditions, the daytime canopy temperature may consistently exceed the surrounding air temperature, reaching up to 8 °C above ambient levels in wheat.
- The Nobel Prize in Physiology or Medicine was awarded in 1979 to Allan M. Cormack and Godfrey N. Hounsfield for their development of computed tomography (CT). In 2003, the prize was presented to Paul C. Lauterbur and Sir Peter Mansfield for their pioneering discoveries concerning magnetic resonance imaging (MRI).

Learning Objectives

After reading this chapter, students will be able to:

- Define the classification of crop traits
- Understand the physiological meaning of canopy temperature
- Comprehend the principles of physiological breeding
- Illustrate the basics of plant phenotyping
- Name the major phenotyping platforms for crop traits
- Appreciate the significance of mining new physiological traits

Introductory Remarks

Both basic science and applied science have distinct objectives. The primary objective of basic research is to unveil the inherent laws governing the functioning of the natural world, providing theoretical guidance and intellectual inspiration for human endeavours aimed at transforming nature. Conversely, applied research aims to take the discoveries from fundamental research and put them into practical use. This helps us innovate and develop new solutions that meet humanity's changing needs in a more tailored way.

Currently, driven by cutting-edge science and technology, such as genomics and artificial intelligence, fundamental research in plant physiology continues to make new breakthroughs, deepening our understanding of the essence of plant life and its interactions with the environment. Compared to plant physiology, crop physiology can be viewed as an applied discipline, with the primary goal of applying the principles of molecular biology, physiology, and ecology to enhance crop production. It addresses the urgent needs of intelligent crop breeding and precision management by providing guidelines and methodologies. To achieve this, crop physiology must elucidate the formation and regulatory mechanisms of crop yield, quality, and stress resistance across multiple levels of biological organization while developing practical tools and technologies.

The preceding six chapters have covered the basics of crop physiology. Focusing on the practical requirements of students and researchers in crop science, we have systematically delineated key physiological questions, such as carbon and nitrogen metabolism, embryo–endosperm interactions, source–sink relationships, plant architecture, crop phenology, as well as canopy structure and function. At the end of this book, we will explore how to apply insights from crop physiology to intelligent breeding and smart management, with an emphasis on physiological breeding approaches as well as on the identification, phenotyping, and application of novel physiological traits.

Concept Map

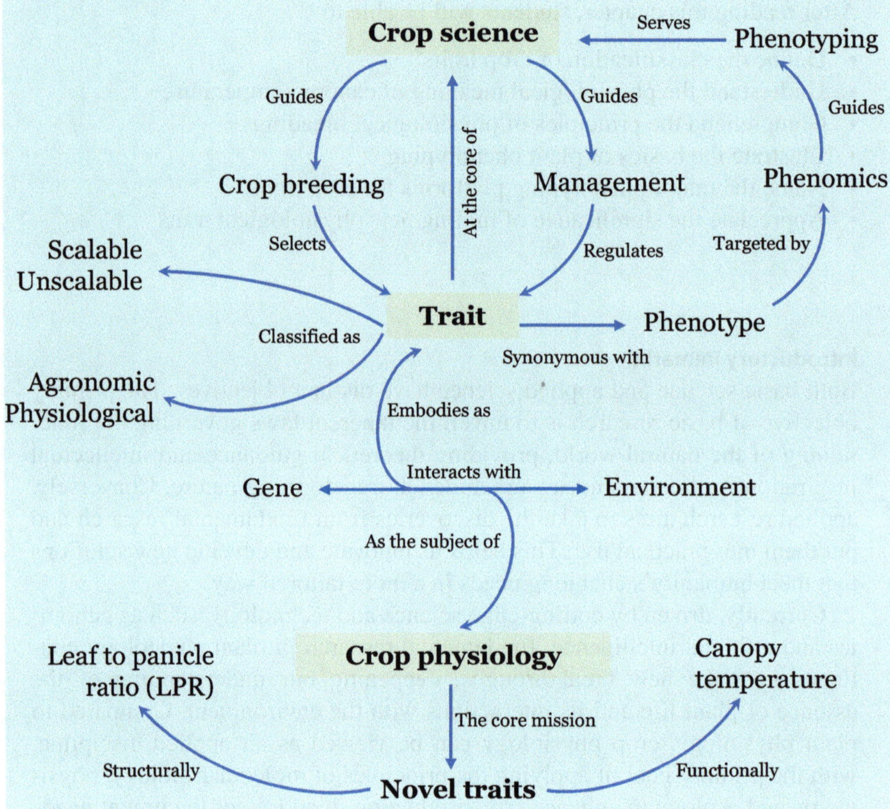

Annotation: Traits, synonymous with phenotypes, reflect the interaction between genes and the environment. The essence of crop science can be viewed as acquiring knowledge related to the quantification (phenotyping), selection (breeding), and regulation (management) of traits that serve as critical indicators for evaluating crop performance. Therefore, mining and phenotyping novel traits should be a central focus of crop physiology. Traits in crop science can be categorized into agronomic and physiological traits. Physiological traits reflect the structural and functional characteristics of crop plants and their populations, such as the architectural trait of leaf to panicle ratio (LPR) and the functional trait of canopy temperature. While some of these traits can be scaled from lower to higher levels of biological organization, most cannot due to the emergent properties inherent in biological systems.

1 Traits in Crop Science

Crop science is a dynamic and multidisciplinary field that integrates genomics, physiology, phenomics, ecology, informatics, breeding, and agronomy to understand how crops grow, develop, and respond to the shifting environments. By leveraging state-of-the-art tools ranging from high-throughput phenotyping to genome editing, it endeavours to propose sustainable solutions for enhancing crop yield, quality, resilience, and adaptability to climate change. From an application-oriented perspective, the essence of crop science lies in acquiring knowledge about the quantification, selection, and regulation of traits suggestive of crop productivity and resilience (Fig. 8.1). Identifying novel, high-value traits that directly predict crop performance has become one of the central priorities in modern crop science.

1.1 Trait and Phenotype

A trait, literally, refers to a specific characteristic, quality, or tendency possessed by someone or something. The term *trait* is synonymous with *phenotype* in the

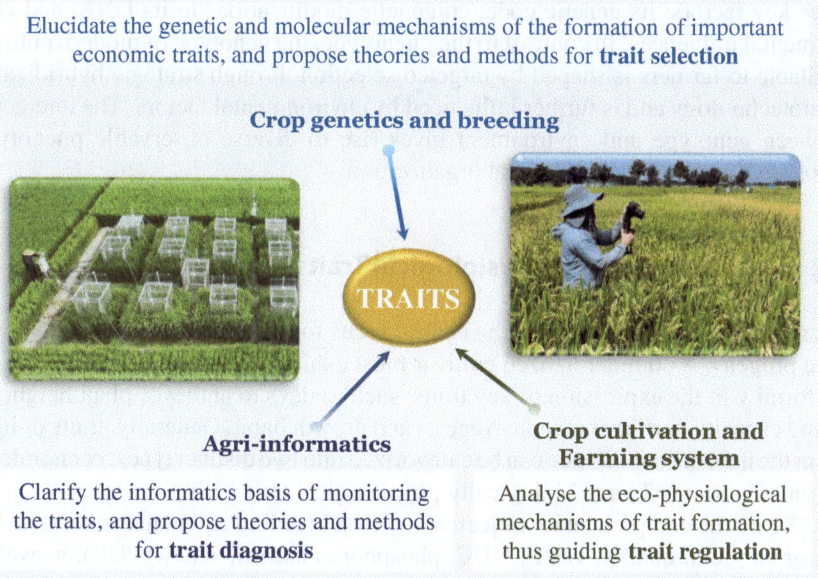

Fig. 8.1 Crop science: all about the theory and knowledge of traits. Crop science is an applied discipline, and its subdivisions differ across scientific institutions. For agricultural universities in China, it is generally categorized into three primary subdisciplines: crop genetics and breeding, crop cultivation and farming systems, and agricultural informatics or crop informatics (phenomics). Collectively, these subdisciplines aim to address a common objective: the study and optimization of plant and canopy traits. In this context, crop science can be defined as the body of knowledge concerned with the quantification, selection, and regulation of agronomically significant traits

scientific terminology of crop science, while its counterpart is the *gene*; similarly, the *genotype* serves as the counterpart of the *phenotype*.

Genetically, a trait is defined as the characteristics or attributes of an organism that are determined by genes and influenced by the gene–environment interactions. It represents a phenotypically evident, observable, and quantifiable outcome of gene expression, which is influenced by multiple factors, including genetic composition, environmental conditions, ontogenetic development, and epigenetic mechanisms. At cellular level, the formation and expression of a trait involve a series of interconnected processes that sequentially alter the composition of transcribed genes (*transcriptome*) and the resultant proteins (*proteome*) and ultimately shape the *metabolome*. These physiological processes collectively drive plant growth and development, thereby resulting in observable variations in structure, function, and phenology at crop level. Although specific key crop traits display discrete genetic control, as exemplified by the well-known pea phenotypes in Mendel's experiments or major disease-resistance loci, the majority are highly quantitative, such as those related to plant architecture and phenology. Consequently, the principles of *quantitative genetics* have been integral to plant breeding practices throughout history (Wallace et al. 2018).

The phenotype is characterized by a multitude of structures, processes, and functions that undergo changes during plant growth and development (Walter et al. 2015). Similar to the trait, the phenotype of a crop plant is primarily determined by three key factors: its genetic code, epigenetic modifications to its DNA, and environmental influences. In contrast to the phenotype, the genotype of modern cultivars available to farmers is shaped by targeted selection through strategic hybridization or biotechnology and is further influenced by environmental factors. The interaction between genotype and environment gives rise to diverse observable phenotypes across various levels of biological organization.

1.2 Agronomic and Physiological Traits

In crop breeding, traits serve as the central focus for the selection and evaluation of elite progeny. A commercialized cultivar must exhibit distinctiveness, stability, and uniformity in the expression of key traits, such as days to anthesis, plant height, tillering capacity, fertilizer responsiveness, and growth habit. Generally, traits of interest in the field of crop science can be categorized into two distinct types: economically agronomic trait (AT) and biologically physiological trait (PT).

ATs are the comprehensive objectives of crop breeding and management, including grain yield, quality, WUE, NUE, phosphorus use efficiency (PUE), as well as resistance to biotic and abiotic stresses. It can be quantified either through biological identification, such as disease or pest resistance, or via mathematical calculation, such as resource use efficiency, quality grading, and yield components. ATs are integrative indices and thus can be decomposed into different components. For instance, the yield of rice is determined by factors such as the number of panicles per unit area, grains per panicle, and grain weight. Similarly, the quantification of WUE can involve factors such as the volume of irrigation and precipitation, the amount of water uptake by roots, and the leaf transpiration coefficient.

PTs are the biophysical and biochemical properties of the crop plant. They reveal the structural and functional characteristics of a crop plant or stand, which are essential for determining crop performance across diverse environments. PTs are the primary focus in physiology research, acting as a critical bridge between plant fundamental science and its applications in crop breeding and management. In contrast to the integrative nature of AT, PT is a simple characteristic and can be assessed through a single chemical or physical analysis.

At the cellular level, PT mainly includes the indicators of metabolic processes, like the expression profile or pattern of transcriptome, proteome, and metabolome, as well as enzyme activity. Recent breakthroughs in structural genomics and advanced physicochemical analytical platforms, including high-throughput sequencing, mass spectrometry-based proteomics, and metabolomics, empower researchers to concurrently capture the entire spectrum of expressed mRNAs, peptides, and small-molecule metabolites from specific cells within a tissue or organ. By integrating these multi-omics scales, scientists can attain a more accurate, systems-level understanding of cellular function. Furthermore, enhancements in instrumentation sensitivity, automation, and data-analysis pipelines have significantly reduced per-sample costs, rendering comprehensive molecular profiling more readily accessible for both basic and applied researches.

At organ and plant level, PT refers to the parameters regarding the ontogenetic or developmental events, including (i) morphological traits of plant organs such as plant height, leaf angle, and tillering angle; (ii) anatomical traits consisting of developmental events of embryo and endosperm, histochemical properties of vascular and parenchyma in the internode, and so on; and (iii) functional traits that indicate the physiological processes within the organ, like enzyme activity, stomatal conductance (Gs), and carbon isotope discrimination (CID; Box 8.1). Modern microscopy, particularly confocal laser scanning microscopy and X-ray computed tomography (CT), improves the visualization and resolution of anatomical features at sub-organ scales. Traditional histochemical stains have been improved and integrated with immunolabeling and fluorescence antibodies, thereby enhancing the sensitivity and specificity for detecting cell wall components, enzymes, and signalling molecules. Parallel advancements in biochemical analysis establishes a connection between structural observations and physiological responses. Collectively, these tools are revolutionizing physiological research by shifting it from descriptive anatomy to integrative and systematic investigations of the relationships between plant form and function.

At the population or canopy level, PT includes (i) structural traits of the crop stand such as LAI, normalized difference vegetation index (NDVI), and leaf-to-panicle ratio (LPR; Yang et al. 2020) and (ii) functional traits indicating photosynthesis, respiration, and evapotranspiration. Detailed information regarding the structural and functional traits of the canopy is presented in Chap. 6. The significant and rapid advancement in plant phenomics has led to a revolutionary enhancement in quantifying crop populations. Through the seamless integration of cutting-edge imaging technologies, hyperspectral sensors, and machine-learning algorithms, scientists are now capable of non-invasively capturing detailed spectral signatures of individual plants and plant canopies. This allows for high-throughput measurement of both structural traits, including LAI, PAI, and LPR, and functional traits, such as canopy carbon assimilation and respiration rate, evapotranspiration, and GHG emission. As a result, large-scale phenotypic data can be collected rapidly and precisely, which in turn speeds up breeding programs, improves crop management, and deepens our understanding of crop physiology in diverse environments.

Box 8.1: Carbon Isotope Discrimination in Plants

Isotopes represent distinct forms of an element with different nuclear composition. The number of protons remains constant across different isotopes, as this characteristic defines the elemental identity, while the number of neutrons varies among isotopes. Radioactive isotopes undergo decay over time, transforming into different elements. By contrast, stable isotopes retain their composition indefinitely without experiencing decay. The two stable isotopes of carbon are ^{12}C and ^{13}C, differing in composition only by the addition of a neutron in ^{13}C. ^{11}C and ^{14}C are radioactive isotopes of carbon that are commonly utilized in biological tracer studies.

Atmospheric CO_2 is primarily composed of three types of carbon isotopes: primarily $^{12}CO_2$, with about 1% being $^{13}CO_2$ and a small fraction as the radioactive $^{14}CO_2$. The isotope composition of plant biomass differs from that of the atmosphere, partially due to the slower diffusion of $^{13}CO_2$ in air compared to that of the lighter isotope $^{12}CO_2$ (which is 1.0044 times slower). Thus, leaves have a more negative ^{13}C value due to this diffusion effect.

Carbon isotope composition or ratio ($\delta^{13}C$; usually a negative value) varies between C3 and C4 plants. In C3 plants, Rubisco exhibit a higher affinity for $^{12}CO_2$ compared to $^{13}CO_2$, leading to discrimination against the heavier isotope. For instance, Rubisco of spinach (*Spinacia oleracea*) demonstrates a discrimination factor ($\delta^{13}C$) of −30.3‰. By contrast, the discrimination against $^{13}CO_2$ in C4 plants is impeded by the diffusion barrier between the vascular bundle sheath and the mesophyll cells. In addition, the mesophyll cells contain PEP carboxylase, which scavenges most of the CO_2 escaping from the bundle sheath. Consequently, C3 plants have a mean $\delta^{13}C$ value of about −28‰, while C4 plants display a significantly higher mean value of around −14‰.

Carbon-isotope discrimination (CID, $\Delta^{13}C$), the positive value of $\delta^{13}C$, serves as a useful tool for estimating time-integrated measures of photosynthetic activities and WUE in individual plant or population, which is difficult to be assessed through direct measurements (Lambers et al. 2008). In C3 plants, the potential for isotope discrimination by Rubisco is not achieved due to limited CO_2 availability at the carboxylation site. Greater discrimination against $^{13}CO_2$ occurs when the intercellular CO_2 concentration (c_i) is elevated, which typically happens when stomata are open. However, the opening of stomata also results in a greater amount of water loss via transpiration. Generally, lower ratios of photosynthesis to transpiration (lower WUE at leaf level) are associated with higher CID values.

The relationship between CID and grain yield varies depending on genotype and growing environment (Jiang et al. 2021). In wheat, a high CTD is positively correlated with grain yield under favourable conditions. In contrast, a low CTD is considered an effective indicator for selecting high-yielding genotypes under drought stress. This large disparity poses the primary obstacle to applying CTD in wheat breeding programs aimed at improving WUE.

1.3 Scalability of Crop Traits

Scalability refers to the ability of a system, network, process, or organization to effectively handle an increase in workload, demand, or scale. It denotes the capacity of a system or organization to expand its performance, functionality, or resources in response to growing demands, such as an increase in users, data volume, or processing requirements. Antonyms of scalability include non-scalability, limitation, and inflexibility, terms that reflect an inability to adapt, grow, or respond effectively to changing demands. When a system or organization fails to scale effectively under increased demand, it may experience performance degradation, higher costs, or inefficient resource utilization. Scalability is a widely recognized concept in technology, business, and organizational management, highlighting the importance of adaptability and flexibility in responding to change and supporting growth.

In crop science, Sadras and Richards (2014) proposed an operational definition in the context of crop improvement: a trait scales up if it remains agronomically relevant at higher levels and eventually at the population level where yield is defined. Accordingly, crop traits can be basically classified into two groups:

1. Traits that generally scale up. Traits such as the presence of awns, plant height, leaf angle, tillering ability, low grain amylose content, and resistance to herbicides frequently manifest at the crop level from the cellular, tissue, or organ level because of their high heritability. Most of these traits are governed by single genes, such as the waxy gene responsible for grain amylose content. Consequently, genetic improvement achieved via marker-assisted selection, transgenic technologies, or genome-editing approaches has led to substantial advancements in these traits.
2. Traits that do not scale up. These traits, such as CID, Gs, photosynthesis, and stay-green (see Sect. 2.3 in Chap. 7), serve as indicators of the physiological status of tissues or organs. Most PTs belong to this category, and their relevance to crop yield remains to be validated due to the emergent properties observed at higher levels of organization. Compared with biotechnology, conventional breeding and agronomic management have been proven to be more efficient in improving this type of trait.

When integrating a physiological trait into breeding or management practices, it is crucial to evaluate its scalability. Physiological processes within cropping systems operate across diverse spatiotemporal scales, spanning from the cellular to the canopy level and from seconds to days, or from fine-scale to broad-scale contexts. These processes often exhibit distinct characteristics depending on the scale under consideration. For example, while the small-scale and rapid processes at the cellular level are predominantly governed by chemical and biophysical mechanisms, the large-scale and slower processes at the crop level are primarily regulated by interactions among organs or plants. Nonetheless, even when these processes are identical, for instance, photosynthetic CO_2 fixation, their responses to the same driving force (e.g. light quantity) vary due to the emergent properties of the biological organization. While leaf photosynthesis follows a saturation function influenced by light intensity, canopy photosynthesis exhibits a more linear response (Sect. 3.2 in Chap. 6). Specifically, the canopy attains significantly higher assimilation rates at high irradiance without reaching saturation because the lower leaves in the middle and bottom layers can intercept and utilize additional light.

With the advances in molecular technologies for genome sequencing and functional genomics, genes have evolved from abstract concepts into tangible biological entities. It is generally anticipated that a gene-by-gene engineering approach will substantially improve the efficiency of crop improvement. In fact, remarkable achievements have been made in developing plants with enhanced resistance to pests or tolerance to herbicides. These cases typically involved single-gene transformations, in which the phenotypic response was directly proportional to the level of molecular intervention. Nevertheless, it is not suitable for most complex ATs and PTs, which are regulated by multiple genes and influenced by growing environments. This complexity presents significant challenges to accurately predict phenotypes using a gene-by-gene approach (Hammer et al. 2006).

2 Canopy Temperature as an Exemplar Physiological Trait

Canopy temperature (CT), defined as the surface temperature of the plant canopy, serves as a comprehensive indicator for assessing the energy exchange between the entire crop stand and its surrounding environment. It is widely recognized as a reliable indicator of physiological response to heat and drought conditions and has been incorporated into breeding programs aimed at enhancing abiotic stress resilience.

2.1 Eco-physiological Aspect of Canopy Temperature

Temperature is a critical environmental factor that dictates the distribution of plants, influencing nearly all aspects of physiological processes. The air temperature within a habitat provides a general proxy of leaf temperature. Nevertheless, the actual temperature of a plant organ (e.g. a leaf or spike) frequently diverges remarkedly from that of the atmospheric environment. Leaves efficiently absorb short-wave radiation (SR) from the sun. Only a relatively small portion of the incident solar radiation is reflected, transmitted, or used for metabolic processes other than heating. Under intense sunlight, the net absorption of solar radiation (SR_{net}) serves as the primary energy input to a leaf. If the leaf lacked an effective mechanism to dissipate this energy, its temperature would exceed 100 °C in less than 1 min. Consequently, regulatory processes governing leaf heat loss are essential for maintaining temperature homeostasis to ensure normal physiological functioning.

The plant undergoes heat loss via multiple biophysical and biochemical processes (Fig. 8.2). The leaf itself emits long-wave radiation (LR). Simultaneously, it absorbs infrared LR from the ambient environment. The net effect of emission and

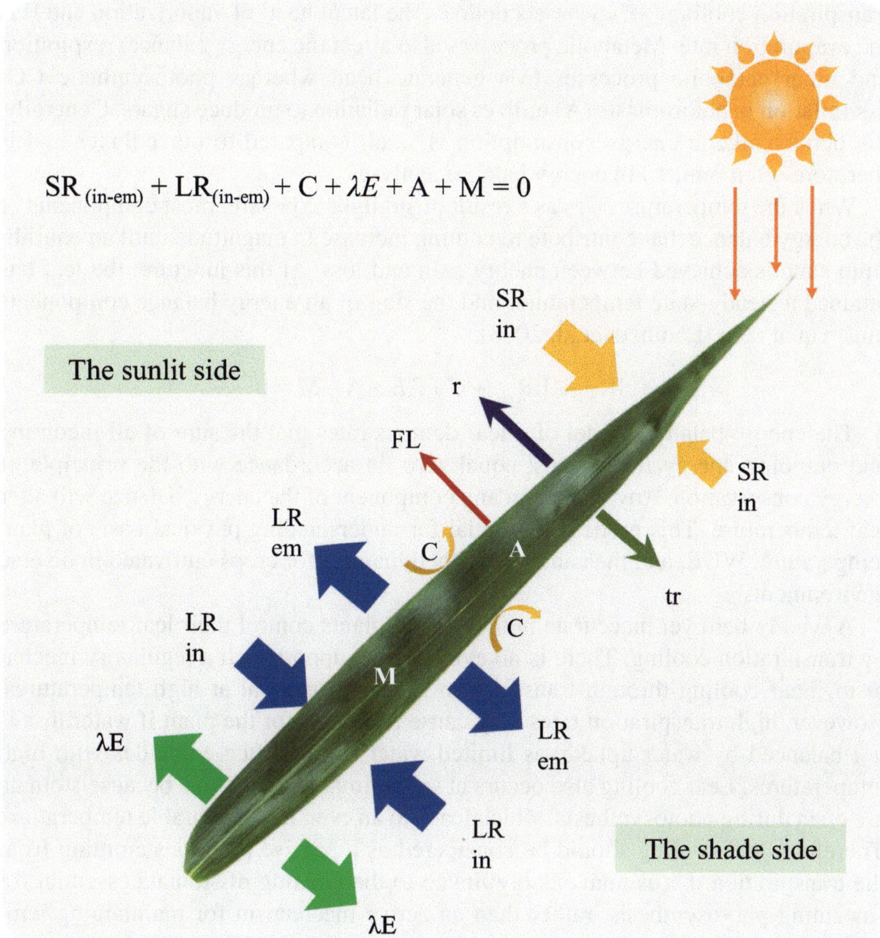

Fig. 8.2 Schematic representation of the leaf energy balance. Two types of electromagnetic radiation, which are distinguished by their sources and spectral distributions, play a crucial role in crop environments. Solar radiation, originating from a very hot thermal radiator, the Sun, is referred to as short-wave radiation (SR), ranging from 0.3 to 3 μm. In contrast, thermal radiation emitted by objects on our planet, such as soils, plants, and the atmosphere, occurs at longer wavelengths because these radiating bodies are at significantly lower temperatures. Such long-wave radiation (LR) is predominantly found between 5 and 100 μm. The leaf energy balance equation comprises SR, LR, convective heat transfer (C), and evaporative heat loss (λE). SR_{em}, including reflection (r), transmission (tr), and fluorescent emission (FL), is only given for one side of the leaf. A and M represent CO_2 assimilation and heat-producing metabolic processes, respectively. em emitted, in incident. (Redrawn from Lambers et al. (2008))

absorption (LR_{net}) can be either negative or positive, depending on whether the leaf is exporting or importing radiation, respectively. When a temperature gradient exists between the leaf and the air, convective heat transfer (*C*) occurs in the direction of descending temperature. Another major component of the energy balance is transpiration cooling (λE), where λ denotes the latent heat[1] of vaporization and E is the evaporation rate. Metabolic processes also affect the energy balance: respiration and other catabolic processes (M) generate heat, whereas photosynthetic CO_2 assimilation in chloroplasts (A) utilizes solar radiation to produce sugars. Generally, the photosynthetic energy consumption is small compared to other fluxes and is therefore often omitted in energy balance analyses.

When the temperature rises as a result of sunlight exposure, most components of the energy balance that contribute to cooling increase in magnitude until an equilibrium state is achieved between energy gain and loss. At this juncture, the leaf has attained a steady-state temperature, and the sum of all energy balance components must equal zero (Lambers et al. 2008):

$$\mathrm{SR}_{\mathrm{net}} + \mathrm{LR}_{\mathrm{net}} + C + \lambda E + \mathrm{A} + \mathrm{M} = 0$$

The energy-balance model of a leaf demonstrates that the sum of all incoming and outgoing energy fluxes must equal zero, in accordance with the principle of energy conservation. Any change in any component of the energy balance will alter leaf temperature. This model is essential for understanding physical basis of plant temperature, WUE, and the canopy energy dynamics for crops cultivated in diverse environments.

A widely held yet inaccurate notion is that plants control their leaf temperature by transpiration cooling. There is no evidence to support such a regulatory mechanism. Leaf cooling through transpiration is only beneficial at high temperatures. However, high transpiration rates may cause problems for the plant if water loss is not balanced by water uptake, as limited water supply often coincides with high temperatures. Leaf cooling also occurs at suboptimal temperatures because stomata are open during photosynthesis, which leads to an even less favourable temperature. Therefore, leaf cooling should be considered as a passive process stemming from the transpiration that is unavoidably linked to the opening of stomata essential for sustaining photosynthesis, rather than an active mechanism for maintaining temperature homeostasis.

2.2 Factors Influencing Canopy Temperature

Multiple physiological and environmental factors jointly regulate the energy balance of crop canopies, ultimately shaping their temperature dynamics (Fig. 8.3).

The primary environmental factors affecting CT are as follows (Reynolds et al. 2012): (i) incident solar radiation, which directly heats the plants and is modulated by cloud cover; (ii) soil moisture, which governs transpiration potential and thereby affects transpiration cooling; (iii) wind velocity, which enhances transpiration by

[1] Convective heat transfer is commonly termed sensible heat because it directly changes leaf temperature and can be measured with a thermometer. In contrast, the heat involved in transpiration is called latent heat: it is absorbed during evaporation (or released during condensation) without causing an immediate change in temperature.

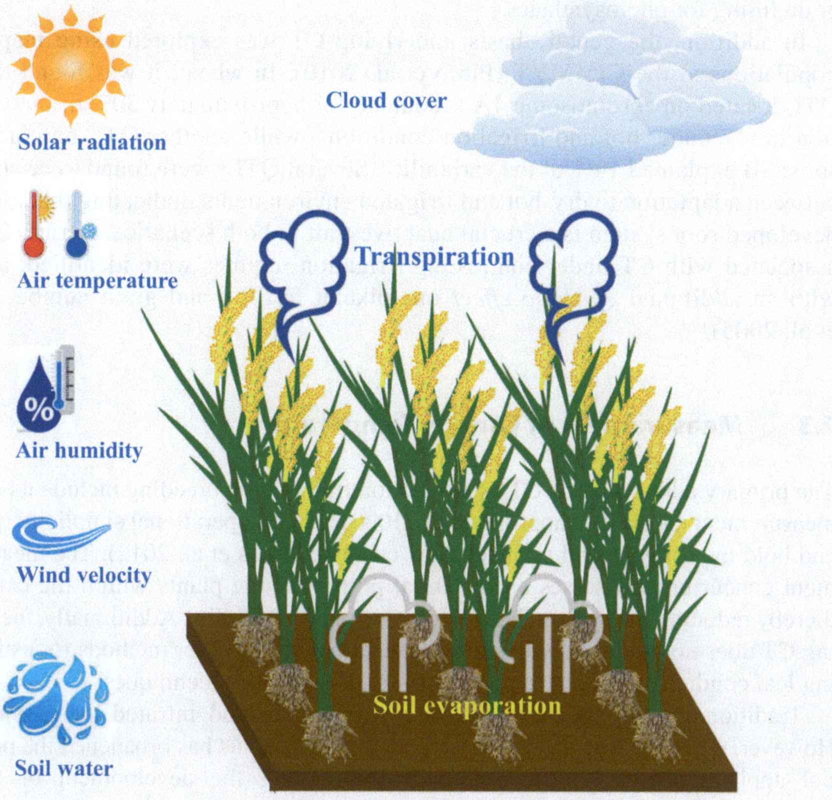

Fig. 8.3 Physiological and environmental factors controlling canopy temperature. Environmental factors encompass solar radiation, air temperature, humidity, wind velocity, soil moisture, and evaporation. The physiological process primarily involves transpiration, driven by a continuous water potential gradient extending from the soil, through the root–xylem continuum and into the atmosphere. Any constraint along this pathway, such as limited root hydraulic conductivity, elevated xylem resistance, a thick leaf boundary layer, or stomatal closure, can reduce the overall water flux. (Redrawn from Reynolds et al. (2012))

disturbing boundary layers between the leaf and its surrounding air; and (iv) air humidity, which determines the vapor pressure deficit (VPD) between plant organs and the atmospheric environment.

The main physiological factors affecting CT within a given environment are as follows (Reynolds et al. 2012): (i) xylem system connecting leaves, shoots, and roots, which governs transpiration capacity; (ii) stomatal aperture, which regulates transpiration rate and can be modulated by root-derived hormonal signals, such as ABA and CKs; (iii) root depth, which determines water accessibility, especially under drought conditions; and (iv) source-sink relation, such as the increased leaf

CO_2 uptake through greater stomatal conductance as a result of the high demand of grain filling for photosynthates.

In addition, the genetic basis underlying CT was explored using mapping populations at the CIMMYT (Pinto et al. 2010). In wheat, it was found that a QTL located on chromosome 4A accounted for approximately 30% of the variation in CT under hot and irrigation conditions, while another QTL on chromosome 3B explained 14% of the variability. Several QTLs were found to be shared between adaptation to dry-hot and irrigated environments, indicating that a well-developed root system is a crucial adaptive trait in both scenarios. In rice, QTLs associated with CT under contrasting irrigation regimes were identified, along with an additional *epistatic effect* on spikelet fertility and grain number (Liu et al. 2005).

2.3 Measurement of Canopy Temperature

The primary advantages of CT as a selection trait in crop breeding include its rapid measurement capability (approximately 10 s per plot), operational simplicity (point and hold the trigger), and relatively low cost (Reynolds et al. 2012). The measurement concurrently assesses a substantial portion of the plants within the canopy, thereby reducing errors associated with inter-plant variability. Additionally, measuring CT does not interfere with stomatal function, unlike other methods for estimating leaf conductance, such as porometry or gas exchange techniques.

Traditionally, CT has been measured using handheld infrared thermometers. However, the decreasing cost of infrared imaging systems has broadened the potential applications for CT measurement, facilitating the development of high-throughput phenotyping platforms. Recently, an airborne thermal imaging platform was established to precisely measure CT in large-scale field trials. This platform enables the precise estimation of organ-specific temperatures within individual plants and allows for the simultaneous capture of CT across all plots in extensive trials.

Although CT is widely used, it has several limitations that require careful consideration. First, CT measurements represent a composite of source organs (e.g. leaves) and sink organs (e.g. wheat spikes or rice panicles), which exhibit contrasting eco-physiological activities such as respiration, transpiration, and surface temperatures. Furthermore, CT is highly sensitive to ambient variables, including incident radiation, wind velocity, and humidity. These factors collectively obscure the effects of physiological processes and genetic differences. Consequently, it is essential to disentangle the influences of environmental conditions and physiological activities to comprehensively elucidate the mechanisms underlying CT (Li et al. 2021).

2.4 Canopy Temperature as a Representative Physiological Trait

The surface temperature of the canopy is associated with the transpiration rate, which drives leaf cooling. Physiologically, it reflects the influence of transpiration on leaf cooling and correlates with numerous physiological factors, such as stomatal conductance (Gs), CID, WUE, LAI, and root depth, ultimately affecting crop yield. Moreover, a lower CT is positively correlated with both grain yield and quality under heat and drought stress, indicating the greater adaptability to changing environments.

Lower CT can serve as indicators of improved hydration status (Pask et al. 2012). This approach is commonly employed for stress diagnosis and the selection of stress-tolerant genotypes in breeding: (i) under drought stress, it reflects the capacity of the roots to absorb water from deeper soils; (ii) under irrigated conditions, it may reflect source strength, sink activity, and/or vascular transportation ability; and (iii) under heat stress, it is associated with vascular capacity, transpiration cooling, and heat dissipation. Genotypic differences in CT facilitate the characterization of diverse germplasms in terms of leaf energy balance, stomatal conductance, and transpiration activity (Thapa et al. 2018).

Canopy temperature depression (CTD), defined as the difference between air temperature and canopy temperature (CTD = $T_{air} - T_{canopy}$), is commonly used in breeding and management practices. When the canopy is cooled below air temperature through transpiration cooling, CTD assumes a positive value. Conversely, when stomata close and T_{canopy} rises above air temperature, CTD becomes negative. CTD typically exhibits a higher (more positive) magnitude in well-irrigated plants, whereas it is lower (more negative) under drought stress. Under water-deficit conditions, CTD shows a strong correlation with key physiological parameters such as Gs and leaf water potential, as well as with grain yield, suggesting its utility as an effective selection criterion for drought tolerance (Thapa et al. 2018).

Caution is undoubtedly necessary when interpreting CT data in crop breeding (Reynolds et al. 2012). For instance, as observed in numerous Mediterranean regions, under drought conditions where water availability at soil depth is limited, there may be no correlation between CT and yield. Additionally, although a low CT value may indicate the genetic potential of root systems to utilize soil moisture, confounding factors such as micronutrient deficiency or soil-borne diseases that inhibit root growth could complicate this relationship.

3 Physiological Breeding

3.1 The Concept of Physiological Breeding

According to the historical stage classification of plant breeding by Wallace et al. (2018), approximately three decades ago, we transitioned into the Breeding 3.0 era. This period was marked by the integration of molecular markers and genomic data alongside phenotypic data. We currently stand at the threshold of Breeding 4.0, an era marked by revolutionary advancements in genetics and information technology, enabling the rational combination of functional genetic variants with unprecedented speed and precision. For instance, *genome resequencing* is now more cost-effective than time-consuming and laborious field trials, while *genome editing* enables precise, simultaneous modifications at multiple loci within a single generation. Meanwhile, high-throughput phenotyping platforms provide spatiotemporal measurements of numerous traits with high resolution, and AI-driven approaches facilitate the processing and interpretation of agronomic big data far beyond human capacity.

Physiological breeding involves deliberately crossing parents with distinct yet complementary traits to harness cumulative gene action for enhanced yield while screening progeny through phenotyping platforms in conjunction with *genomic selection* (Reynolds and Langridge 2016). Key steps in a physiological breeding pipeline are illustrated in Fig. 8.4. In practice, it deviates from conventional breeding by considering a broader range of traits, including genetically complex physiological characteristics; it also differs from molecular breeding in that it integrates both phenotypic and genotypic information. Overall, physiological breeding synthesizes knowledge spanning from genes to field applications, integrating multiple disciplines to establish a robust linkage between genomics and phenomics. Therefore, it should be regarded as an essential approach in the era of Breeding 4.0.

The approaches of physiological breeding also serve as valuable platforms for improving breeding methodologies. Recent examples include screening genetic resources to identify novel expression patterns of CBC enzymes (Prins et al. 2016), identifying common genetic bases for heat and drought adaptation (Pinto et al. 2010), and analysing genetic trade-offs among yield components (Lopes et al. 2015). This substantial amount of information, when integrated with the findings from physiological strategic crosses, contributes to the development of models that explore genotype-by-environment interactions. As a result, it aids in the creation of new cultivars and test environments that are suitable for future climatic scenarios.

3.2 Application in Crop Breeding

CT is a versatile phenotyping tool that can support breeding efforts, as it captures multiple physiological traits critical for adaptation to specific environments. CT was first reported for detecting genotypic variations in wheat by Blum and co-workers in 1982 (Blum et al. 1982) and subsequently accepted as an excellent predictor of yield in hot, irrigated environments. The use of CT in CIMMYT's breeding research dates back to the early 1990s, when it was first adopted as a tool for parental selection in strategic crosses and early-generation screening under drought stress. A synthetic analysis was conducted on multi-location yield trials in Asia, North Africa, and Mexico to evaluate the *genetic gains* achieved through CT-based breeding at CIMMYT for various target environments (Reynolds and Langridge 2016). The results indicated that the most promising new lines exhibited the highest average

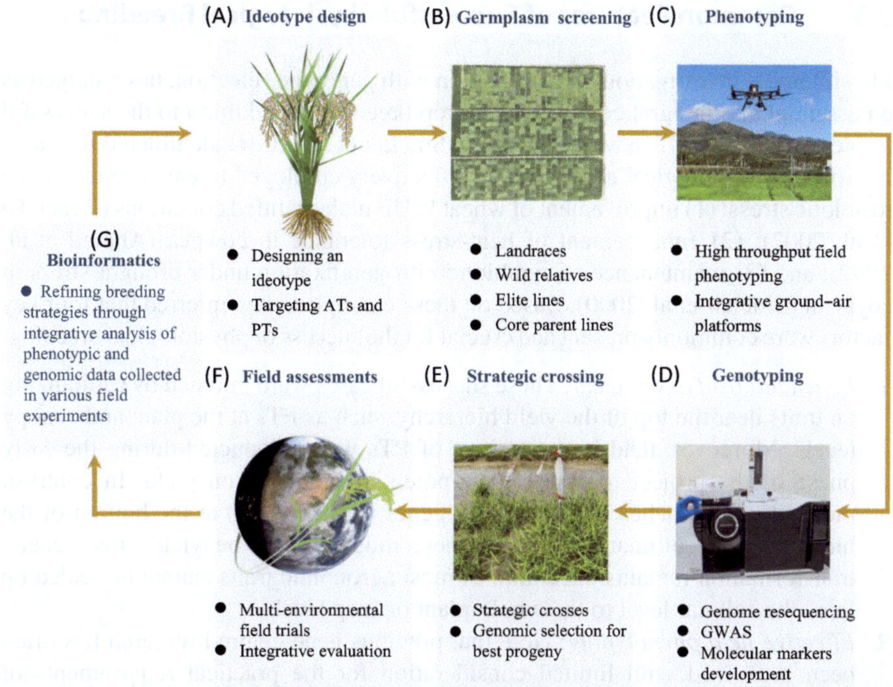

Fig. 8.4 Physiological breeding pipeline. (**A**) designing of crop ideotype, formulating a plant with high-yield potential and strong adaptation; (**B**) germplasm screening, identifying genetic resources with complementary allelic variation; (**C**) field phenotyping, maximizing resolution of physiological trait expression to select parents; (**D**) genotyping, including GWAS, genome resequencing, and development of molecular markers; (**E**) strategic crossing and progeny selection, achieving cumulative gene action and selecting elite progeny by the combination of phenotyping and genotyping; (**F**) assessment of genetic gains, through multi-location testing systems, such as the International Wheat Improvement Network (IWIN), which facilitates the evaluation of wheat genotypes across approximately 700 field sites in over 90 countries; (**G**) informatics services, underpinning the iterative refinement of breeding strategies. ATs agronomic traits, GWAS genome-wide association study, PTs physiological traits. (Redrawn from Reynolds and Langridge (2016))

yield across environments stressed by heat and drought, with an increase ranging from 8% to 34%. In China, Zhang and Wang (2008) proposed a novel physiological breeding strategy that harnessed the *germplasm* of low-CT wheat, resulting in the development of a series of superior wheat cultivars or lines. Notably, Xinong-805 achieved an unprecedented yield of 11.0 t/ha in the Guanzhong Plain in 2020. The significance of CT in enhancing rice drought resistance was demonstrated by Caine et al. (2019), who genetically modified the high-yielding rice cultivar IR64 to reduce stomatal density through overexpression of the epidermal patterning factor *OsEPF1*. This modification led to lower canopy temperatures under water-deficit conditions, thereby enhancing WUE.

3.3 Common Features of Successful Physiological Breeding

Physiological investigation, in conjunction with genotype selection, has emerged as an essential and integral component of crop breeding. In addition to the successful implementation of CT in wheat breeding, Sinclair et al. (2004) identified three cases in which a physiological approach was effectively employed to enhance resilience to abiotic stress: (1) improvement of wheat WUE under rainfed conditions (Rebetzke et al. 2002); (2) enhancement of heat-stress tolerance in cowpea (Ahmed et al. 1992); and (3) maintenance of symbiotic nitrogen fixation under drought stress in soybean (Sinclair et al. 2000). Based on these examples, they inferred that four key factors were commonly present and crucial for the success of physiological breeding.

1. *Beneficial trait assessment*: These successful cases were initiated by capitalizing on traits near the top of the yield hierarchy, such as PTs at the plant and canopy levels. Moreover, field investigations of PTs were conducted during the early phase of the project to assess their potential influence on yield. In contrast, molecular approaches aim to modify genes (or pathways) at the bottom of the hierarchy. The ultimate impact of these modifications on yield after genetic transformation remains uncertain, as most agronomic traits cannot be scaled up from the cellular level to the whole-plant or crop level.
2. *Effective field phenotyping*: Note that previous gene-centred research has often been organized with limited consideration for the practical requirements of enhancing crop performance under applied conditions. Although practical benefits were frequently claimed, the fundamental assumptions underlying much of this research and the experimental protocols used often have little relevance to actual crop production systems. Therefore, a comprehensive evaluation of phenotypic expression is essential. This evaluation encompasses assessments of crop performance and yield under field conditions, as well as under different cultivation methods, such as direct seeding and manual or mechanical transplanting in rice cultivation.
3. *Multi-disciplinary effort*: It should integrate scientists from multiple disciplines, including crop geneticists, physiologists, agronomists, and breeders, throughout the entire research process. These large-scale projects ought to be headed by breeders and crop physiologists, who can guide the project towards the intended objective without being sidetracked, even by attractive but irrelevant endeavours. Nevertheless, fulfilling this requirement may become progressively more challenging because of the decreasing funding for whole-plant physiology.
4. *Long-term commitment*: A minimum of 15 years is required from the identification of a specific problem to the release of a cultivar. There is still a long way to go for molecular scientists to translate the discovery of novel genes into practical applications in crop improvement, owing to the complex emergent properties that arise when scaling up from the cellular level to the whole-plant or field level. Nevertheless, this remains a valuable endeavour, as it is the only pathway through which fundamental plant science can meaningfully impact the agricultural sector.

4 Phenotyping of Physiological Traits

A prerequisite for physiological breeding is the establishment of a connection between genotype and phenotype. Since the construction of the first plant *genome* for *Arabidopsis* in 2000, high-quality reference genomes and extensive *genotyping* data have substantially facilitated the genetic dissection of complex traits via quantitative genetic approaches. In contrast to the remarkable advancements in genotyping, traditional *phenotyping*, which mainly relies on manual field operations, has notably lagged.

Conventional crop phenotyping generally relies on manual methods, such as visual inspection, handheld sensors, and destructive sampling. These methods are labour-intensive, time-consuming, and prone to operator bias due to subjective judgement. Moreover, repeated physical contact with plants may disrupt the canopy, thereby compromising data accuracy. In contrast, modern phenomics employs high-throughput imaging, remote sensing, and automated data analysis to enable rapid, non-invasive measurement of plant structure and function. By minimizing human involvement, shortening data collection timelines, and reducing canopy interference, these advanced technologies effectively overcome the major bottlenecks in large-scale phenotyping.

4.1 Crop Phenomics

Phenomics conducts a systematic investigation of phenotypes via multidisciplinary approaches that integrate biology, data science, engineering, and other disciplines. It focuses on quantifying the *phenome*, the collection of physical and chemical characteristics exhibited by an organism. In the field of plant sciences, phenomics research is carried out in both natural and artificial environments, such as glasshouses or growth chambers, where growth conditions can be precisely controlled.

Crop phenomics is defined as a comprehensive exploration of crop phenotypes (traits) across multiple levels of biological organization. Owing to the rapid advances in sensor technology, machine vision, automation, wireless communication, cloud-based technologies, and artificial intelligence (AI), crop phenotyping has emerged as a powerful tool for quantifying physiological and agronomic traits. It facilitates the quantification and monitoring of large, genetically diverse populations in field experiments and breeding nurseries at various temporal and spatial scales. The integration of phenomic, genomic, and transcriptomic data has created new opportunities for understanding the genetic and molecular mechanisms underlying phenotypic differences and plant responses to environmental cues. Combining phenotypic data with genome-wide association study (GWAS) and expression quantitative trait locus (eQTL) analyses helps uncover key gene regulatory networks, enabling precise, biology-informed strategies for crop improvement.

The abundance of phenotypic data, offering detailed insights into crop performance and resource utilization at high spatial and temporal resolutions, holds immense potential for advancing precision agriculture. Remote sensing technologies, along with ground-based sensors and UAVs, monitor crop growth, health, and stress in field environments. This data, in conjunction with environmental information and crop models, is crucial for optimizing practices such as seeding, irrigation, fertilization, and protection. This knowledge is essential for developing sustainable agricultural practices, managing natural resources, and mitigating the effects of climate change.

4.2 Sensors for Crop Phenotyping

Within a plant body or crop canopy, light propagation is a complex process governed by interactions between photons and structural components, i.e. reflection, transmission, and absorption (Sun et al. 2022). The quantity of light reflected from or transmitted through plant body can be measured using optical sensors to quantify light attenuation. Electromagnetic waves are absorbed, reflected, emitted, or transmitted by plant tissues in different ways, depending on their biochemical composition and physiological state. These differences provide valuable information for estimating parameters such as physical and chemical properties, biotic and abiotic stress status, plant architecture, canopy structure and microclimate, and yield components.

Sensors capable of detecting the electromagnetic wavebands emitted by plants serve as the key component in crop phenotyping. A variety of optical sensors have been developed to quantify the distinct electromagnetic responses of plants. Each sensor has specific advantages in measuring a particular category of traits, depending on its operating principle (Fig. 8.5).

Red-green-blue (RGB) is the most prevalent sensor due to its advantages in spatial resolution, signal-to-noise ratio (SNR), throughput, environmental adaptability, and high repeatability. Hyperspectral imaging, which offers high spectral resolution, can capture a wide range of phenotypes. Chlorophyll fluorescence (ChlF) can indicate photosynthetic activities when equipped with appropriate sensor specifications. X-ray computed tomography (X-ray CT) can obtain detailed internal 3D information with good spatial resolution and repeatability; however, it remains costly, has low throughput, and exhibits poor environmental adaptability. Light detection and ranging (LiDAR) provides distinct advantages in characterizing canopy structure, boasting higher throughput and better environmental adaptability. Thermal imaging is commonly used to acquire physiological traits related to surface temperature. Nevertheless, it demonstrates relatively low spatial resolution and poor repeatability. Raman spectroscopy offers excellent spectral resolution, along with a reasonable SNR and good repeatability for biochemical components. Nuclear magnetic resonance (NMR) has been proposed as an indicator of seed viability (Kano et al. 1990) and has been applied to investigate the water status of developing caryopses of cereals (Ishimaru et al. 2009). Magnetic resonance imaging (MRI), a technique based on the NMR phenomenon, has helped elucidate the precise water distribution in developing cereal grains (Horigane et al. 2001).

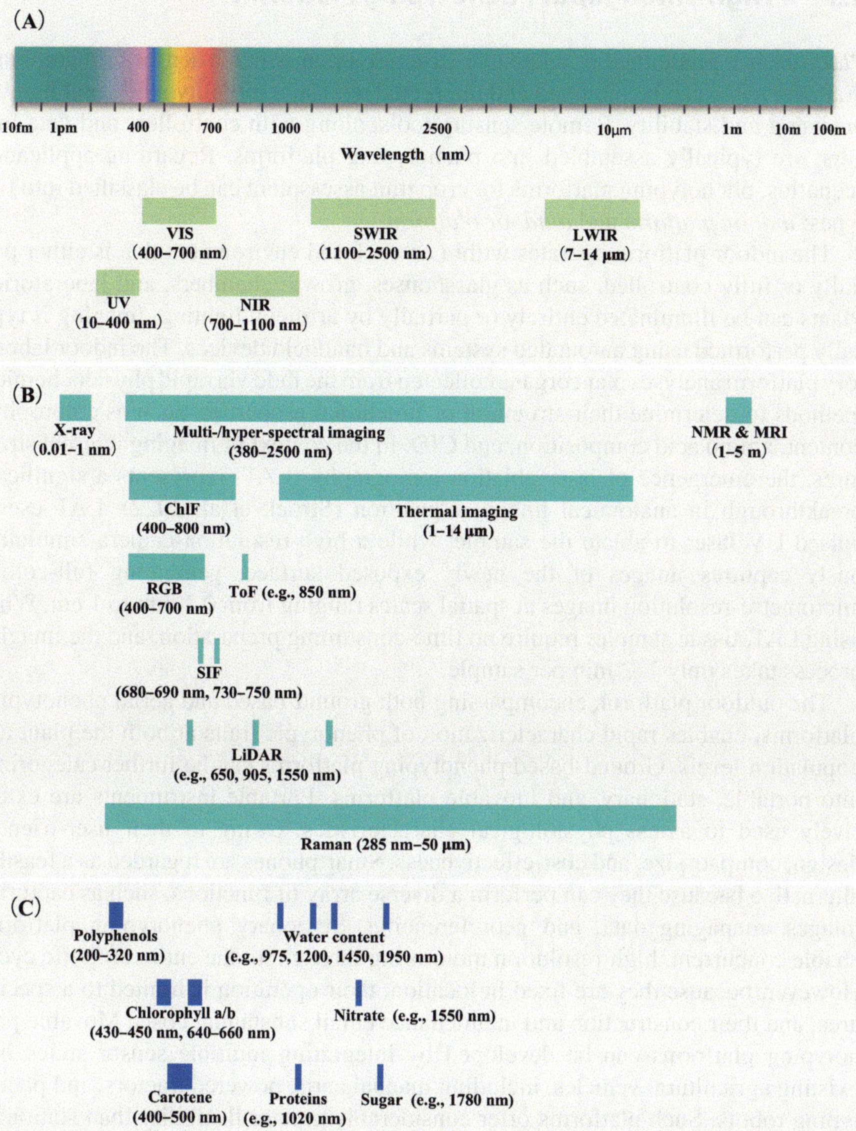

Fig. 8.5 Spectral-based phenotyping of the biochemical state of plants. (**A**) Electromagnetic spectrum depicting all wavebands of interest in imaging techniques for plant phenotyping. UV ultraviolet, VIS visible, NIR near-infrared, SWIR short-wave infrared, LWIR long-wave infrared. (**B**) Imaging techniques and their wavelength ranges. NMR nuclear magnetic resonance, MRI magnetic resonance imaging, ChlF chlorophyll fluorescence, RGB red-green-blue, ToF time-of-flight, SIF solar-induced fluorescence, LiDAR light detection and ranging. (**C**) Plant biochemical components and their detectable wavelengths. (Redrawn from Sarić et al. (2022) and Sun et al. (2022))

4.3 High-Throughput Phenotyping Platforms

Platform is a predefined or adaptable structure upon which a sensor or a combination of sensors can be mounted and integrated to ensure mobility while maintaining precision and stability. Remote sensing tools, along with controllers and data loggers, are typically assembled into phenotyping platforms. Regarding application scenarios, phenotyping platforms for crop trait assessment can be classified into two types: *indoor platform* and *outdoor platform.*

The indoor platform operates within an enclosed environment that is either partially or fully controlled, such as glasshouses, growth chambers, and laboratories. Plants can be illuminated entirely or partially by artificial lighting. Imaging is typically performed using automated systems and handheld devices. The indoor laboratory platform analyses plant organs collected from the field via rapid physicochemical methods to determine their structural or functional properties, such as chlorophyll content, amino acid composition, and CID. In the context of imaging internal structures, the emergence of laser ablation tomography (LAT) represents a significant breakthrough in anatomical image acquisition (Strock et al. 2022). LAT uses a pulsed UV laser to ablate the sample, while a high-resolution camera simultaneously captures images of the newly exposed surface, generating full-colour, micrometre-resolution images at spatial scales ranging from 0.1 mm to 1 cm. When using LAT, tissue samples require no time-consuming preparation, and the imaging process takes only 1–2 min per sample.

The outdoor platform, encompassing both ground-based and aerial phenotyping platforms, enables rapid characterization of phenotypic traits at both the plant and population levels. Ground-based phenotyping platforms can be further categorized into portable, stationary, and movable platforms. Portable instruments are extensively used to assess physiological characteristics, owing to their user-friendly design, compact size, and cost-effectiveness. Smartphones are regarded as a feasible alternative because they can perform a diverse array of functions, such as capturing images, managing data, and georeferencing. Stationary phenotyping platforms enable concurrent, high-resolution monitoring throughout the entire crop life cycle. However, because they are fixed in location, their operation is limited to a specific area, and their construction and maintenance entail substantial costs. Movable phenotyping platforms can be developed by integrating multiple sensor suites into existing agricultural vehicles, including manual carts, powered tractors, and phenotyping robots. Such platforms offer considerably greater flexibility than stationary ones and higher efficiency than portable instruments. However, their application is limited to specific circumstances. For example, they are not suitable for phenotyping rice canopies in waterlogged paddy fields.

Aerial imaging, whether conducted by manned or unmanned low-flying vehicles, is revolutionizing field phenotyping. It offers two primary advantages that enhance throughput and precision compared to ground-based systems. Firstly, the ability to capture hundreds of field plots in a single image eliminates the confounding effects of *environmental drift* associated with time-consuming plot-to-plot measurements. Secondly, image analysis enables data curation by removing outlying pixels within each plot. Both factors strengthen the associations between remotely sensed traits and yield relative to ground-based measurements. As sensor and image analysis technologies advance, the scope and precision of remote sensing are expected to improve. However, many important phenotypic traits still cannot be effectively measured using remote sensing. For example, detailed growth analysis requires precise destructive harvests, and it remains difficult to measure partially or fully obscured stem or root characteristics via remote sensing.

High-throughput phenotyping (*HTP*) is advancing at a rapid pace and has already begun to narrow the genotype–phenotype gap. Currently, however, existing HTP platforms primarily capture macroscopic traits such as plant height and leaf area, or spectral signatures like ChlF, NIR, and RGB reflectance. To gain a more in-depth physiological understanding, next-generation phenotyping should extend beyond these physical and spectral proxies and directly probe subcellular functional states through targeted metabolite profiling. Integrating complementary high-resolution sensors, such as optical molecular spectroscopy, hyperspectral imaging, and ambient mass spectrometry, into automated phenotyping pipelines will enable dynamic, multi-scale monitoring of plant metabolism (Hall et al. 2022).

5 LPR: A Deep-Learning-Based Physiological Trait

To address the slowdown in genetic gains for crop yield, it is crucial to identify and accurately characterize novel physiological traits grounded in a solid scientific foundation, thereby advancing crop breeding and management. Deep learning, a subfield of machine learning, has emerged as a powerful tool for image processing. It enables the extraction of spatial and temporal patterns in crop growth and development, thereby enhancing the accuracy of physiological trait identification and characterization.

One such trait is the leaf to panicle ratio (LPR), which reflects the light interception pattern within the rice canopy (Yang et al. 2020). The LPR value is calculated as the leaf pixel number (*L*) divided by panicle pixel number (*P*) in the RGB image of rice canopy (LPR = *L*/*P*), after leaf-panicle segmentation by FPN-mask Network. Physiologically, LPR indicates the distribution of light interception within the canopy between the source organ leaf and the sink organ panicle, which is critical for the RUE of the crop canopy. With the rapid advancement of plant phenomics, LPR can be efficiently quantified using RGB images acquired from aerial platforms. It is regarded as a novel PT indicative of canopy structure and function in response to genotype and nitrogen fertilization (Xiao et al. 2021; Yang et al. 2023).

One of the substantial challenges in rice breeding is to develop an ideotype or ideal plant architecture (IPA) to improve RUE and thereby increase grain yield. The hybrid japonica rice Jiayou Zhongke 6 (JYZK-6) is one of the representative cultivars exhibiting IPA characteristics. However, the eco-physiological basis underlying its high-yield potential remains incompletely understood. Xiao et al. (2021) conducted 2-year field experiments on four japonica rice cultivars, including JYZK-6, that differed in canopy structure. A dynamic canopy light interception simulation system was developed to capture time-series canopy images representing the diurnal variation in solar angle. Subsequently, LPR was calculated to quantify the spatial and temporal pattern of light distribution within the rice canopy (Fig. 8.6). The conventional cultivars, such as NJ-5055, exhibited a unimodal pattern in LPR, with peak values occurring at noon. In contrast, the IPA cultivar JYZK-6 displayed a V-shaped diurnal pattern of LPR, characterized by relatively higher values at sunrise and sunset and a pronounced decline to the lowest level at noon, during the mid- to late grain-filling stages. Morphological measurements revealed that this V-shaped pattern was associated with the height difference between the flag leaf and the panicle, the curvature of the panicle, the ratio of leaf area to panicle area, and the changing solar angles as the growth stage progressed.

This distinctive characteristic of the light pattern suggests that, within the canopy of IPA cultivars, leaves receive greater solar radiation during the morning and afternoon, whereas panicles intercept more solar radiation at noon. It can thus be inferred that this unique trait may confer a physiological advantage to IPA cultivars over conventional ones in alleviating midday photosynthetic suppression, as the panicle is more resistant to high radiation at noon (PAR, 1575.4 μmol m^{-2} s^{-1}; Fig. 8.6).

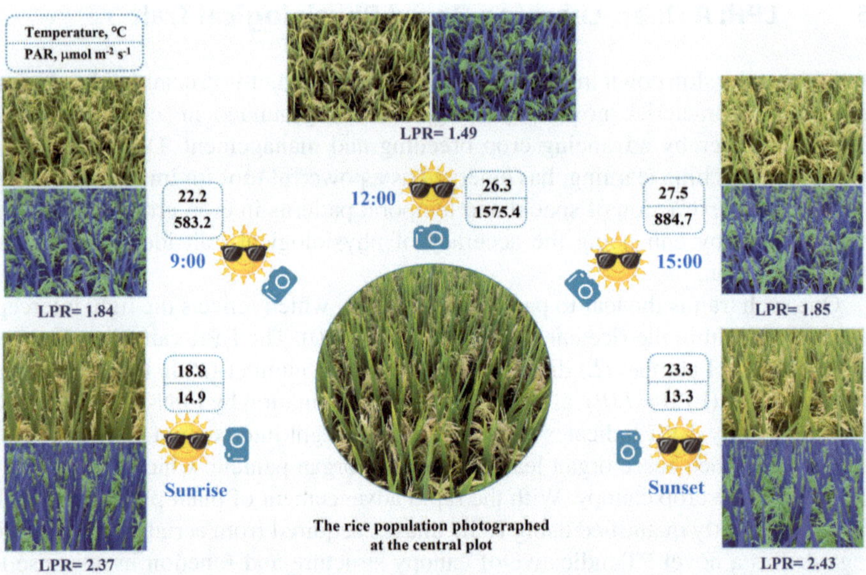

Fig. 8.6 Diurnal variations of LPR for IPA cultivar JYZK-6. Each RGB image is accompanied by a segmentation map of leaves and panicles, generated through the deep learning model FPN–Mask Network (Xiao et al. 2021). RGB images were obtained using the dynamic canopy light interception simulation device (DCLISD). By simulating the sun's trajectory, DCLISD portrays the "camera" as the "sun" directed towards the rice canopy at the centre of the field plot. The RGB images were captured by simulating five time points of a day, namely, sunrise, 9 a.m., 12 p.m., 3 p.m., and sunset at 30 DAA in 2019. The embedded box displays the temperature and PAR on the day of photography. DAA days after anthesis, IPA ideal plant architecture, LPR leaf to panicle ratio, PAR photosynthetically active radiation, RGB red-green-blue

However, the underlying mechanism still needs to be clarified. Nevertheless, finding from the perspective of LPR offers a novel insight into the eco-physiological mechanisms underpinning RUE at canopy level and hence is valuable for breeding programs for ideotype cultivars.

6 Concluding Remarks and Perspectives

The Green Revolution that took place in the mid-twentieth century led to a substantial increase in global cereal yields, primarily due to the development of dwarf, high-yielding cultivars with an enhanced harvest index (HI). Recent studies indicate that HI in high-performing rice and wheat cultivars has nearly reached its theoretical maximum of approximately 0.6, leaving minimal genetic potential for further improvement (Sanchez-Bragado et al. 2020). As a result, traditional breeding strategies based on HI optimization and dwarfing genes have largely plateaued. To sustain future yield gains and enhance resilience to climate change, crop breeders urgently need novel, heritable traits that can serve as reliable, early-stage indicators of field performance.

Physiological breeding, employing PTs like CT as selection criteria, has been recognized as a helpful complement to conventional and molecular breeding, in particular for the areas that suffer drought and heat stresses. Despite the existing set of PTs serving as breeding targets, there is a need for additional traits with clearer indicative value to enhance the efficiency of germplasm screening and early-generation selection. Recent advancements in genomics and phenomics have substantially improved our understanding of the biochemical and biophysical processes that govern crop yield and quality. Consequently, this has facilitated the mining and phenotyping of novel PTs for modern crop breeding and management.

The rapid advancements in computer science and mathematics are ushering in an AI-driven era with the potential to fundamentally transform crop science and technology. The major bottleneck in the crop breeding pipeline is the limited capacity to link genotype with phenotype, owing to the lack of efficient, high-throughput phenotyping methodologies and platforms. Machine learning (e.g. deep learning) algorithms now enable more efficient analysis of large-scale genomic, phenotypic, and environmental datasets than traditional methods, facilitating the identification of novel gene functions, optimization of breeding strategies, as well as discovery of new PTs, such as COV (Liu et al. 2021) and LPR (Yang et al. 2023). Furthermore, AI-enhanced simulation models can integrate climatic conditions with physiological traits to accurately forecast crop productivity under future climate scenarios, thereby supporting more resilient and well-informed decision-making by all stakeholders involved in crop science and technology. Collectively, these advances are expected to accelerate the development and deployment of climate-smart, high-yielding, and resource-efficient crops, marking a transformative shift in agricultural research and practice.

Summary

1. In the field of crop science, traits of interest can be broadly categorized into two distinct types: agronomic and physiological traits. The agronomic trait encompasses the comprehensive objectives of crop breeding and management, while the physiological trait reveals the structural and functional aspects of a crop plant or stand.
2. Canopy temperature is the surface temperature of the canopy, serving as a reliable indicator of plant response to high temperatures and water scarcity.
3. Physiological breeding involves crossing parents with different complex yet complementary traits to achieve cumulative gene action for yield while selecting progeny through phenotyping platforms in combination with genomic selection.
4. In terms of application scenarios, phenotyping platforms for crop traits can be categorized into the indoor laboratory platform and the outdoor field platform. The former analyses plant organs using rapid physical or chemical methods to determine their structural or functional properties. The latter consists of ground-based and aerial phenotyping platforms.
5. In the future, crop science should prioritize the mining and phenotyping of new physiological traits for crop breeding and management, with the aid of AI technologies.

Important Terms

Agronomic trait; Canopy temperature; CTD; Crop phenomics; Phenomics; Phenotype; Physiological breeding; Physiological traits; Platform; Scalability; Trait.

Review Questions

Multiple Choice

1. Among the traits listed below, (   ) is different from the other three according to the classification of TRAIT.

   (a) Gs
   (b) CT
   (c) LPR
   (d) LAI

2. Among the traits listed below, (   ) is different from the other three according to the classification of TRAIT.

   (a) Grain yield
   (b) WUE
   (c) NUE
   (d) CT

3. Among the traits listed below, (   ) is scalable from the tissue/organ level to the crop level.

   (a) Leaf photosynthetic capacity
   (b) Herbicide resistance
   (c) Gs
   (d) Leaf surface temperature

4. Among the traits conferred by genes (shown in the brackets), (   ) is NOT scalable from the tissue or organ level to the crop level.

   (a) Powdery mildew resistance of wheat leaf (*Pm21*)
   (b) Low amylose content in rice endosperm (*Waxy*)
   (c) High sink activity in maize endosperm (*O2*)
   (d) Narrow leaf angle of maize (*lg1*)

5. Concerning the typical physiological trait of canopy temperature (CT), (   ) is NOT right.

   (a) CT integrates many physiological functions necessary to ensure adaptation to a given environment.
   (b) There might be no correlation between CT and yield under drought conditions where water availability at soil depth is limited.
   (c) CT reflects the amount of transpiration that influences evaporative cooling and is correlated with many physiological factors, including Gs, CID, WUE, LAI, and eventually crop yield.
   (d) The value of canopy temperature depression (CTD), defined as the difference between air temperature and canopy temperature, is typically negative during the daytime.

True or False

1. The term "trait" is used interchangeably with "phenotype" in scientific terminology, while its counterpart is the gene; similarly, the genotype serves as the counterpart of phenotype (   )
2. Transpiration cooling is an active mechanism for controlling leaf temperature (   )

Blank Filling

1. The phenotype is characterized by a multitude of __________, __________, and __________ that undergo changes during growth and development.
2. The essence of crop science lies in acquiring knowledge related to the selection (breeding) and regulation (management) of __________ that serve as indicators for crop performance.

3. In the realm of crop science, traits of interest can be broadly categorized into two distinct types: __________ traits and __________ traits.

Short Answer

1. List three cases of successful physiological breeding, and summarize their common features.
2. Classify traits in crop science and interpret the meaning of each classification.
3. State the key steps in a physiological breeding pipeline.
4. Elucidate the physiological significance of canopy temperature.
5. Summarize the influencing factors of canopy temperature.
6. Interpret the eco-physiological foundation for canopy temperature in terms of energy balance.
7. List the major types of crop phenotyping platforms.

Essay Items

1. Do you think it's important for crop physiology to focus on exploring new physiological traits that can indicate how well crops perform in real field conditions? If not, please suggest an alternative statement.
2. Physiological trait (PT) reflects the structural and functional aspects of a crop plant or stand. Please select one PT, and list its pros and cons in crop breeding or management.
3. Evaluate the application of plant phenomics in crop physiology by comprehensively delineating its strengths, weaknesses, and the challenges it faces.

References

Ahmed FE, Hall AE, DeMason DA (1992) Heat injury during floral development in cowpea (*Vigna unguiculata, Fabaceae*). Am J Bot 79(7):784–791. https://doi.org/10.2307/2444945

Blum A, Mayer J, Gozlan G (1982) Infrared thermal sensing of plant canopies as a screening technique for dehydration avoidance in wheat. Field Crop Res 5:137–146. https://doi.org/10.1016/0378-4290(82)90014-4

Caine RS, Yin X, Sloan J et al (2019) Rice with reduced stomatal density conserves water and has improved drought tolerance under future climate conditions. New Phytol 221(1):371–384. https://doi.org/10.1111/nph.15344

Hall RD, D'Auria JC, Silva Ferreira AC et al (2022) High-throughput plant phenotyping: a role for metabolomics? Trends Plant Sci 27(6):549–563. https://doi.org/10.1016/j.tplants.2022.02.001

Hammer G, Cooper M, Tardieu F et al (2006) Models for navigating biological complexity in breeding improved crop plants. Trends Plant Sci 11(12):587–593. https://doi.org/10.1016/j.tplants.2006.10.006

Horigane AK, Engelaar WMHG, Maruyama S et al (2001) Visualisation of moisture distribution during development of rice caryopses (*Oryza sativa* L.) by nuclear magnetic resonance microimaging. J Cereal Sci 33(1):105–114. https://doi.org/10.1006/jcrs.2000.0348

Ishimaru T, Horigane AK, Ida M et al (2009) Formation of grain chalkiness and changes in water distribution in developing rice caryopses grown under high-temperature stress. J Cereal Sci 50(2):166–174. https://doi.org/10.1016/j.jcs.2009.04.011

Jiang MJ, Xu HF, Yang NF et al (2021) Temporal and spatial variations of carbon isotope signature reveal substantial contribution of bracts and internode assimilates to grain filling of japonica rice. Crop J 9(2):271–281. https://doi.org/10.1016/j.cj.2020.06.011

Kano H, Ishida N, Kobayashi T et al (1990) ^{1}H-NMR imaging analysis of changes of free water distribution in barley and soybean seeds during maturation. Jpn J Crop Sci 59:503–509. https://doi.org/10.1626/jcs.59.503

Lambers H, Chapin FS III, Pons TL (2008) Plant physiological ecology, 2nd edn. Springer

Li Y, Zhou QL, He MJ et al (2021) Dissection of environmental and physiological effects on the temperature difference between superior and inferior spikelets within a rice panicle. Crop J 9(5):1098–1107. https://doi.org/10.1016/j.cj.2020.10.016

Liu HY, Zou GH, Liu GL et al (2005) Correlation analysis and QTL identification for canopy temperature, leaf water potential and spikelet fertility in rice under contrasting moisture regimes. Chin Sci Bull 50:317–326. https://doi.org/10.1007/BF02897572

Liu FS, Song QF, Zhao JK et al (2021) Canopy occupation volume as an indicator of canopy photosynthetic capacity. New Phytol 232(2):941–956. https://doi.org/10.1111/nph.17611

Lopes MS, Dreisigacker S, Peña RJ et al (2015) Genetic characterization of the wheat association mapping initiative (WAMI) panel for dissection of complex traits in spring wheat. Theor Appl Genet 128(3):453–464. https://doi.org/10.1007/s00122-014-2444-2

Pask A, Pietragalla J, Mullan DM et al (2012) Physiological breeding II: a field guide to wheat phenotyping. CIMMYT, Mexico

Pinto RS, Reynolds MP, Mathews KL et al (2010) Heat and drought adaptive QTL in a wheat population designed to minimize confounding agronomic effects. Theor Appl Genet 121(6):1001–1021. https://doi.org/10.1007/s00122-010-1351-4

Prins A, Orr DJ, Andralojc PJ et al (2016) Rubisco catalytic properties of wild and domesticated relatives provide scope for improving wheat photosynthesis. J Exp Bot 67(6):1827–1838. https://doi.org/10.1093/jxb/erv574

Rebetzke G, Condon AG, Richards R et al (2002) Selection for reduced carbon isotope discrimination increases aerial biomass and grain yield of rainfed bread wheat. Crop Sci 42(3):739–745. https://doi.org/10.2135/cropsci2002.0739

Reynolds M, Langridge P (2016) Physiological breeding. Curr Opin Plant Biol 31:162–171. https://doi.org/10.1016/j.pbi.2016.04.005

Reynolds MP, Pask A, Mullan DM (2012) Physiological breeding I: interdisciplinary approaches to improve crop adaptation. CIMMYT, Mexico

Sadras VO, Richards RA (2014) Improvement of crop yield in dry environments: benchmarks, levels of organisation and the role of nitrogen. J Exp Bot 65(8):1981–1995. https://doi.org/10.1093/jxb/eru061

Sanchez-Bragado R, Vicente R, Molero G et al (2020) New avenues for increasing yield and stability in C3 cereals: exploring ear photosynthesis. Curr Opin Plant Biol 56:223–234. https://doi.org/10.1016/j.pbi.2020.01.001

Sarić R, Nguyen VD, Burge T et al (2022) Applications of hyperspectral imaging in plant phenotyping. Trends Plant Sci 27(3):301–315. https://doi.org/10.1016/j.tplants.2021.12.003

Sinclair TR, Purcell LC, Vadez V et al (2000) Identification of soybean genotypes with N_2 fixation tolerance to water deficits. Crop Sci 40(6):1803–1809. https://doi.org/10.2135/cropsci2000.4061803x

Sinclair TR, Purcell LC, Sneller CH (2004) Crop transformation and the challenge to increase yield potential. Trends Plant Sci 9(2):70–75. https://doi.org/10.1016/j.tplants.2003.12.008

Strock CF, Schneider HM, Lynch JP (2022) *Anatomics*: high-throughput phenotyping of plant anatomy. Trends Plant Sci 27(6):520–523. https://doi.org/10.1016/j.tplants.2022.02.009

Sun DW, Robbins K, Morales N et al (2022) Advances in optical phenotyping of cereal crops. Trends Plant Sci 27(2):191–208. https://doi.org/10.1016/j.tplants.2021.07.015

Thapa S, Jessup KE, Pradhan GP et al (2018) Canopy temperature depression at grain filling correlates to winter wheat yield in the U.S. Southern High Plains. Field Crops Res 217:11–19. https://doi.org/10.1016/j.fcr.2017.12.005

Wallace JG, Rodgers-Melnick E, Buckler ES (2018) On the road to breeding 4.0: unraveling the good, the bad, and the boring of crop quantitative genomics. Annu Rev Genet 52:421–444. https://doi.org/10.1146/annurev-genet-120116-024846

Walter A, Liebisch F, Hund A (2015) Plant phenotyping: from bean weighing to image analysis. Plant Methods 11(1):14. https://doi.org/10.1186/s13007-015-0056-8

Xiao F, Li WW, Xiao MH et al (2021) A novel light interception trait of a hybrid rice ideotype indicative of leaf to panicle ratio. Field Crop Res 274:108338. https://doi.org/10.1016/j.fcr.2021.108338

Yang ZF, Gao S, Xiao F et al (2020) Leaf to panicle ratio (LPR): a new physiological trait indicative of source and sink relation in japonica rice based on deep learning. Plant Methods 16(1):117. https://doi.org/10.1186/s13007-020-00660-y

Yang ZF, Qi XT, Dai Y et al (2023) Nitrogen fertilization produces divergent effects on canopy structure between indica and japonica rice reflected in leaf to panicle ratio based on deep learning. Field Crop Res 304:109184. https://doi.org/10.1016/j.fcr.2023.109184

Zhang SW, Wang CF (2008) Research progress and tendency in wheat low temperature genotypes. Sci Agric Sin 41(9):2573–2580. https://www.chinaagrisci.com/CN/10.3864/j.issn.0578-1752.2008.09.002

Appendixes

Appendix 1: Scientific Name of Crop Species

African rice	*Oryza glaberrima*
Alfalfa	*Medicago sativa*
Arabidopsis	*Arabidopsis thaliana*
Avocado	*Persea americana*
Barley	*Hordeum vulgare*
Bean (adzuki)	*Vigna angularis*
Bean (common, dry, pinto)	*Phaseolus vulgaris*
Bean (faba)	*Vicia faba*
Bean (lima)	*Phaseolus lunatus*
Bean (mung)	*Vigna radiata*
Bean (urd/black gram)	*Vigna mungo*
Buckwheat (common)	*Fagopyrum esculentum*
Buckwheat (tartary)	*Fagopyrum tataricum*
Cassava	*Manihot esculenta*
Chayote	*Sechium edule*
Chickpea	*Cicer arietinum*
Chili pepper	*Capsicum annuum*
Clover (subterranean)	*Trifolium subterraneum*
Clover (arrow leaf)	*Trifolium vesiculosum*
Cocoa	*Theobroma cacao*
Coffee	*Coffea arabica*
Cotton	*Gossypium hirsutum*
Cowpea	*Vigna sinensis*
Cucumber	*Cucumis sativus*
Lentil	*Lens culinaris*
Lupin (white)	*Lupinus albus*
Maize	*Zea mays*
Milk vetch	*Astragalus glycyphyllos*
Millet (broomcorn)	*Panicum miliaceum*
Millet (finger)	*Eleusine coracana*
Millet (foxtail)	*Setaria italica*
Millet (pearl)	*Pennisetum glaucum*
Mustard (white)	*Sinapis alba*
Oat	*Avena sativa*
Oil palm	*Elaeis guineensis*
Pea	*Pisum sativum*

Z. Liu, *Crop Physiology*, https://doi.org/10.1007/978-981-92-2960-4

Peanut	*Arachis hypogaea*
Pineapple	*Ananas comosus*
Potato	*Solanum tuberosum*
Quinoa	*Chenopodium quinoa*
Rape (canola)	*Brassica napus*
Rice	*Oryza sativa*
Rye	*Secale cereale*
Ryegrass (perennial)	*Lolium perenne*
Sesame	*Sesamum indicum*
Sorghum	*Sorghum bicolor*
Soybean	*Glycine max*
Sugar beet	*Beta vulgaris* ssp. *vulgaris*
Sugarcane	*Saccharum officinarum*
Sunflower	*Helianthus annuus*
Sweet potato	*Ipomoea batatas*
Taro	*Colocasia esculenta*
Teosinte	*Zea mays* ssp. *parviglumis*
Tomato	*Solanum lycopersicum*
Turnip	*Brassica rapa*
Vetch (common)	*Vicia sativa*
Watermelon	*Citrullus lanatus*
Wheat	*Triticum aestivum*
Yam	*Dioscorea alata*

Appendix 2: Idea-Based Course Design Document

Teaching success is rooted in careful preparation long before students enter the classroom. At the planning phase, establishing clear, measurable learning objectives is the foundational step; these objectives define precisely what students should know, understand, and be able to demonstrate by the end of the course. Serving as a guiding framework, they inform decisions about content selection, instructional strategies (e.g. lectures, discussions, problem-based learning, or collaborative projects) and the design of assessments that accurately measure mastery. By aligning objectives with learning activities and evaluations in advance, instructors can organize topics in a logical sequence, anticipate potential difficulties, allocate class time effectively, and construct a coherent learning experience that promotes progressive knowledge development. This intentional, forward-looking approach reduces the need for improvisation during instruction, enabling teachers to focus on fostering student engagement and deep conceptual understanding. Ultimately, comprehensive pre-semester planning ensures alignment among learning goals, instructional methods, and assessment practices, thereby supporting meaningful learning, enhancing student motivation, and promoting sustained academic achievement.

A course design functions as a comprehensive and systematic framework that not only delivers essential content to students but also enables effective coordination among instructors. It outlines the learning structure for each class session, clearly defining the objectives of every component of the curriculum. By dividing the syllabus into distinct units and learning outcomes, the design specifies corresponding teaching activities and instructional resources tailored to support these goals.

Furthermore, it delineates assessment strategies used to evaluate whether students have achieved the intended learning outcomes by the end of the course. By explicitly outlining both the instructional sequence and the criteria for success, course design ensures that all stakeholders, including students, teachers, and administrators, share a common understanding of educational objectives, the pedagogical methods employed, and the standards for measuring progress. As a result, this structured blueprint enhances consistency, transparency, and communication across the entire learning process.

Designing a course is akin to conducting an orchestra, where each instrument possesses a distinct voice that contributes to the overall harmony of the performance. Similarly, every instructional component, such as lectures, activities, assessments, and resources, must function in seamless coordination to create a cohesive and engaging learning environment that resonates with students. When a course's blueprint is thoroughly developed, teaching becomes the day-to-day enactment of this design, transforming planned structures into dynamic interactions, real-time adjustments, and effective delivery. This continuous implementation ensures that the theoretical framework evolves into a meaningful and memorable educational experience, just as a conductor brings a musical score to life through precise and unified execution.

In today's information-saturated environment, the constant influx of data often undermines deep thinking and long-term planning, making thoughtful guidance more critical than ever. While search engines, social media platforms, and online learning resources provide immediate access to vast amounts of knowledge, they frequently promote fragmented learning, selective exposure, and a preference for instant answers. Consequently, many individuals resort to skimming headlines or accepting superficial explanations without questioning underlying assumptions or considering broader contexts. As the brain adapts to rapid rewards, essential cognitive habits, such as critical thinking, genuine curiosity, and the ability to synthesize disparate ideas, are increasingly at risk of erosion. Over time, this trend may impair the capacity for sustained contemplation and the development of nuanced, interconnected understanding, underscoring the importance of deliberate, reflective practices to counteract the dominance of instant gratification.

Idea-based learning (IBL), first introduced by Hansen in 2011, is a pedagogical approach centred on a learning orientation designed to address the challenges posed by today's rapid information explosion. Hansen (2011) identifies several core characteristics that define IBL. First, it prioritizes conceptual understanding by placing ideas, rather than isolated facts, at the centre of instruction, enabling learners to explore and integrate concepts across disciplines. Second, it supports student autonomy by encouraging learners to formulate their own questions, locate relevant resources, and construct meaning independently. Third, IBL emphasizes relevance by connecting classroom content to real-world contexts and learners' personal experiences. Fourth, assessment in IBL focuses on the development of critical thinking and problem-solving skills, moving beyond rote memorization. Finally, IBL positions teachers as facilitators who guide inquiry, provide instructional scaffolding, and cultivate a collaborative environment in which ideas can be examined, challenged, and refined. Collectively, these features aim to equip students with the cognitive and metacognitive skills required to navigate and make sense of an ever-expanding knowledge landscape.

Idea-based course design adheres to a set of principles that closely align with student-centred, outcomes-focused learning environments. In contrast to traditional curricula, which primarily emphasize the transmission of factual knowledge, often resulting in a syllabus composed of loosely connected topics and activities, idea-based design is grounded in broad conceptual frameworks that provide both individual courses and entire programs with a clear sense of coherence and unity. A typical idea-based course outline comprises nine key components: big ideas, enduring understandings, learning outcomes, common misconceptions, essential questions, guiding concepts, authentic performance tasks, criteria for evaluating task performance, and core competencies (see Table A.1). These elements function cohesively to ensure that each learning activity is intentionally aligned with overarching conceptual goals, thereby promoting deeper understanding and the

Table A.1 The nine essential elements for idea-based course design (Hansen 2011)

Elements	Description
Big ideas	Big ideas function as the fundamental pillars that unify a field or as the cohesive element that binds it, thereby defining a discipline. They encompass the genuinely significant meta-concepts and theories that serve as "conceptual lenses" for entire knowledge domains. It is essential to focus solely on a few key big ideas (typically 2–4), as each usually generates several enduring understandings, which in turn can yield numerous learning outcomes.
Enduring understanding	They are more precise conceptual derivations stemming from big ideas, including the key definitions, applications, and implications. These understandings serve as transferable generalizations central to a discipline that is adaptable to previously unencountered scenarios. They represent the core knowledge and skills that students should retain and be able to apply several years after completing the course.
Learning outcomes	They reflect the key aspects of enduring understandings, specifically, the competencies students are expected to acquire by the end of this course. They should embody the following characteristics: (1) student-centred; (2) focused on high-order thinking skills; (3) quantifiable; (4) tangible.
Common misconceptions	The primary obstacles to students' conceptual understanding are as follows: (1) insufficient reasoning abilities, specifically the capacity to recognize problems, assumptions, and implications within intricate arguments; (2) intellectual habits that impede perseverance, awareness of bias, or tolerance of ambiguity;[a] (3) systemic misconceptions, such as overly simplistic models of complex natural processes or stereotypes that obstruct in-depth understandings in social science or humanities.
Essential questions	Essential questions, often rooted in misconceptions, serve as the cornerstone of the course. They guide the exploration of significant concepts and core content, prompt a re-evaluation of prior learning, and establish the course's logical progression. A well-structured course is organized around a coherent sequence of such questions, aiding learners in grasping the meaning and relevance of the course content.

(Continued)

Table A.1 (Continued)

Elements	Description
Guiding concepts	They establish a more direct connection between course content and enduring understandings than learning outcomes and big ideas do. Moreover, they are not merely course-specific topics or isolated facts; rather, they span course segments, entire courses, and even disciplines. Whereas facts contribute to the acquisition of knowledge, concepts facilitate the development of understanding.
Authentic performance tasks	In-depth understanding cannot be effectively evaluated through multiple-choice tests or even essay tests. True comprehension is most effectively demonstrated when students engage in realistic, discipline-specific tasks. A task is considered authentic if it: (1) is realistically contextualized; (2) requires judgment and innovation; (3) involves active engagement with the subject matter; (4) mimics challenging professional situations; (5) assesses the application of a comprehensive set of knowledge and skills.
Performance criteria	Authentic performance tasks simulate real-world problems that professionals may encounter. The evaluation criteria for these tasks focus on four key aspects: (1) cognitive skills (critical and creative thinking, problem solving); (2) aesthetic appreciation; (3) social interaction; (4) oral and written communication.
Needed competencies	Complex and authentic tasks require a diverse range of latent skills that must be identified and explicitly taught. Competent instructors break down these tasks into specific cognitive processes and provide targeted practice opportunities. Scoring rubrics are valuable tools for identifying the necessary abilities.

[a]Tolerance of ambiguity refers to how individuals or groups perceive and process unfamiliar, complex, or inconsistent information. In foreign language learning, it specifically describes learners' ability to handle thoughts and ideas that challenge their existing knowledge. It reflects the psychological tolerance of learners when encountering uncertain language phenomena

development of transferable skills. For a comprehensive, step-by-step guide to implementing this approach, readers are directed to Hansen's 2011 publication, which offers detailed procedures and illustrative examples. By adopting this framework, educators can shift from fragmented content delivery to a more integrated and meaningful educational experience that reflects contemporary pedagogical standards.

The book is intentionally designed to serve as a reference book for training the next generation of crop physiologists. To support this objective, Table A.2 provides a comprehensive reference guide for structuring a Crop Physiology course, reflecting the book's central themes and foundational philosophy. This guide has been carefully developed and successfully implemented in the Crop Physiology course launched by the author. By offering a structured framework and practical teaching materials, the supplement aims to enhance instructional effectiveness and inspire innovative research, thereby contributing to the advancement of the field of crop physiology.

Table A.2 Course design document for crop physiology

Big ideas	Interdisciplinary approaches		Structure and function		Interaction	
Enduring understanding	Crop science is the knowledge and theory about the traits	Crop physiology is the key interface between genomics and phenomics	The function of a biological system reflects its structure	Optimizing structure is an effective way to improve function	Interactions among components exist at every organizational level	Interaction leads to emergent property
Learning outcomes	Place TRAIT at the core of crop science; understand the mission of crop physiology is to mining novel traits	Construct a knowledge bridge connecting molecular-level gene expression with field-based crop performance	Analyse the structure of a biological system and interpret its functional mechanisms	Design an optimal architectural framework for maximizing yield in a specific crop	Dissect a biological system, clarify the interactions within it, and relate them with system function	Recognize the emergent properties of a system and cultivate a holistic perspective
Common misconceptions	Most of traits are scaled up from the molecular level to crop level	Crop yield is determined at the molecular level and can be increased by modifying just one or a few genes	The structure of a biological system is an intricate black box that defies quantification	There exists just one ideal plant type that can achieve the maximum yield	The biological systems are isolated and quiescent, not interrelated and dynamic	A system is merely the aggregate of its individual components
Essential questions	Why do crop scientists need a shared linguistic foundation?	How does crop physiology bridge the gap between gene activity and crop performance?	How can the structure–function relationship be applied to crop science?	How can we design an optimal plant type that maximizes crop yield?	What's the difference between the interactions at various levels of biological organization?	Why cannot the behaviour of a system be interpreted solely based on its individual components?

Big ideas	Interdisciplinary approaches		Structure and function		Interaction	
Guiding concepts	Agronomic traits; gene; physiological breeding; physiological traits; traits	Biological organization; complexity; genomics; hierarchy; phenomics; scalability	Anatomy; function; histochemistry; meristem; modularity; ontogenesis; phytomer; structure	Domestication; ideotype; micro-climate; plant architecture; weak competitor	C–N; crop–environment; embryo–endosperm; source–sink; systematic thinking	Emergent property; heterogeneity; holism; information; interface; mechanism; reductionism; signal
Authentic performance tasks	Assemble a multidisciplinary team comprising experts in crop breeding, genetics, physiology, informatics, and agronomy to develop an integrated and site-specific strategy aimed at enhancing high-yielding crop production		Analyse the structure–function relationship within a specific biological system, identify the pivotal structural factors that govern its functionality, and devise a strategy to enhance its performance through altering the structure		Analyse a specific biological system, such as a seed, an individual plant, or a complex crop stand, by deconstructing it into its constituent components to identify the critical interactions that influence its functionality	
Performance criteria	Detect concepts in real situation; look at things with new eyes; draw concept maps; integrate divergent perspectives into a comprehensive framework		Gain in-depth understanding of the target system; develop a systematic solution; present arguments logically and clearly		Provide a logical description of the system; identify the critical bottleneck points; deliver a clear and concise presentation	
Needed competencies	Appropriate turn-taking in discussions; active listening; brainstorming with others; giving and receiving criticism constructively; writing up group discussions and decisions		Visualizing the structure of the system; formulating questions; developing hypotheses; generating novel alternatives; intellectual autonomy		Visualizing the structure of the system; developing hypotheses; reflecting upon one's own thinking; intellectual autonomy	

Glossary

14-3-3 proteins A protein family consisting of highly conserved regulatory molecules that modulate the function of a wide array of client proteins through phosphoserine or phosphothreonine binding. By influencing protein stability, subcellular localization, and enzymatic activity, they play crucial roles in signal transduction, cell cycle control, and apoptosis regulation.

Abscisic acid (ABA) A plant hormone that promotes leaf abscission, induces seed and bud dormancy, and inhibits germination.

ADP-glucose pyrophosphorylase (AGPase) The enzyme responsible for catalysing the synthesis of ADP-glucose from glucose 1-phosphate and ATP.

Agrichemical A general term denoting various chemical substances utilized in agricultural practices. In most cases, it refers to a broad category of substances that includes various types of pesticides such as insecticides, herbicides, fungicides, and nematicides. It also includes synthetic fertilizers, plant growth regulators like hormones and other chemical agents, as well as concentrated forms of raw animal manure. Also *agrochemical*.

Agronomy A branch of agriculture focusing on field crop production and soil management practices. Agronomy is the application of knowledge and technologies from various disciplines such as biology, chemistry, economics, ecology, soil science, hydrology, pest management, and genetics, aimed at managing field crops.

Aleurone The outermost layer of the cereal endosperm responsible for storing proteins, lipids, vitamins, and minerals.

Allosteric effector A small molecule that interacts with either an allosteric site of an enzyme or a protein molecule, thereby inducing a functional change in the target molecule. Also *allosteric modulator*.

Allotetraploid Polyploid organisms with four sets of chromosomes, derived from the hybridization of two distinct species.

Amylase An enzyme that catalyses the hydrolysis of starch into simple sugars, which belongs to the class of glycoside hydrolases and specifically targets **α**-1,4 glycosidic bonds.

Amylopectin The major component of starch (~80%) consisting of branched chains of glucose units. It is water-insoluble and exhibits a red-brown coloration when reacted with iodine.

Z. Liu, *Crop Physiology*, https://doi.org/10.1007/978-981-92-2960-4

Amylose The minor component of starch (~20%), consisting of long, unbranched chains of glucose units. It is water-soluble and produces an intense blue colour when reacted with iodine.

Anabolism One of the two subcategories of metabolism, the formation of complex organic molecules from simpler precursor compounds. Compare with *catabolism.*

Anatomy A branch of biology that examines the form and structure of living organisms. It includes the study of both external (gross) and internal (microscopic) features, the spatial relationships among anatomical components, and how these structures contribute to physiological function.

Angiosperm Any seed-bearing plant belonging to the phylum Angiospermophyta, in which the ovules are enclosed within an ovary that develops into a fruit following fertilization.

Annuals Plants that complete their entire life cycle, from germination to reproduction and death, within a single year.

Antiflorigen A plant hormone that originates from the same gene family as florigen but functions as a flowering suppressor. The antiflorigen in Arabidopsis is TERMINAL FLOWER 1 (TFL1), whereas in tomato, it is SELF PRUNING (SP). Compare with *florigen.*

Antiporter An integral membrane protein that mediates secondary active transport by translocating two or more distinct molecules or ions in opposite directions across a phospholipid bilayer. The movement of one substrate along its electrochemical gradient provides the energy required to drive the uphill transport of the other substrate against its electrochemical gradient. Compare with *symporter.*

Apical dominance Growth inhibition of axillary buds by the growing shoot apex.

Apoplast The region of plant tissue located outside the plasma membranes, primarily composed of cell walls, constitutes a tissue-level compartment that facilitates the transport of water and solutes across a tissue or organ. Compare with *symplast.*

Apoplastic pathway Nutrients exchange between two plant cells through the diffusion of chemical signals across the extracellular space.

Artificial intelligence (AI) The capacity of a digital computer or a computer-controlled robot to perform tasks commonly associated with human cognition, such as reasoning, interpreting meaning, formulating generalizations, and learning from experience.

Asparagine synthetase (AS) An enzyme primarily located in the cytoplasm that catalyses the conversion of aspartate into asparagine.

Aspartate aminotransferase (AspAT) An enzyme that catalyses the reversible transfer of an amino group from glutamate to oxaloacetate, resulting in the formation of 2-oxoglutarate (**α**-ketoglutarate) and aspartate.

Assimilation The process by which complex substances that constitute an organism are synthesized from simpler materials and elements present in the external environment.

Auxins Hormones involved in cell elongation, apical dominance, and rooting, among other processes.

Axillary meristem A group of rapidly dividing (meristematic) cells situated at the base of a leaf where it attaches to the stem, found in growing shoots.

Azimuthal canopy orientation The arrangement of leaves around the stem of an individual plant in relation to the planting axis, as observed from an overhead perspective.

Binding protein A protein that reversibly binds to a variety of small molecules, such as amino acids, sugars, inorganic ions, vitamins, and other similar compounds.

Biological organization The hierarchical organization of complex biological structures and systems that define life through a reductionist approach. Also *hierarchy of life* or *levels of organization*.

Bloom's taxonomy Taxonomy of educational objectives, developed in the 1950s by the American educational psychologist Benjamin Bloom, who established a standardized framework for aligning educational objectives, curricula, and assessments used in schools.

Branch An axillary shoot or inflorescence that originates from an axillary bud.

Brassinosteroids (BRs) Endogenous polyhydroxylated plant steroid hormones that play a crucial role in regulating various aspects of growth and development, including brassinolide, castasterone, and typhasterol. BRs function by promoting cell elongation and division, stimulating shoot and root growth, facilitating vascular tissue differentiation, supporting pollen tube formation, and regulating flowering time.

Breeding The scientific discipline focused on modifying the agronomic traits of crop plants to develop desired cultivars.

C3 carbon fixation The standard form of carbon fixation in plants, in which carbon dioxide is directly fixed into sugars through the enzymatic activity of Rubisco and the metabolic processes of the Calvin-Benson cycle.

C4 carbon fixation A specialized form of carbon fixation that enhances the effective concentration of carbon dioxide within the cells exhibiting Rubisco activity, thereby reducing carbon loss via photorespiration. C4 plants chemically fix carbon dioxide in mesophyll cells by catalysing its addition to the three-carbon compound phosphoenolpyruvate (PEP) via PEP carboxylase, yielding the four-carbon compound oxaloacetate.

Calvin–Benson cycle (CBC) The biochemical pathway responsible for carbon fixation during photosynthesis in plants, algae, and cyanobacteria. Also *Calvin cycle* or *C3 cycle*.

Canalization A self-organizing pattern of polar auxin transport, in which an initially broad domain of auxin-transporting cells gradually narrows to form a highly focused “canal” of directional transport.

Canopy The aboveground part of a plant community, formed by the collective crowns of individual plants.

Canopy function The capacity of the canopy to exchange energy (e.g. short and long-wave radiations) and substances (e.g. CO_2, H_2O, O_2, CH_4, N_2O) with the surrounding environment.

Canopy gap fraction The proportion of the sky visible through openings in the vegetation canopy, serving as an indicator of canopy openness or porosity.

Canopy structure or architecture The spatial arrangement of aboveground plant organs within the canopy.

Carbohydrates Organic compounds composed of carbon, hydrogen, and oxygen, typically with a hydrogen-to-oxygen atom ratio of 2:1. They serve as a primary energy source for living organisms and are found in foods such as grains, fruits, and vegetables. Chemically, carbohydrates include sugars, starches, and dietary fibre, and they are essential nutrients for human health and cellular function.

Caryopsis A single-seeded fruit (e.g. cereal grains) in which the pericarp completely encloses the seed and is firmly adhered to the seed coat.

Catabolism One of the two major subcategories of metabolism, involving the breakdown of complex substances into simpler molecules, typically accompanied by the release of energy. Compare with *anabolism*.

Cell differentiation A process in which cells that originated mitotically from a common ancestor differentiate from one another, leading to distinct variations in structural and functional characteristics.

Cell division A process in which cells duplicate their components, including chromosomal DNA and organelles, and distribute them equally to produce daughter cells that are genetically identical to the mother cell.

Cell expansion The increase in cell volume results from the expansion of vacuolar volume, accompanied by a corresponding increase in cell surface area. This process is regulated by turgor pressure, cell wall properties, and the cytoskeleton.

Cell lineage The developmental history of a tissue or organ within an organism, from its origin in specific cells of the fertilized egg or embryo, through a series of progressive stages, to its attainment of a fully differentiated state.

Cell wall A rigid, non-living outer layer that surrounds the plasma membrane of plant, fungal, bacterial, and algal cells. It is mainly composed of structural polymers such as cellulose in plants, chitin in fungi, and peptidoglycan in bacteria.

Cellular endosperm An endosperm that is divided into multiple cells as a result of cytokinesis occurring after mitosis. Compare with *nuclear endosperm*.

Cellularization The subdivision of a coenocyte or syncytium into distinct cellular units.

Central Dogma A fundamental principle of modern molecular biology that describes the fundamental process of genetic information transfer within biological systems. It posits a unidirectional flow of information: DNA is transcribed into RNA, which is subsequently translated into protein.

Chemical property A substance's capacity to undergo chemical transformations that result in the formation of new substances. Compare with *physical property*.

Circadian clock A biochemical mechanism that enables organisms to synchronize their physiological processes and behavioural patterns with the daily and seasonal variations of the day–night cycle. Also *circadian oscillator*.

Circadian rhythm Rhythm occurring in living organisms within a cycle of approximately 24 h, maintained even in the absence of external environmental cues, such as light–dark cycles.

Co-chaperone A protein that assists molecular chaperone in the process of protein folding.

Coenocyte A multinucleate mass of protoplasm formed by repeated nucleus divisions without cell fission. This contrasts with a *syncytium*, which forms through the fusion of multiple cells followed by the dissolution of internal cell membranes within the cellular mass.

Collenchyma A type of plant tissue characterized by elongated cells with unevenly thickened cell walls composed primarily of cellulose and pectin; it provides structural support, especially in young, growing shoots and leaves.

Community An interacting group of various species coexisting in a shared geographical area.

Companion cells (CCs) Specialized plant cells that directly support the function of phloem sieve tube elements. Originating from the same mother cell, they remain connected via plasmodesmata, facilitating the transport of nutrients into the sieve tubes for distribution throughout the plant.

Comparative genomics The study of similarities and differences in the genome sequences of different species, to understand evolutionary relationships, identify functional genomic elements, and uncover the genetic mechanisms underlying biological traits.

Compartmentalization The formation of cellular compartments that are surrounded by a single or double lipid bilayer membrane.

Compatible osmolyte A small organic solute that cells accumulate to counteract external osmotic stress without exerting significant adverse effects on macromolecular structure and function. Representative examples include proline, glycine betaine, trehalose, glycerol, and various polyols, which serve as chemical chaperones, reactive oxygen species (ROS) scavengers, and osmotic volume regulators.

Competition The interaction between organisms or species where the fitness of one is reduced due to the presence of another, typically competing for limited resources such as nutrients, space, water, or light.

Concept A mental representation or idea that enables individuals to understand, organize, and classify information about the world. It functions as an abstract generalization of objects, events, or relationships, forming the foundation for thinking, reasoning, and communication across diverse disciplines such as philosophy, psychology, education, and natural sciences.

CONSTANS (*CO*) A circadian clock-regulated gene encoding a nuclear protein that contains zinc finger domains implicated in protein-protein interactions, as well as a second plant-specific domain exhibiting sequence similarity to the DNA-binding domain of the yeast protein HAP2.

Cooperation Symbiotic interactions among crop species or individual plants that improve nutrient absorption, light capture, stress resilience, and overall yield of the entire population.

Cotransporter A subcategory of transporters that couples the favourable movement of one molecule along its concentration gradient with the unfavourable movement of another molecule against its concentration gradient, including *antiporters* and *symporters*.

Critical period The period during the crop cycle in which the number of seeds per square meter is highly sensitive to growing environments.

Crop A plant or plant-derived product that can be widely cultivated and harvested for commercial profit or subsistence purposes in the agricultural sector.

Crop development The sequence of processes and events that occur during the development of new organs throughout the entire life cycle of a crop plant.

Crop phenomics A comprehensive study of crop phenotypes (traits) across multiple levels of biological organization.

Crop physiology An applied basic science that examines the structure–function relationships underlying the biochemical and biophysical processes involved in yield and quality formation within crop stands, as well as their responses to environmental signals and management practices.

Crop quality The distinguishing characteristic, property, or attribute of a crop product with respect to its end use, such as food, feed, or industrial applications.

Crop yield A measurement of the quantity of a crop produced per unit area of farmland.

Crosstalk The interaction among different signalling pathways or cellular components, which facilitates the integration of multiple signals and enhance cellular responses to environmental changes.

Cryptochrome A flavoprotein that exhibits sensitivity to blue light and serves functional roles in attracting pollinators as well as regulating circadian rhythms.

Culm The hollow stem of a grass or cereal plant, especially the one that bears the flower.

Cultivar A plant variety that has been intentionally selected and cultivated by humans due to desirable traits, such as enhanced yield, disease resistance, or aesthetic features. It is distinguished by its capacity to retain consistent and distinct characteristics when propagated through cultivation.

Cuticle A protective, waxy, or hardened layer that covers the plant epidermis.

Cytoplasm The protoplasm of a cell that lies within the cell membrane but excludes the nucleus, containing organelles, vesicles, cytosol, and other inclusions.

Cytosol The aqueous solution of proteins, metabolites, and other soluble molecules within a biological cell, in which organelles are suspended.

Cytokinins (CKs) A class of plant hormones that promote cell division and growth, and delay plant senescence, such as kinetin and zeatin.

Cytoskeleton A highly intricate and dynamic network of protein filaments that interacts with a diverse array of molecular motors and cross-linking proteins.

Day-neutral plant A plant that blooms following a phase of vegetative growth, irrespective of the photoperiod, such as tomato, cucumber, strawberry, and sunflower.

De novo Starting from the beginning.

Deep learning A subset of machine learning, the utilization of multiple, interconnected layers within a neural network.

DELLA protein A master growth repressor that belongs to the GRAS family of putative transcriptional regulator.

Dermal tissue The outermost tissue of a plant, such as the epidermis or the periderm (of woody plants).

Desiccation The maturation-drying process that enables seeds to downregulate the transcription of genes associated with seed development, thereby facilitating an appropriate transition to the germination.

Determinate growth A growth pattern in which the apical meristem of a plant loses or transforms its division function, resulting in the cessation of the elongation growth of the stem axis. Eventually, the apical meristem develops into a reproductive structure of a fixed size (such as a terminal inflorescence or flower) or forms an apical lignified structure. Compare with *indeterminate growth*.

Development The initiation and differentiation of organs, as well as the sequential progression through various stages that plants undergo throughout their life cycle.

Dextrin Any of a group of adhesive substances that serve as intermediate products in the process of starch hydrolysis.

Differentiation A series of changes that take place in cells and tissues during development, leading to their specialization.

Diffuse radiation The portion of solar radiation that is scattered by atmospheric molecules and particulate matter in the air.

Digital twin A virtual replica of a physical object or system, continuously updated with real-time data to simulate, analyse, and optimize performance throughout its lifecycle.

Diphosphate Two phosphate groups linked directly through a P-O-P bond, forming a single functional unit attached to the parent molecule, as exemplified by adenosine diphosphate (ADP) or inorganic pyrophosphate. In contrast, *bisphosphate* denotes two distinct phosphate groups that are covalently bound to different positions on the same molecule but are not directly bonded to each other, as observed in molecules such as fructose 1,6-bisphosphate and ribulose 1,5-bisphosphate. Also *pyrophosphate*.

Diploid Cells or organisms that possess pairs of homologous chromosomes, resulting in a chromosome number that is twice the haploid count.

Direct radiation The portion of solar radiation that originates directly from the Sun.

Disproportionating enzyme A 4-α-glucanotransferase that transfers 1,4-α-D-glucan segments, such as maltodextrin, to the 4′ position of another molecule, thus disproportionating the substrate into a longer chain (e.g. amylose) and a shorter chain (typically glucose).

Domestication The process by which wild animals and plants undergo hereditary reorganization to become domesticated or cultivated forms, in accordance with human interests.

Domestication syndrome A set of traits that differentiates most domesticated crops from their wild ancestors.

Dormancy A physiological condition in which a viable seed does not germinate despite the presence of favourable environmental conditions.

Double fertilization The process that involves two simultaneous nuclear fusions within an embryo sac: the fusion of the egg cell with a sperm cell and the fusion of a second male gamete with the two polar nuclei.

Dry matter (DM) The mass of a substance determined after complete drying, predominantly representing the solid content that encompasses all constituent elements except water.

Ecology A branch of biology that investigates the spatial and temporal patterns in the distribution and abundance of organisms, along with the underlying causes and the resulting consequences.

Ecosystem The complex of living organisms, their physical environment, and the intricate interrelationships among them within a defined spatial unit.

Ecosystem services Services that natural ecosystems provide humans with various economic and social values. These services include, but are not restricted to, water source protection, soil conservation, climate regulation, species protection, food supply, and tourism.

Ectopic expression The synthesis of a gene or protein in cells, tissues, or developmental stages where it is not typically expressed. This phenomenon often stems from experimental manipulation or disease, and it can uncover functional roles or lead to abnormal phenotypes.

Embryo An immature, multicellular plant individual enclosed within a seed, which develops into a mature plant after germination.

Emergent property A property or phenomenon of complex systems that arises from the interactions among components within a system, yet is not exhibited by the individual components in isolation.

Endoreduplication DNA replication that occurs in the absence of subsequent mitosis, resulting in the formation of polyploid nuclei with elevated DNA content.

Endosperm A triploid tissue found in the seeds of angiosperms, formed through the fusion of a sperm nucleus with two polar nuclei within the embryo sac.

Energy The capacity of matter or radiation to perform work, which is derived from its mass, motion, electric charge, and other related properties. In physics, energy is defined as the quantitative property that must be transferred to an object in order to perform work on the object or to increase its thermal energy.

Entity Something that exists independently of others and possesses a distinct and well-defined identity.

Environmental drift A phenomenon where external environmental conditions diverge from their projected course. In machine learning, environmental drift occurs when changes in the operating environment violate model assumptions, causing predictive performance to decline over time.

Enzyme A protein that functions as biochemical catalysts and plays a key role in regulating the rates of biological processes that occur within cells.

Epidermis The outermost layer of the plant's primary body, serving as a protective covering for leaves, floral parts, young stems, and roots.

Epistatic effect A genetic interaction in which the expression of one gene masks or modifies that of another, resulting in a phenotype that deviates from the simple sum of individual gene effects.

Essential genes Genes that participate in fundamental metabolic and cellular processes essential for organismal growth and survival, including DNA replication, transcription, and translation.

Ethylene (ETH) A plant hormone that plays a regulatory role in fruit ripening, the formation of aerenchyma in submerged roots and stems, and various other developmental processes.

Euchromatin A lightly packed form of chromatin that is rich in gene concentration and transcriptionally active, allowing easier access for RNA polymerase and transcription factors during gene expression. Compare with *heterochromatin*.

Eudicot A wide range of flowering plants, typically possess two cotyledons in their seeds, pollen grains with three pores, the ability to undergo secondary growth, leaves with reticulate venation, and floral parts arranged in multiples of four or five. This clade of angiosperms includes the majority of species that were previously classified as dicots.

Evapotranspiration (ET) The combined loss of water through evaporation from the soil surface and transpiration from the leaves.

Expansin Non-enzymatic protein that facilitates cell wall loosening and cell expansion. The protein consists of two domains: a cellulose-binding domain and a domain homologous to glycoside hydrolase family 45, which collectively disrupt hydrogen bonds between cellulose microfibrils and matrix polysaccharides.

Expression quantitative trait loci (eQTLs) Regions in the genome where genetic variations (like SNPs or indels) influence the expression level of one or multiple genes, thereby linking genotype to gene expression phenotypes.

F-Box proteins A family of regulatory proteins defined by the presence of an F-Box domain, a conserved structural motif of about 40 to 50 amino acid residues. This domain was first identified in cyclin F, encoded by the *FBXO1* gene, from which it derives its name.

Floret A small flower located within a densely packed cluster. This terminology is commonly used to describe individual florets within spikelets and spikes in cereal crops.

Florigen The chemical signal that is transmitted from the leaves to the shoot apical meristem to convey the flowering information. Compare with *antiflorigen*.

Flow The translocation of nutrients from source organs to sink organs via vascular tissues.

Flower The reproductive structure of a plant, typically short-lived, that contains the sexual organs, including stamens and carpels.

Fructan A polymer composed of fructose units, commonly found in certain types of fruits.

Function The system's ability to engage in the reciprocal exchange of matter, energy, and information with its surrounding environment.

Gametophyte The plant body that produces gametes in species with alternation of generations. Compare with *sporophyte*.

Gene The fundamental unit of heredity, typically composed of a segment of DNA (or RNA in certain cases), that encodes genetic information and governs the development of specific traits or characteristics in an organism.

Gene expression The process by which the information stored in a gene is utilized to synthesize a functional product, such as a protein or RNA molecule, through transcription and translation.

Gene loci Specific, fixed positions on chromosomes where particular genes or genetic markers reside.

Genetic gain The enhancement of phenotypic value per generation resulting from genetic improvement through artificial selection in breeding programs.

Genetic modification The process that involves the deliberate alteration of an organism's genetic material, typically through laboratory-based techniques, in order to introduce novel traits or modify existing traits. Also *genetic engineering*.

Genetic regulation The precise cellular mechanism that governs the timing, location, and extent of gene expression, thereby ensuring accurate transcription into RNA and subsequent translation into functional proteins. This highly coordinated process depends on multiple interdependent components, including sequence-specific transcription factors that bind to regulatory DNA elements, core promoters that recruit RNA polymerase, distal enhancers that precisely regulate spatial and temporal expression, and epigenetic modifications, such as DNA methylation and histone acetylation, that establish heritable, context-dependent gene activity states. Also *gene regulation*.

Genetics A branch of biology that focuses on the study of genes, heredity, and variation in living organisms. It investigates the mechanisms by which traits are transmitted from parents to offspring, the molecular structure and function of genes, and the processes underlying genetic variation and evolutionary change within populations.

Genome The complete genetic blueprint of an organism, encompassing its DNA (or RNA in certain viruses), which contains all the heritable information required for its development, functioning, and reproduction.

Genome editing A set of techniques that enable highly specific modifications to the genome of an organism by targeting particular DNA sequences using engineered enzymes or nucleases. This process permits the insertion, deletion, or substitution of DNA sequences at defined genomic locations, thereby allowing precise alterations to genetic material.

Genome resequencing The methodology for sequencing the genomes of multiple individuals within species that already have established reference genomes.

Genome-wide association study (GWAS) A research approach that scans the entire genome of many individuals to identify genetic variants, typically SNPs, that are statistically associated with specific traits.

Genomic selection (GS) A method of marker-assisted selection that employs genome-wide genetic markers to estimate breeding values and facilitate the selection of individuals based on traits of interest.

Genomics The scientific study of genomes, encompassing the structure, function, and analysis of an organism's genetic material. As a branch of molecular genetics, this field involves processes such as sequencing, mapping, and functionally analysing genes and entire genomes.

Genotype The genetic makeup or constitution of an organism, which determines its hereditary characteristics, representing the specific combination of genes or alleles that an organism inherits from its parents. Compare with *phenotype*.

Genotyping The process of determining the genetic constitution (genotype) of an individual or organism by analysing its DNA sequence and comparing it to a reference sequence or the sequence of another individual. This procedure enables the identification of variations in genetic markers, such as SNPs, insertions, deletions, and copy number variations (CNVs).

Germplasm The fundamental genetic resources, including seeds and tissues, that are maintained in a living state for the purposes of breeding, conservation, and scientific research.

Gibberellins (GAs) A class of hormones that play a role in promoting stem elongation, seed germination, flowering, and various other physiological processes.

Globulins A group of proteins that are insoluble in pure water, yet soluble in dilute salt solutions and capable of being coagulated by heat.

Glucan water dikinase (GWD) A starch hydrolase that binds to the surface of starch granules and catalyses the incorporation of phosphate groups at the C-6 position of glucosyl residues.

Gluconeogenesis The metabolic pathway through which glucose is synthesized from non-carbohydrate precursors, such as amino acids, pyruvate, or glycerol.

Glutamate dehydrogenase (GDH) An enzyme catalysing the reaction of glutamate, H_2O, and NAD^+ producing **α**-ketoglutarate (2-oxoglutarate), ammonia, and NADH.

Glutamine 2-oxoglutarate aminotransferase (GOGAT) An enzyme that catalyses the synthesis of glutamate from glutamine and 2-oxogluatrate (**α**-ketoglutarate) and, thereby, in conjunction with glutamine synthetase, plays a central role in regulating nitrogen assimilation. Also *glutamate synthase*.

Glutamine synthetase (GS) An enzyme that plays a crucial role in nitrogen metabolism by catalysing the condensation of glutamate and ammonia to synthesize glutamine.

Glycogen A polysaccharide composed of glucose units, serving as the primary form in which carbohydrates are stored in the liver and muscles of humans and animals.

Grain A fruit type containing a seed that cannot be easily separated, scientifically referred to as *caryopsis*.

Grain chalkiness The opaque part of rice grain because of loose packing of starch granules and protein bodies.

Grain crops Field crops whose harvestable organs consist of dry grains (e.g. cereals) or seeds (e.g. legumes, oilseed crops).

Grain filling A metabolic process occurring in endosperms, wherein sugars transported from the source tissues are utilized for the synthesis of starch and other storage compounds.

Granule-bound starch synthase (GBSS) The key enzyme encoded by the waxy genes, which is responsible for amylose synthesis in the endosperm of cereals. It associates tightly with starch granules and catalyses glucose addition via **α**-1,4-glycosidic bonds to extend amylose chains within the pre-existing amylopectin matrix. Also *waxy protein*.

Gravitropism The directional growth or movement of a plant or plant organ in response to gravitational stimuli, as exemplified by roots growing downward and shoots growing upward.

Green Revolution The profound transformation in agriculture that primarily took place between the 1940s and the late 1970s. Key features of this revolution included the development of high-yielding crop varieties (such as semidwarf wheat and rice), the application of synthetic fertilizers and pesticides, improvements in irrigation infrastructure, the implementation of agricultural mechanization, and the adoption of modern management practices.

Ground tissue A plant tissue that occupies the interior of organs, fulfilling functions such as photosynthesis, storage, transport, and support. It encompasses parenchyma, collenchyma, and sclerenchyma cells, each varying in cell wall thickness, viability, and mechanical properties.

Growth The change in size or weight of the initiated organs.

Growth habit The external morphology, structural features, and growth characteristics of a plant, such as prostrate or erect growth habit, and determinate or indeterminate growth pattern.

Growth respiration The respiration serving as the energy source for the synthesis of new tissues.

Gymnosperm Any seed-bearing plant in which the ovules are exposed and not enclosed within an ovary, typically borne on the surface of megasporophylls that are often organized into cones. Compare with *angiosperm*.

Harvest index (HI) The biomass of the harvested product expressed as a percentage of the total plant biomass.

Heliotropism The growth of a plant towards the Sun, such as the phenomenon of solar tracking exhibited by the leaves and flowers of certain plants (e.g. sunflower, soybean).

Heterochromatin A tightly packed form of DNA that is transcriptionally inactive. It remains condensed throughout the cell cycle and is typically associated with gene silencing, playing crucial roles in chromosome structure and epigenetic regulation. Compare with *euchromatin*.

Heterofertilization A kind of double fertilization that the egg and central cell of a female gametophyte within a single ovule are fertilized by sperm cells from two different pollen tubes.

Heterogeneity In environmental science, the spatiotemporal variability of environmental conditions. It reflects the disparities in physical, chemical, and biological characteristics, such as microclimate, topography, vegetation, and soil properties, which collectively form a mosaic of distinct patches and ecological gradients.

Heterosis The phenomenon where the hybrid offspring display superior biological qualities or enhanced performance compared to their parents. This includes increased growth rate, higher fertility, improved disease resistance, or greater overall productivity. Also *hybrid vigour*.

Hierarchy A systematic organization or categorization of elements based on their relative significance or degree of inclusiveness.

High throughput An approach that involves the execution of multiple repeated measurements within a relatively short timeframe.

High-throughput phenotyping A technology-driven methodology that employs advanced tools such as imaging systems, sensors, and automated platforms to efficiently and accurately quantify and analyse crop traits.

Histochemistry A branch of biology that examines the chemical composition of cells and tissues. By bridging histology and biochemistry, it generates a "chemical map" of tissue architecture.

Holism A philosophical and systemic concept that emphasizes the principle that the whole is greater than the sum of its parts, advocating for the understanding of systems as integrated entities rather than mere aggregations of individual components. Compare with *reductionism.*

Homeostasis The maintenance of a stable internal environment within biological systems to preserve optimal physiological functioning, regardless of external changes.

Homeotic genes A class of genes that regulate the development of anatomical structures across a wide range of organisms, including insects, mammals, and plants. Mutations in these genes can result in substantial morphological alterations, such as the transformation of one body structure into another (e.g. the development of legs in place of antennae).

Homologs Chromosomes, genes, nucleic acids, or proteins that exhibit similar sequences, structural features, and/or functional characteristics.

Hormone The signalling molecule in multicellular organisms that is transported to distant organs to regulate physiological processes or behavioural responses.

Hyponasty The increased growth of the lower surface of a plant organ (e.g. the leaf), resulting in an upward bending of the part. Its counterpart is *epinasty*, downward curvature due to differential expansion between the upper and lower tissues.

Ideotype A biological model that is designed to exhibit predictable performance or behaviour within a specified environment.

Imprinted gene A gene that is expressed in a parent-of-origin-specific manner. Due to epigenetic marking during gametogenesis, only one allele is active, and the other allele remains silenced.

In vitro Biological processes or experiments that occur outside of a living organism, typically conducted within controlled laboratory settings or scientific apparatus.

In vivo Biological processes or experiments that occur or are conducted within a living organism.

Indeterminate growth A continuous growth pattern that persists indefinitely due to the ongoing activity of the shoot apical meristem. Compare with *determinate growth.*

Indoor platform The phenotyping conducted in enclosed and partially or fully controlled environments, such as glasshouses, growth chambers, and laboratories. Compare with *outdoor phenotyping*.

Inflorescence The arrangement and structural organization of flowers along the stem axis of a plant, such as racemes and panicles.

Inflorescence meristem A shoot meristem that generates an inflorescence or part of an inflorescence.

Information A measure that mitigates uncertainty and randomness, converting raw data into meaningful and structured knowledge. This knowledge empowers systems to differentiate states and propels adaptive evolution across the fields of physics, biology, and computer science. Information can take on diverse forms, including signals, messages, data, or knowledge, which are perceptible, recordable, transmittable, or processable.

Interaction The reciprocal relationship between two or more biological entities, such as genes, proteins, cells, organisms, or ecological partners, in which they mutually influence each other's structure, function, or fitness.

Intercalary meristem A meristematic region located within the internode that enables its elongation.

Intercropping The practice of growing multiple crops simultaneously in the same area, with the objective of increasing yield and reducing input requirements through improving the biodiversity of agroecosystem.

Interface The physical or functional boundary at which a living cell, tissue, organism, or biological material interacts with another material, be it solid, liquid, gas, or a different biological entity. This boundary serves as a surface facilitating the exchange of matter, energy, and signals between the two adjacent domains.

Intermediate organ A plant organ that can function as either a source or sink, depending on its physiological status and developmental stage.

Internode A segment of a plant stem located between two adjacent nodes.

Intragenic promoter A DNA sequence located within the coding region of gene, as opposed to upstream regulatory region. This promoter can initiate transcription in both the sense and antisense directions, frequently giving rise to alternative transcript or long non-coding RNA that regulates the expression of its host gene.

Invertase An enzyme that catalyses the hydrolysis of sucrose into glucose and fructose.

Irradiance The amount of solar energy received per unit time and per unit area, expressed in watts per square meter (W/m^2). It should be noted that *radiation* is the general phenomenon in which energy is emitted or propagated in the form of electromagnetic waves or particles. It may refer to the source, the propagation process, or the emitted radiation field itself. *Irradiance* is a measure of power per unit area at a given instant, representing the intensity of radiation at that moment. *Irradiation*, expressed in joules per square meter (J/m^2), refers to the total accumulated energy delivered over a specified time interval, obtained by integrating irradiance over time.

Isoamylase A hydrolase enzyme that catalyses the hydrolysis of 1,6-**α**-glycosidic branch linkages present in glycogen and amylopectin.

Jasmonates (JAs) The lipid-based plant hormones that regulate a diverse array of physiological processes in plants, including the defence against herbivory and the responses to abiotic stresses.

Krebs cycle A series of chemical reactions in most aerobic organisms in which cells break down glucose and other molecules in the presence of oxygen to produce carbon dioxide and energy in the form of ATP. The cycle is named after Hans Adolf Krebs, who identified its key components and was honoured with the Nobel Prize in Physiology or Medicine in 1953. Also *citric acid cycle* or *tricarboxylic acid (TCA) cycle*.

Landrace A locally adapted strain of a species that has been selectively developed and acclimatized by farmers.

Lateral meristem A cylindrical band of undifferentiated cells positioned parallel to the longitudinal axis of stems or roots. It is accountable for secondary growth, which enhances the plant's girth through the production of vascular cambium and cork cambium.

Leaf age The developmental stage of a crop plant, defined by the number of fully expanded leaves on the main stem.

Leaf angle The angle between the leaf blade and the main stem, or the angle between the leaf normal and the zenith (vertical direction). Also *leaf tilt, insertion*, or *inclination angle*. Note that leaf inclination angle is sometimes used to denote the angle formed between the leaf blade and the horizontal plane.

Leaf area index (LAI) The ratio of the leaf surface area to the ground area covered by a crop stand.

Level A position within an actual or conceptual hierarchy.

LiDAR An acronym for Light Detection and Ranging, a remote sensing technology that measures distances by emitting pulsed laser light and generates three-dimensional data.

Light extinction coefficient A constant that characterizes the exponential rate of light attenuation with increasing depth in a homogeneous medium.

Limit dextrinase A debranching enzyme that catalyses the hydrolysis of **α**-1,6-glucosidic bonds in branched dextrins, amylopectin, pullulan, and related polysaccharides. Also *limit-dextrin debranching enzyme*, *R-enzyme*, *pullulanase*, or *isoamylase*.

Linkage drag The co-inheritance of undesirable alleles along with a desirable trait due to the physical proximity of their corresponding genes on the same chromosome.

Lodging The permanent displacement of cereal stems from their originally vertical position.

Long-day plant A plant that requires a day length greater than a specific minimum threshold to initiate flowering.

Long noncoding RNA (lncRNA) A class of non-coding RNA molecule longer than 200 nucleotides that does not encode protein, with multiple regulatory functions such as the regulation of gene expression, chromatin remodelling, and epigenetic modifications.

Maintenance respiration The respiration serving as the energy source for all processes that involve in maintaining cellular structure, such as the turnover of cellular components and the maintenance of intracellular ion gradients.

Maltase A glucohydrolase that catalyses the hydrolysis of terminal, non-reducing 1,4-linked **α**-glucose residues, resulting in the release of **α**-glucose. Also **α**-*glucosidase*.

Mass Both a property of a physical body and a measure of its resistance to acceleration when a net force is applied, reflecting its tendency to maintain its current state of motion.

Mass flow The movement of material substances. In physics, mass flow occurs within open systems and is typically quantified as the movement of mass across a defined boundary, characterized by its cross-sectional area and flow velocity. Also *mass transfer* or *bulk flow*.

Matter The physical substance that constitutes the entities present in the world. In classical physics and general chemistry, matter is defined as any substance that possesses mass and occupies space by virtue of having volume. Compare with *substance*, which refers to a specific, tangible, or chemically defined material or entity.

Mechanism The fundamental process or structure that gives rise to a specific outcome or function. It may be a physical system (e.g. the gears of a watch), a biological pathway (e.g. DNA replication), or a conceptual model elucidating the occurrence of a phenomenon.

Meristem A type of plant tissue consisting of undifferentiated cells known as meristematic cells. These cells possess the capacity for continuous or periodic division and can differentiate into various specialized cell types, thereby facilitating plant growth and the development of new tissues and organs.

Metabolic pathway A series of enzyme-catalysed biochemical reactions that take place within a cell to convert specific substrates into corresponding products. This process may involve either the release or consumption of energy and is typically organized into linear or branched routes.

Metabolic pool The reservoir of biomolecules within the cell upon which enzymes can act to generate useful products in accordance with the cell's needs.

Metabolism The sum of intracellular chemical reactions occurring within a living organism, serving as a vital source of energy for essential biological processes and playing a key role in the synthesis of new organic compounds.

Metabolome The complete set of metabolites present within a cell, tissue, or organism under a specific physiological or developmental condition.

Metabolomics The large-scale investigation of metabolites present in cells, tissues, or organs, to offer a functional assessment of cellular processes and biological states.

Microclimate A localized climate or set of atmospheric conditions that differ from those of the surrounding larger area. It describes small-scale variations in climate within a specific environment, such as a garden, forest, or urban area, where factors like temperature, humidity, wind, and other weather elements can significantly deviate from the broader regional climate. Also *microenvironment*.

microRNA (miRNA) A small single-stranded non-coding RNA molecule consisting of approximately 22 nucleotides. It is present in plants, animals, and certain viruses and plays a key role in RNA silencing and the post-transcriptional regulation of gene expression. Compare with *siRNA*.

Mitosis A biological process in which the nucleus undergoes division to form daughter nuclei, each possessing an identical number of chromosomes as the original parent nucleus.

Model plant A plant species that is widely studied to elucidate specific biological phenomena, with the expectation that findings in the model plant will provide insights into the mechanisms of other plant species. Examples include Arabidopsis, rice, and pea.

Modular construction The process through which the plant body is formed via repeated arrangement of small, modular units.

Modularity The extent to which the components of a system can be separated and recombined, often providing the advantage of enhanced flexibility and versatility in application. In biology, it denotes the repetitive occurrence of morphological components or architectural units, such as phytomers, during the processes of growth and development.

Monocarpic plant A plant that flowers, produces seeds, and then dies.

Monocot A flowering plant characterized by an embryo containing a single cotyledon (seed leaf), typically exhibiting elongated, sessile leaves with parallel venation. Short form for *monocotyledon*.

Monosaccharide A type of sugar, such as glucose and fructose, that cannot be hydrolysed into simpler sugar molecules.

Morphogen A diffusible chemical substance that establishes a gradient of activity to guide cells towards appropriate fates over long distances, thereby contributing to tissue patterning.

Morphogenesis A biological process through which a cell, tissue, or organism establishes its structural organization and morphological characteristics.

Morphological plasticity The ability of a plant to alter its architectural structure in response to environmental changes.

Neural network A computational architecture inspired by the structure and function of the human brain. It consists of interconnected nodes, or neurons, that process information in a hierarchical manner. By learning from data through iterative adjustments, neural networks refine the weights assigned to input signals to enhance their performance in tasks such as pattern recognition, prediction, and decision-making. As a foundational element of artificial intelligence (AI), neural networks can adapt to new information and generating outputs based on previously learned patterns.

Nikolai Vavilov (1887–1943) A distinguished Russian and Soviet agronomist, botanist, and geneticist, widely recognized for identifying the centres of origin of cultivated plants.

Non-structural carbohydrates (NSCs) Carbohydrates that are not incorporated into structural components like cell walls. They are primarily composed of soluble sugars (e.g. glucose, fructose, sucrose) and starch and serve as a buffer for carbon allocation.

Normalized difference vegetation index (NDVI) A remote sensing index of ground surface greenness that utilizes the relative brightness of the surface in the red and near-infrared spectral bands.

Nucellus The central region or the nucleus within the ovule of seed plants.

Nuclear endosperm A multinucleate endosperm that remains undivided into distinct cellular compartments as a result of the absence of cytokinesis following mitotic divisions. Compare *cellular endosperm.*

Nucleic acids Complex macromolecules composed of nucleotide units that encode the genetic information essential for cellular functions and hereditary transmission.

Nucleoside A compound consisting of a purine or pyrimidine base covalently bonded to a sugar moiety, typically ribose or deoxyribose.

Nucleotide A compound composed of a nucleoside chemically bonded to phosphoric acid.

Nucleus A membrane-bound organelle that houses the cell's genetic material.

Nutrients Chemical compounds that are essential for the growth and development of plants, including sugars, amino acids, nitrate, nitrite, ammonium, as well as minerals. Plants require at least 13 essential mineral nutrients, which are classified into macrominerals and microminerals. The macrominerals include nitrogen (N), phosphorus (P), potassium (K), calcium (Ca), sulphur (S), and magnesium (Mg). The microminerals consist of iron (Fe), boron (B), chlorine (Cl), manganese (Mn), zinc (Zn), copper (Cu), and molybdenum (Mo).

Oligosaccharides A class of carbohydrates composed of a small number of monosaccharide units connected by glycosidic linkages.

Omics Any of several biological disciplines defined by the comprehensive analysis of the complete set of a specific type of biomolecule or an entire molecular process within an organism, typically named with the suffix -omics, such as genomics, proteomics, metabolomics, metagenomics, and transcriptomics. The omics sciences share the overarching goal of identifying, characterizing, and quantifying the biomolecules and molecular processes that underlie the structure and function of cells and tissues.

Ontogenesis The origin and development of an organism, typically from the time of egg fertilization to the attainment of its mature form. Also *ontogeny*.

Organ A functional unit composed of tissues that work together to carry out a specific role, such as nutrient absorption, photosynthesis, or reproduction.

Organelles The "little organs" of a cell, such as the plastids, mitochondria, and ribosomes, performing specialized functions essential to cellular operation.

Organogenesis The process by which undifferentiated plant cells develop into structurally distinct and functionally specialized organs, such as roots, shoots, leaves, and flowers, through coordinated cell division, differentiation, and patterning.

Orthologs Genes found in different species that originate from a common ancestral gene through the evolutionary divergence of species. Although their functions may vary, orthologs are generally regarded as homologs, meaning they share a common evolutionary origin. They are clearly distinguished from *paralogs*, which result from gene duplication events occurring within a single species.

Osmosis The passive process by which a solvent, typically water, diffuses through a semipermeable membrane from a region of lower solute concentration to a region of higher solute concentration, with the aim of equalizing the solute concentrations on both sides of the membrane.

Osmotic potential A measure of the tendency of a solution to draw water from pure water through osmosis across a selectively permeable membrane.

Outdoor platform The phenotyping conducted in outdoor environments, typically within agricultural fields or natural ecosystems, utilizing solely natural light sources. Compare with *indoor platform*.

Parenchyma A fundamental, thin-walled tissue composed of living cells that are primarily responsible for essential physiological functions such as photosynthesis, nutrient storage, and substance transport in plants. Typical parenchyma tissues include the cortex and pith of stems, the root cortex, the mesophyll of leaves, the pulp of fruits, and the endosperm of seeds.

Patterning The deliberate design and implementation of recurring structures or configurations, whether in nature, technology, or human-made systems. It reflects a systematic, rule-driven approach to organizing elements into coherent, scalable, and often functional arrangements.

Pentose phosphate pathway A metabolic pathway composed of irreversible oxidative reactions responsible for the generation of NADPH, as well as reversible reactions that facilitate the equilibration of intermediates within the pentose phosphate and triose phosphate pool.

Perennials Plants that typically require more than one growing season to reach maturity, live for more than 2 years, and can produce seeds multiple times throughout their lifespan.

Pericarp The portion of a fruit that surrounds the seeds and originates from the development of the ovary wall.

Periderm The corky outer layer of a plant stem formed during secondary thickening or as a protective response to injury or infection.

Permease A class of multipass transmembrane proteins that facilitate the diffusion of a specific molecule in or out of the cell by passive transport.

Phase The interval between development stages. Compare with *stage*.

Phenology The timing of periodic biological phenomena, such as flowering, breeding, and migration, in relation to climatic conditions. In the field of science, phenology refers to the systematic study of how climate influences the periodic occurrence of biological phenomena, particularly the timing of seasonal events in plant and animal species.

Phenome The complete set of phenotypes exhibited by a cell, tissue, organ, organism, or species under specific environmental and genetic backgrounds.

Phenomics A systematic and interdisciplinary research field dedicated to the comprehensive study of phenotypes, including the measurement, analysis, and integration of phenotypic data across multiple biological scales.

Phenotype The complete set of observable characteristics of an organism, which are shaped by the interactions between genetic and environmental factors. Synonymous with *trait*. Compare with *genotype*.

Phenotyping The systematic observation, measurement, and analysis of observable traits of an organism.

Phloem loading The process by which sucrose synthesized in leaf mesophyll cells is loaded into the phloem for export from the leaf.

Phloem unloading The process by which sucrose transported through the phloem is delivered into sink organs.

Phloem A specialized tissue in vascular plants that is responsible for transporting organic compounds, such as sugars and amino acids, from photosynthetic organs (e.g. leaves) to the sinks. It is a complex tissue composed of sieve tubes, companion cells, parenchyma cells, and fibres. Compare with *xylem*.

Phosphoenolpyruvate carboxylase (PEPC) A kind of carboxy-lyases that catalyses the addition of bicarbonate (HCO_3^-) to phosphoenolpyruvate (PEP) to form the four-carbon oxaloacetate and inorganic phosphate.

Phosphoglucan water dikinase (PWD) A hydrolase that acts on starch by binding to the prephosphorylated starch substrate mediated by glucan water dikinase (GWD), thereby introducing phosphate groups at the C-3 position of the glucosyl residue.

Phosphoglucomutase An enzyme that catalyses the interconversion between glucose 1-phosphorate and glucose 6-phosphorate.

Phosphoglucose isomerase An enzyme that catalyses the interconversion between glucose 6-phosphate and fructose 6-phosphate.

Phosphorylation The introduction of a phosphoryl group (a trivalent phosphorus-oxygen radical) into a compound through chemical or enzymatic means.

Photosynthates The energy-storing carbohydrates generated as the product of photosynthesis. Also *photoassimilates*.

Photoperiod The number of daylight hours, varying throughout the seasons and with latitude.

Photoperiodism The plant response to day length that allow adaptation to the changing seasons.

Photoreceptor A light-sensitive protein in plants that absorbs specific wavelengths of light and triggers signalling cascades regulating key developmental processes, such as seed germination, stem elongation, leaf expansion, and flowering.

Photorespiration A metabolic process that occurs in plants during photosynthesis, characterized by the uptake of oxygen and the release of carbon dioxide. Also *the oxidative photosynthetic carbon cycle*, or **C2** *photosynthesis*.

Photosynthesis The biochemical process by which plants, algae, and certain bacteria convert light energy into chemical energy. During this process, carbon dioxide and water are utilized to synthesize organic compounds, such as sugars.

Photosynthetically active radiation (PAR) The solar radiation within the wavelength range of 400–700 nm.

Phototropin The plant-specific, light-activated receptor kinase that regulates a wide range of blue light-induced physiological responses.

Phototropism The growth response of plant parts to light stimulation, resulting in a curvature towards the light source.

Phyllochron The interval between the emergence of successive leaves on a shoot. Compare with *plastochron*.

Physical property The intrinsic structural characteristic of a substance that can be observed or measured without altering its chemical composition. Compare with *chemical property*.

Physiology The scientific discipline that investigates the functional processes of living organisms. As a subfield of biology, it focuses on the mechanisms by which organisms, organ systems, organs, cells, and biomolecules execute chemical or physical functions within a living system.

Physiological maturity The stage where a seed attains its maximum dry weight, signifying the conclusion of dry matter accumulation and seed filling.

Phytochrome A light-sensitive chromoprotein in plants that detects red and far-red light. It switches between two forms: Pr, which absorbs red light, and Pfr, which absorbs far-red light and is the active form. Red light converts phytochrome to Pfr, triggering signalling pathways that regulate seed germination, shade avoidance, flowering time, and circadian rhythms.

Phytomer A structural and developmental unit in botany, specifically defined as a repeating segment along the growth axis of a plant. It typically comprises four primary components: a node, an attached leaf, an internode, and an axillary bud.

PIN auxin efflux carriers (PINs) A family of proteins that are widely recognized as auxin efflux carriers. The family is named after its founding member, PIN-FORMED 1 (PIN1), which was originally named based on the mutant phenotype characterized by defective organ initiation at the shoot meristem, a consequence of abnormal auxin distribution.

Plant architecture The three-dimensional structural organization of the plant body, including the branching pattern, as well as the size, shape, and spatial positioning of vegetative and reproductive organs.

Plant growth regulators Organic substances that influence plant growth and development at low concentrations, including both naturally occurring hormones and synthetic compounds.

Plant hormone The endogenous organic substance synthesized by plants that regulates a wide range of physiological functions at very low concentrations. It is generally produced in specific tissues or organs and subsequently transported to target site where it exerts its biological effects. Also *phytohormone*.

Plant physiology The scientific study of the functional processes and mechanisms that occur within plants at multiple levels, including the cellular, molecular, and whole-organism levels.

Plants Multicellular, predominantly photosynthetic organisms that are classified within the kingdom Plantae. They are characterized by the presence of cell walls composed of cellulose, chloroplasts utilized for photosynthesis, and a lifecycle marked by alternating diploid (sporophyte) and haploid (gametophyte) generations.

Plasmalemma A thin membrane composed of lipids and proteins that surrounds the cytoplasm of a cell and regulates the transport of substances into and out of the cell. Also *cell membrane* or *plasma membrane*.

Plastids A diverse family of specialized organelles found exclusively in plant cells and some protists, playing essential roles in photosynthesis, pigment synthesis, and nutrient storage. For example, *proplastids* are undifferentiated plastids typically found in meristematic cells. *Chloroplasts* are chlorophyll-containing plastids responsible for photosynthesis. *Amyloplasts* are specialized for the storage of starch. *Chromoplasts* contain carotenoid pigments and are commonly found in flowers and fruits. *Leucoplasts* are non-pigmented plastids. *Gerontoplasts* are derived from chloroplasts during the senescence of leaf tissues.

Plastochron The time interval between the initiation of successive leaf primordia at the shoot apex. Compare with *phyllochron.*

Platform A predefined or adaptable framework upon which a sensor or a combination of sensors can be mounted and integrated to ensure mobility while preserving precision and stability.

Pleiotropy The phenomenon whereby a single gene exerts an influence on multiple distinct and seemingly unrelated phenotypic traits, typically due to the gene's participation in various biochemical pathways or developmental processes.

Polar auxin transport (PAT) A directional and active transport mechanism in plants that facilitates the movement of auxin between cells. This mechanism involves specific transporters, such as the efflux carrier PIN and influx carrier AUX1/LAX, which play a crucial role in regulating auxin distribution and establishing gradients that are essential for plant development.

Population A group of individuals of the same species that live in a defined geographic area and can interbreed under natural conditions.

Primary nitrogen assimilation The initial conversion of inorganic nitrogen (predominantly nitrate or ammonium), which is absorbed by roots, into organic molecules (amino acids) within a cell in either roots or shoots. Compare with *secondary nitrogen assimilation.*

Primordium An organ at its earliest stage of development, serving as the foundation for subsequent growth and differentiation.

Process A sequence of events, particularly those that lead to natural transformations or developments.

Programmed cell death (PCD) A genetically regulated, active mechanism of cellular suicide wherein cells undergo self-destruction in response to specific signals or stimuli. This process is crucial for maintaining cellular homeostasis, supporting organismal development, and ensuring survival.

Promoter A specific DNA sequence located upstream of a gene that serves as the binding site for RNA polymerase and transcription factors to initiate transcription. Key regulatory elements (e.g. TATA box, CAAT box) within the promoter determine not only whether a gene is transcribed but also when, where, and at what level.

Protein disulfide isomerase An enzyme located in the endoplasmic reticulum of eukaryotic cells, which catalyses the formation and reduction of disulfide bonds between cysteine residues during protein folding.

Protein storage vacuole (PSV) The vacuole that stores protein bodies in seeds, tubers, or roots.

Protein turnover The dynamic, ongoing cycle of protein synthesis and degradation within living cells, which removes damaged or dysfunctional proteins and replaces them with newly synthesized ones.

Proteinase Any enzyme capable of performing proteolysis, by catalysing the hydrolysis of peptide bonds. Also *peptidase*.

Proteome The complete set of proteins present within a cell, tissue, or organism.

Proteomics The large-scale study of proteins, particularly regarding their structures, functions, interactions, and patterns of expression within a cell, tissue, or organism.

Protoplasm The living material within cells, which is a semi-fluid, gel-like substance that constitutes the fundamental structure of cellular organisms and facilitates essential life processes.

Protoplast A plant or bacterial cell that has its rigid cell wall enzymatically removed, leaving behind a spherical, membrane-bound structure containing the plasma membrane, cytoplasm, and nucleus (in eukaryotes), all essential living components necessary for regeneration, fusion, and genetic manipulation.

Quantitative trait locus (QTL) A defined region of the genome that contributes to variation in a quantitative trait, such as plant height, grain yield, or disease resistance, characterized by a continuous phenotypic distribution across a population and shaped by the combined effects of multiple genes and environmental factors.

Quantitative genetics A branch of genetics that focuses on the inheritance of complex, continuously varying traits that are influenced by multiple genes and environmental factors.

Quiescence The state of a cell that is not actively engaged in the cell division cycle at a given time.

Radiation use efficiency (RUE) The ratio of biomass produced per unit of intercepted radiation.

Raffinose A trisaccharide found in cottonseed and beet root molasses, consisting of D-galactose, D-glucose, and D-fructose.

Reactive oxygen species (ROS) The highly reactive chemical molecule formed due to the electron acceptability of O_2, such as peroxides, superoxide, hydroxyl radical, singlet oxygen, and alpha-oxygen.

Reducing agent A substance capable of reducing another substance by itself being oxidized.

Reductionism A philosophical and methodological approach that seeks to understand complex systems, phenomena, or concepts by breaking them down into their constituent parts or simpler, more fundamental elements. This approach emphasizes the analysis of individual components to gain insight into the functioning or nature of the whole system. However, it is frequently criticized for potentially oversimplifying complex systems and overlooking emergent properties that emerge from the interactions among their components. Compare with *holism*.

Resistant starch The portion of starch and its degradation products that resist digestion in the small intestine of healthy individuals. It is predominantly composed of retrograded amylose, which is characterized by a relatively low molecular weight and a linear, densely packed molecular structure that endows it with resistance to enzymatic hydrolysis.

Respiration The metabolic process through which plants convert the energy stored in carbohydrates into usable chemical energy required to support growth, maintenance, and other essential physiological processes.

Ribosome A small, granular cellular structure located in the cytoplasm, composed of ribosomal RNA and associated proteins. It plays a crucial role in protein synthesis by translating genetic information encoded in mRNA into specific polypeptide chains.

Rubisco Ribulose 1,5-bisphosphate carboxylase/oxygenase, a crucial enzyme involved in the process of photosynthesis.

Salicylic acid (SA) A phenolic phytohormone that plays essential roles in various physiological processes, including defence against pathogens, photosynthesis, transpiration, as well as ion uptake and transport.

Scalability The capacity of a system, network, process, or organization to effectively manage an increase in workload, demand, or scale without compromising performance.

Sclereid A type of thick-walled, lignified plant cell commonly found in sclerenchyma tissue, providing mechanical support and structural integrity to various plant organs.

Sclerenchyma A specialized plant tissue composed of cells with thick, lignified secondary cell walls. These highly rigid walls provide mechanical support, enabling plants to resist compression, bending, and other physical stresses.

Scutellum An abaxial embryonic tissue in cereals, involving in nutrient uptake from the endosperm, particularly during germination.

Secondary nitrogen assimilation A set of enzymatic reactions in plant cells that detoxify internally generated ammonium, primarily from photorespiration and protein turnover, by incorporating it into organic nitrogen compounds such as glutamine and glutamate. Compare with *primary nitrogen assimilation*.

Seed The reproductive structure of a plant, specifically a mature ovule containing an embryo and essential nutrients, which has the capacity to develop into a new plant under appropriate conditions.

Self-thinning A phenomenon in ecology and plant biology that occurs when individuals within a densely populated plant or animal community compete for limited resources, such as light, nutrients, and space, resulting in the mortality or suppression of less competitive individuals.

Senescence The aging process in plants, characterized by progressive degeneration and developmental changes that ultimately lead to the functional decline and death of plant tissues or organs.

Sensor A device that detects and measures physical properties, such as temperature, light, motion, or pressure, by converting them into signals.

Shade-avoidance syndrome (SAS) The response of shade avoidance that encompasses various physiological and morphological changes, such as stem elongation, altered flowering time, increased apical dominance, and a reallocation of resources.

Shoot apical meristem (SAM) A dynamic, self-renewing stem cell niche located at the tip of plant shoots, comprising undifferentiated cells organized into four functionally distinct zones: the central zone (CZ), the peripheral zone (PZ), the rib zone (RZ), and the organizing zone (OZ).

Short-day plant A plant that flowers when the day length falls below a critical threshold.

Sieve tube elements (STEs) Specialized cells within the phloem tissue of angiosperms that are responsible for the translocation of sugars and amino acids throughout the plant. These cells rely on *companion cells* to perform certain metabolic functions and should be distinguished from sieve cells, which are present in gymnosperms.

Signal A chemical or physical message that varies over time or space and carries information. Signals may include physical quantities such as mechanical pressure, pH, temperature, and light, as well as chemical substances like peptides, sugars, and hormones.

Signal peptide A peptide segment comprising approximately 20 to 40 amino acids, fused to a protein to facilitate its secretion and subsequent transport across the cell membrane.

Signalling The transmission of information between two physical entities. Analogous to *communication*.

Sink The physical structure of the plant that imports and utilizes resources produced by source organs.

Sink strength The capacity of sink organs to import nutrients from source organs, which is determined by the size, metabolic activity, and storage capacity of the sink organs.

Small interfering RNA (siRNA) A class of double-stranded, non-coding RNA molecule, 20–25 nucleotides in length, that is critically involved in regulating and suppressing gene expression. It is an externally introduced or synthetically produced RNA, whereas *microRNA (miRNA)* is naturally produced within the cell as a hairpin primary transcript.

Soluble starch synthase A starch synthase enzyme that exists in a soluble form within the plastid stroma and is also partially associated with the starch granule, playing a role in amylopectin biosynthesis by preferentially elongating glucan chains of varying lengths.

Source The physical structure of the plant responsible for the production and export of resources.

Source–sink interaction The mutual influence between source organs like leaves and sink organs such as seeds, tubers, and roots, regulated by complex physiological and biochemical signalling pathways.

Source strength The capacity of source organs to assimilate CO_2, convert it into sugars, and transport these sugars to sink organs.

Specific leaf area The ratio of leaf surface area to leaf dry mass. "*Specific*" denotes a property per unit mass, enabling standardized comparisons of material properties regardless of sample quantity.

Spherosome A small cell organelle bounded by a single membrane that participates in the storage and synthesis of lipids.

Spikelet A small and secondary inflorescence unit typical of grasses that consists of one or more florets enclosed by a pair of glumes and is often surrounded by lemmas and paleas.

Sporophyte The diploid generation in plants exhibiting alternation of generations, which originates from a zygote and is capable of producing asexual spores. Compare with *gametophyte*.

Stachyose A tetrasaccharide composed of two **α**-D-galactose residues, one **α**-D-glucose residue, and one **β**-D-fructose residue. Stachyose, along with related oligosaccharides such as raffinose, naturally occurs in a wide variety of vegetables, such as green beans, soybeans, and other legumes.

Stage The time points when distinct changes in morphology, dimensions, function, and anatomy occur during the progression of crop development. Compare with *phase*.

Starch A complex carbohydrate, specifically a polysaccharide, predominantly found in plants, where it functions as a primary storage form of energy. It is characterized as a white, tasteless, and odourless substance that is insoluble in cold water and alcohol.

Starch branching enzyme An enzyme that catalyses the formation of **α**-1,6 glycosidic bonds within **α**-1,4-linked glucose chains during amylopectin biosynthesis.

Starch debranching enzyme An enzyme that catalyses the breakdown of branched structures in starch, consisting of a combination of transferases (4-**α**-D-glucanotransferase) and hydrolases (amylo-1,6-glucosidase).

Starch hydrolases Enzymes that facilitate the hydrolysis of starch by transferring an -OH group from water to one moiety and an -H group from water to the other. In plants, these enzymes include amylase, maltase, and dextrinase, among others.

Starch phosphorylase An enzyme that catalyses the conversion of a glucose residue at the non-reducing end of an **α**-1,4-linked glucan into glucose 1-phosphate, which constitutes a key step in the enzymatic degradation of starch.

Starch synthase The enzyme responsible for synthesizing amylose and amylopectin from ADP-glucose. It comprises two types: granule-bound starch synthase (GBSS) and soluble starch synthase (SSS).

Starchy endosperm The inner and main body of the cereal endosperm responsible for storing starch and certain storage proteins.

Stay-green The delayed senescence of leaves.

Stem cell A general term denoting any undifferentiated cell that possesses the ability to undergo long-term self-renewal via cell division and can be induced to differentiate into a specialized and functional cell type.

Stem cell niche The specialized microenvironment that harbours stem cells, offering both structural support and a diverse array of biochemical signals that regulate key cellular processes such as self-renewal, quiescence, differentiation, and survival.

Storage proteins Specialized proteins that accumulate in plant storage organs, such as seeds, tubers, and bulbs, and serve as reservoirs of nitrogen and amino acids. During germination or sprouting, these proteins are enzymatically hydrolysed to release amino acids, which are then reused for de novo protein synthesis essential for seedling growth.

Strigolactones (SLs) Carotenoid-derived phytohormones, functioning as key regulators of lateral root development and bud growth, as well as important triggers for the establishment of symbiotic relationships between plants and mycorrhizal fungi.

Structure The organized arrangement of molecules, cells, tissues, or organs that constitutes a functional entity.

Structure–function relationship The connection between the physical structure of a biological system and its biological function.

Sucrose A disaccharide composed of fructose and glucose, serving as the primary transport form of carbon in plants.

Sucrose phosphate phosphatase The enzyme that catalyses the hydrolysis of sucrose 6-phosphate, resulting in the formation of sucrose and inorganic phosphate (Pi).

Sucrose phosphate synthase The enzyme that catalyses synthesis of sucrose phosphate from fructose 6-phosphate and UDP-glucose.

Sucrose synthase The enzyme that catalyses cleavage of sucrose into UDP-glucose and fructose.

Sucrose transporter (SUT) A type of transmembrane protein that mediates the movement of sucrose across cellular membranes, predominantly in plants. It functions as a proton-sucrose symporter, coupling the transport of sucrose with proton gradients. Also *sucrose carrier (SUC)*.

Symplast A continuous network of protoplasts that are interconnected by plasmodesmata, forming cytoplasmic bridges that enable the efficient transport of water, solutes, and nutrients across cells. Compare with *apoplast*.

Symplastic continuity The cytoplasmic units (cells) connected by plasmodesmata with a high degree of functional and structural integration.

Symplastic isolation The impaired symplastic communication between two cells resulting from the absence, removal, or occlusion of plasmodesmata.

Symporter An integral membrane protein that facilitates secondary active transport by simultaneously translocating two distinct solutes across a biological membrane in the same direction. Compare with *antiporter*.

Synchronization The coordination of events involved in operating a system.

Synteny The presence of large sequence blocks that are conserved across two or more genomes, as indicated by highly similar patterns among the species.

Synthase An ATP-independent enzyme catalysing biosynthetic reactions, e.g. glutamate synthase.

Synthetase An ATP-dependent enzyme catalysing biosynthetic reactions, e.g. glutamine synthetase.

Synthetic biology An interdisciplinary domain that applies engineering principles to the field of biology, including the design and construction of novel biological components, apparatuses, and systems, or the redesign of existing natural biological systems for practical applications.

System A collection of interacting and interrelated components that operate in accordance with a defined set of rules to function as a cohesive and unified entity. Every system is constrained by spatial and temporal boundaries, shaped by its environment, defined by its structure and function, and manifested through its operations.

Systematic thinking A cognitive approach that views a problem or concept as an integrated whole, employing a structured and methodical framework to analyse, integrate, and evaluate interrelated components, thereby facilitating a deeper, more comprehensive understanding and enabling more informed decision-making.

Systems theory An interdisciplinary framework that examines the intrinsic complexity of diverse systems, such as individual organisms, organizations, societies, and electro-mechanical or informational artifacts.

Testa A hard, protective outer layer that covers the seeds of flowering plants. Also *seed coat*.

Thioredoxin A class of small redox proteins that are universally present in all living organisms, characterized at the amino acid sequence level by the presence of two vicinal cysteine residues arranged in a CXXC motif.

Thermal time The cumulative heat units used to predict plant development stages. It measures the accumulation of temperature above a base threshold over time and is commonly expressed as Growing Degree Days (GDD). This concept helps farmers and researchers estimate crop growth stages, determine optimal sowing and harvest times, and assess the suitability of growing conditions for specific crops.

Tiller A lateral shoot that originates from an axillary bud located at the basal nodes of cereals and other grass species.

Tillering The production of lateral shoots from the lower nodes of the plant.

Tissue A group of similar cells organized into unit with particular function and intercellular communication.

Tomography A method for generating images of the interior of the human body or of a solid object using X-rays or ultrasound.

Trade-off A situation in which the improvement of one attribute or characteristic requires the reduction or compromise of another.

Trait The morphological and physiological characteristic of interest for further refinement in a crop breeding program. Synonymous with *phenotype*.

Transcription The process by which genetic information stored in a DNA molecule is transcribed into an RNA molecule, involving the synthesis of RNA using DNA as a template.

Transcription factor A protein that controls the rate of transcription of genetic information from DNA to messenger RNA, by binding to a specific DNA sequence (the promoter).

Transcription initiation complex A multiprotein complex composed of transcription factors that is required for recruiting RNA polymerase and initiating the transcription process.

Transcriptome The complete set of RNA molecules expressed by a cell, tissue, or organism at a given point in time, including mRNA, tRNA, rRNA, and non-coding RNA.

Transcriptomics The analysis of transcriptome by generating genome-wide mRNA profiles, thereby enabling a comprehensive characterization of gene expression under specific experimental conditions.

Transition The process or period in which a change occurs between two distinct states, conditions, or forms.

Translation The process by which the information encoded in mRNA is translated into the amino acid sequence of a protein.

Transporter A protein located within the cell or in cellular membranes, such as the plasma membrane, that facilitates the translocation of specific molecules or ions across biological membranes.

Transposon A DNA sequence capable of migrating from one position to another within a genome. It can excise and relocate itself or duplicate and insert at other sites, functioning as the generation of genetic diversity, the regulation of gene expression, and the reshaping of genome architecture. Also *jumping gene*.

Triacylglycerol The glycerol esterified at each of its three hydroxyl groups by a fatty acid.

Turgor pressure The positive hydrostatic pressure in plant cells due to the resistance of cell walls to volume changes caused by water uptake.

Ubiquitination A post-translational modification process in which ubiquitin molecules are covalently linked to target proteins, playing a critical role in regulating protein degradation, DNA repair, and signal transduction. This process is mediated through a sequential enzymatic cascade involving three key classes of enzymes: ubiquitin-activating enzymes (E1), ubiquitin-conjugating enzymes (E2), and ubiquitin ligases (E3).

UDP-glucose pyrophosphorylase (UGPase) The enzyme catalyses reversible conversion of glucose 1-phosphate and UTP to UDP-glucose and pyrophosphate.

Uniporter A membrane transport protein that facilitates the diffusion of a single specific substrate across a biological membrane. It functions by binding one molecule at a time and transporting it along its concentration gradient, without requiring direct energy input such as ATP or coupling to the movement of other ions or molecules.

Upscaling The ascending from a lower level of biological organization, such as the molecular level, to a higher level, such as the cellular level.

Vascular tissue A tissue of higher plants primarily composed of xylem and phloem, forming a continuous network throughout the plant body. It is responsible for the transport of water, minerals, and organic substances, as well as providing structural and mechanical support.

Vavilov centre A region of the world first indicated by *Nikolai Vavilov* to be an original centre for the cultivated crops.

Vernalization A period of cold exposure which is essential for subsequent flowering and reproductive development.

Vivipary The seed germination occurring prior to detachment from the parent plant.

Wild type (WT) The phenotype of the typical form of a species as it occurs in nature.

Xylem A specialized tissue found in vascular plants that is primarily responsible for transporting water and dissolved minerals from the roots to the leaves. It consists of several types of cells, including tracheary elements (e.g. vessels, tracheids), parenchyma cells, and fibres. Compare with *phloem*.

Zygote A eukaryotic cell formed by a fertilization event between two gametes.

α-Amylase A type of amylase that randomly cleaves glycosidic bonds along the starch chain, ultimately yielding maltotriose and maltose from amylose, or maltose, glucose, and limit dextrin from amylopectin.

β-Amylase A type of amylase that catalyses the hydrolysis of the **α**-1,4 glycosidic bond, successively releasing two glucose units in the form of maltose.

Reference

Hansen EJ (2011) Idea-based learning: a course design process to promote conceptual understanding. Routledge, New York

MIX
Papier aus verantwortungsvollen Quellen
Paper from responsible sources
FSC® C105338

If you have any concerns about our products,
you can contact us on
ProductSafety@springernature.com

In case Publisher is established outside the EU,
the EU authorized representative is:
Springer Nature Customer Service Center GmbH
Europaplatz 3, 69115 Heidelberg, Germany

Printed by Libri Plureos GmbH
in Hamburg, Germany